Cosmology Research - Addressing Current Problems with Astrophysics

Edited by Michael L. Smith and Ahmet M. Mecit Öztaş

Published in London, United Kingdom

Cosmology Research - Addressing Current Problems with Astrophysics
http://dx.doi.org/10.5772/intechopen.1005695
Edited by Michael L. Smith and Ahmet M. Mecit Öztaş

Contributors
Abdullah Alsakka, Ahmet M. Öztas, Alain Blanchard, Andrew Walcott Beckwith, Douglas J. Buettner, Firmin Oliveira, Marius Arghirescu, Martin T. Cole, Martin Tamm, Michael L. Smith, Peter D. Morley, Rafael Cañete Mesa, Richard J. Cook, Yves Gaspar

First published in London, United Kingdom, 2025 by IntechOpen
IntechOpen is the global imprint of INTECHOPEN LIMITED, registered in England and Wales, registration number: 11086078, 167-169 Great Portland Street, London, W1W 5PF, United Kingdom

For EU product safety concerns: IN TECH d.o.o., Prolaz Marije Krucifikse Kozulić 3, 51000 Rijeka, Croatia, info@intechopen.com or visit our website at intechopen.com.

British Library Cataloguing-in-Publication Data
A catalogue record for this book is available from the British Library

Cosmology Research - Addressing Current Problems with Astrophysics
Edited by Michael L. Smith and Ahmet M. Mecit Öztaş
p. cm.
Print ISBN 978-1-83634-809-2
Online ISBN 978-1-83634-808-5
eBook (PDF) ISBN 978-1-83634-810-8

Meet the editors

Michael L. Smith obtained his Ph.D. in Colorado and continued as a post-doc in biophysics at The Johns Hopkins University. There, he became familiar with computerized modeling of data. After obtaining his medical degree in Sweden, he worked as a consultant in the pharmaceutical industry both in Sweden and the U.S. In 2005, he became interested in cosmology after reading about claims related to dark matter and dark energy. He began publishing articles in collaboration with Prof. Ahmet Öztas in 2006 and continues to do so to the present. Mike has coauthored over 60 articles in medicine, biophysics, biochemistry, chemistry, and cosmology. He continues to perform research and editing in cosmology since his retirement in 2017.

Prof. Dr. Ahmet Mecit Öztaş received his Ph.D. in 1991 from Hacettepe University (Ankara, Turkey) in the field of string theory. He has conducted extensive research in theoretical physics. Between 2000 and 2001, he studied at Luleå University of Technology in Sweden, focusing on astrophysics, particularly neutron stars, and published papers in this area. Since 2006, he has been working in the field of cosmology and continues to publish in this area. Currently, Dr. Öztaş is teaching at Hacettepe University, where, through theoretical courses, he also offers introductory courses in cosmology and astrophysics.

Contents

Preface **IX**

Section 1
The Origins of Some Subatomic Particles **1**

Chapter 1 **3**
A Preonic Model of Quarks and Particles, Based on a Cold Genesis Theory
by Marius Arghirescu

Chapter 2 **35**
Phasic Structure of the Standard Model
by Rafael Cañete Mesa

Chapter 3 **75**
How Black Holes Remain Neutron Stars
by Martin T. Cole

Chapter 4 **95**
Gravitational Waves, Tokamaks, and the Pre-Planckian to Planckian era Universe
by Andrew Walcott Beckwith

Section 2
New Boundaries for Cosmological Models **113**

Chapter 5 **115**
On the Use of the Principle of Least Action in Cosmology
by Martin Tamm

Chapter 6 **135**
The Fallacies of ΛCDM Falsifications
by Alain Blanchard

Chapter 7 **151**
Predicting the Future Using Pantheon+TRGB Data: Better Analytical Techniques with Friedmann-Lemaître-Robertson-Walker Models
by Michael L. Smith, Abdullah Alsakka and Ahmet M. Öztas

Chapter 8 173
Dark Matter and Dark Energy in Galaxies Via Gravitational Field Self-Interaction
by Firmin Oliveira

Chapter 9 197
The Gravitational Analog of Maxwell's Equations: A Solution to the Vacuum-Energy and Cosmological-Constant Problems within Conventional General Relativity
by Richard J. Cook

Section 3
Examining the Cosmological Constant 215

Chapter 10 217
Atomic-Clock Detection of CNO-Based Dark Matter
by Peter D. Morley and Douglas J. Buettner

Chapter 11 237
The Problems of Modern Cosmology, the Fine Tuning of Constants and Parameters, and the Notion of Time
by Yves Gaspar

Preface

This book presents a diverse array of the latest investigations into cosmology, primarily theoretical, but several chapters delve into model interpretation and checks against astrophysical data. Many conclusions are interesting because the authors make concrete predictions which can be verified or refuted by future observation, some even with available data. Several offer much more than simple toy models, with insights drawn from recent data that successfully explain puzzling results such as the accelerating Universe expansion and the now-famous *Hubble tension*.

We begin with "A Preonic Model of Quarks and Particles, Based on a Cold Genesis Theory". M. Arghirescu puts his new model to the test by successfully calculating the energies/masses of several heavy and semi-light subatomic particles. The results match expectations, in some instances better than 1%, which is about as good as it gets in particle physics. The author of the chapter, "Phasic Structure of the Standard Model", proposes a new idea for particle creation that correctly predicts fermion masses and some puzzling properties of neutrinos. R.C. Mesa also uses his model to predict that a particle with a mass of around 172 eV, yet undiscovered, should exist. This can obviously be checked by a large collider, probably with results already existing on file. The always interesting problem of neutron star and black hole creation is covered in the chapter, "How Black Holes Remain Neutron Stars". The author, M.T. Cole, suggests that neutrons exhibit a heretofore undiscovered property, an absolute compressibility limit. He interprets this property as important at large object masses above the critical mass for a black hole, where the compact neutron star becomes invisible due to the strong gravitational attraction of electromagnetic radiation. While the heavy neutron star/black hole suffers immense gravitational attraction, becoming invisible, it does not collapse into a singularity. This is a very useful idea, making black holes amenable to mathematical investigation rather than having theorists simply call it a singularity and give up. If this theory is correct, one of the first discarded ideas is the wormhole. This proposal may disappoint many science fiction fans. The last chapter of this section is "Gravitational Waves, Tokamaks, and the Pre-Planckian to Planckian era Universe" by A.W. Beckwith. He proposes that the primitive Universe can be modeled as a huge tokamak that confined the incredible energy of creation for long enough to thoroughly mix the developing matter. Here, he provides several new details to this model.

M. Tamm presents a new vision for the roots of cosmology in "On the Use of the Principle of Least Action in Cosmology". He suggests better cosmological models than the FLRW approximation may be derived if one begins with a time-symmetric principle of least action. For instance, a solution allowing an accelerating universe expansion can be derived directly following his new hypothesis. If true, this places the conclusions of the A. Riess and S. Perlmutter groups on a much firmer theoretical foundation. We next tackle the viability of the current standard model of cosmology. In "The Fallacies of ΛCDM Falsifications", the author takes the conservative view that arguments attempting to falsify the ΛCDM model are subjective interpretations of the data rather than outright contradictions from new, better data. Specifically, he suggests that MoND is more wishful thinking rather than a better, realistic model. He also thinks the current tension over the Hubble constant shall simply be resolved over time with better data. The following chapter suggests that the current standard model of cosmology is the result of astronomers simply misanalyzing SNe Ia data. In "Predicting the Future Using Pantheon+TRGB Data: Better Analytical Techniques with Friedmann-Lemaître-Robertson-Walker Models" the authors point out and graphically demonstrate that Riess, Perlmutter, and coworkers use a century-old technique of convenience to analyze data rather than proper physics. Faulty analysis of the SNe Ia data introduces systematic error leading to fatally flawed results. For instance, the analyses by the aforementioned astronomers predict the future position of our Milky Way will be -∞. An incorrect prediction means their analysis is unscientific. These authors suggest a proper technique that avoids systematic error. Their results are consistent with our Universe being primarily spacetime with very little baryonic matter, dark matter, or dark energy. The proper solution of the correlation between SNe I redshifts and luminosity distances (DL) does not require our Universe to have an accelerating Hubble expansion, either. The authors provide a link to a GitHub site where SNe Ia and TRGB data are stored, and their analysis can be verified by any teenager familiar with Python 3. A conclusion consistent with these results is presented in the chapter, "Dark Matter and Dark Energy in Galaxies Via Gravitational Field Self-Interaction" by F. Oliveira. He argues that dark matter and dark energy can be explained simply as the results of baryonic mass in both spiral and elliptical galaxies. He introduces the gravitational redshift theory (GRST), which is defined in terms of the redshifts of self-interacting gravitational fields in galaxies, a very interesting new concept. This mass is sufficient to account for the missing mass required to match the observed rotation velocities without the need for exotic matter. This clever explanation does not require dark matter to exist as microscopic particles, which is why no such particles have ever been detected on Earth. R. J. Cook proposes a very general version of relativity combining the Maxwell and Einstein equations in "The Gravitational Analog of Maxwell's Equations: A Solution to the Vacuum-Energy and Cosmological-Constant Problems within Conventional General Relativity." Here, the cosmological constant appears as an integration constant unrelated to vacuum energy. This provides a straightforward solution to the relationship (or none) between these two concepts.

Dark matter detection as a particle has been an intractable problem for the last several decades. In "Atomic-Clock Detection of CNO-Based Dark Matter", Morley and Buettner suggest a novel method for detecting very large particles. The authors posit that if DM is enormous, it might be detectable by measuring atomic clocks affected by its gravity. Dark matter may be observed using the gravitational time delay, which should occur between two synchronized spacecraft with atomic clocks aboard. With these moving in opposite solar orbits, separated by at least one astronomical unit (au), the time difference will provide evidence for a large particle encompassing our solar system (or larger). Y. Gaspar has contributed the chapter, "The Problems of Modern Cosmology, the Fine Tuning of Constants and Parameters, and the Notion of Time", where he comments on a fundamental problem of cosmology. That problem is whether the tools of mathematics can adequately describe our Universe. This chapter presents arguments at the heuristic level, examining why some aspects of mathematics struggle with the concept of time. Time, time is all we have.

Michael L. Smith
Independent Researcher,
Sweden

Ahmet M. Mecit Öztaş
Hacettepe University,
Ankara, Turkey

Section 1

The Origins of Some Subatomic Particles

Chapter 1

A Preonic Model of Quarks and Particles, Based on a Cold Genesis Theory

Marius Arghirescu

Abstract

The theoretic model explains the generating of heavy quarks and of heavy particles using the known quark s(strange) and two semi-light quarks: $\lambda^{\pm}$(435 MeV) and $v^{\pm}$($\sim$574 MeV) resulting from a cold genesis theory[1] of astro-particles' forming, (CGT), the masses of the quarks $c^{\bullet}$ (charm) and $b^{\bullet}$ (bottom) of the Standard Model but also the values used by de Souza: c = 1.7 GeV and b = 5 GeV, being re-obtained as tri-quark clusters in the form: $[(q\bar{q})q]]$, by a simple de-excitation reaction, with the emission of a preonic boson obtained in CGT: $z^0 = 34m_e$, respective: $z_2 = 4z^0$ and $z_\mu = 6z^0$- for the quarks $b^{\bullet}$ and b. A similar de-excitation reaction, with emission of a preonic boson $z_k = n \cdot z^0$, (n = 1÷7), representing less than 2.7% from the particle's mass, explains the experimentally obtained masses of heavy baryons and mesons and of some multi-quark particles, with discrepancy <1% generally. The heavy current quarks result by lighter current quarks with quasi-crystalline internal structure, with trigonal/hexagonal symmetry of preonic kernels of two preonic bosons: $z_2 = 4z^0$; $z_\pi = 7z^0$, the top-quark resulting as: $t = 17(b\bar{b}) + b = (7x5) \cdot m(b)$, with a kernel of regular hexagonal polyhedron form, given by kernels of b-quarks. The model is compatible with the generation model and explains the quarks pairs forming from relativist e^-- e^+ jets.

Keywords: preons, quarks, heavy baryons, mesons, cold genesis, astro-particles

1. Introduction

In the Standard Model (S.M.), it is known the constituent quark model, with a valence current quark (u-up, d-down, s-strange) or (c-charm, b-bottom, t-top) with a current mass [1]: (1.8÷2.8; 4.3÷5.2; 92÷104) MeV/c^2, respective: (1.3; 4.2÷4.7; 156÷175) GeV/c^2 and a gluonic shell formed by gluons and sea-quarks [1], the resulted effective quark mass being the constituent quark mass: (336, 340, 486) MeV/c^2, respective: (1.55, 4.73, 177) GeV/c^2.

---

[1] This chapter is approved subject to the inclusion of some discussion of possible improvements to the Standard Model and/or the Generation Model arising from the proposed alternative model based on the strange quark and the two semi-light quarks: λ, v.

The electric charge of u-, c-, t- quarks is $+(^2/_3)$e and the electric charge of d-, s-, b- quarks is $-(^1/_3)$e, the strong interaction of quarks being explained by so-named "color charge", the gluons having two opposed color charges, the gluon field between a pair of color charges forming a narrow flux tube (as a 'string') between them, (the Lund string model [2]).

Conform to S.M., at the high-energy gluons the "breaking" of these strings into new quark–antiquark pairs can occurs, as part of the hadronization process, the upper limit for the gluon's mass experimentally determined being 1÷1.3 MeV/c^2 [3].

Also, the S.M. considers approximately the same size order for the maximum radius of the electron- resulted as scattering center determined inside the electron with X-rays: $\sim 10^{-18}$ m [4, 5] with that of the scattering centers experimentally determined inside the nucleon: 0.43×10^{-18} m [6], considered as quarks in the S.M. and the current quarks are considered un-structured, even if they can transform through weak interactions. As consequence, the quarks of S.M. cannot explain the mass hierarchy of the elementary particles by the sum rule, without the Higgs mechanism of mass acquiring by coupling to the Higgs field- which explains also the gluons' masses.

However, the Grand Unified theories (GUT) predict relations among the fermion masses, such as between the electron and the u-, d-quark, the muon and the strange quark and between the tau lepton and the bottom quark, but these relations were only partially obtained, (ex.: Georgi-Jarlskog mass relation, which for m_d, m_s, m_b gives discrepancy up to 21%, [7]).

An older model, proposed by A. O. Barut, proposed the explaining of the mesonic and baryonic particles by leptonic fermions: electrons, muons and neutrins, [8].

Another theoretic model which sustain the existence of a relation between fermions is the Generation Model, (G.M., [9]) which- using the Harari-Shupe rishons model and the rishons conservation rule, provides a unified classification scheme for leptons and quarks, simplified in comparison with the S.M. by using the conservation of only three additive quantum numbers: Q -charge, p- particle number and g -generation number.

The experimentally evidenced possibility of quark-antiquark –pairs obtaining by the interaction of relativist fluxes of negatrons and positrons is explained in the S.M. by the conclusion of $e^+ - e^-$ annihilation, with paired quarks forming from a high energy photon, resulting by the energy of the collided electrons.

It is sustained in the Standard Model, by the Einsteinian relativity, that "an alternative way to achieve pair production is to use an electron and a positron in the initial state thus producing an intermediate virtual photon. This virtual photon then decays into the final state pair. By raising the photon (or initial e^+e^-) energy just above 200 MeV, it becomes possible to produce pairs of muons and if the energy is raised further, to almost 4 GeV, then the threshold for tau pair production is passed. Eventually, as the energy increases further, the process can also be mediated by the intermediate Z^o weak boson. Quark pairs can be produced in exactly the same way, the main difference being that the final state particles materialize in the real world as streams or jets of particles, mainly hadrons, but with some photons and leptons. At lower energies, the light quarks are produced", [10].

In our opinion, this model of quark pairs production has a serious naturalness problem, because the Eisteinian relativistic mass depends on the observer's frame of reference, i.e. an imaginary observer placed in the inertial system of the moving electron (positron or negatron) cannot observe a mass increasing of that electron, so-their common conclusion cannot confirm the possibility of heavy virtual photon generating (with mass $m_\gamma > 2m_e$). Also, a mechanism of a real speed-dependent increasing of the effective mass is missing.

2. A quasi-crystal quark model as solution to the S.M.'s naturalness problem

In report with the previous case of naturalness problem of the S.M., it results as more natural alternative the possibility to explain the constituent quarks and the resulted elementary particles as clusters of negatron-positron pairs, named 'gammons' ($\gamma(e^-e^+)$) in a Cold Genesis pre-quantum theory of particles and fields, (C.G.T., [11–14]), based on the Galilean relativity, which argued that preonic bosons and quarks can be formed also 'at cold', as Bose-Einstein condensate of 'gammons' which form quasi-stable basic preons z^0 of mass $\sim 34\ m_e$, forming constituent quarks, (M. Arghirescu, 2006, [11, 12], p. 58).

This z^0 -preon was deduced by calibrating the value: $m_k = m_e/2\alpha = 68.5m_e$ obtained by Olavi Hellman [15], by using the masses of the proton and of the Σ-baryon, [11, 12], the proton' mass being given as ~27x68 = 1836 m_e and the Σ-baryon' mass being given as ~69x34 = 2346 m_e.

The Olavi Hellman's relation was deduced by a system of non-linear classical field equations having particle-like solutions, i.e. solutions which differ significantly from zero only in a region whose dimensions are of the same order of magnitude as the elementary length implied by the theory, this relation being obtained in the form:

$$M = (k\text{-}1)\left(\hbar/2l_0c\right) = (k-1)\cdot m_e/2\alpha;\ \left(k \geq 2;\ l_0 = e^2/m_ec^2 = a_c = 2.82\text{fm};\ \alpha = 1/137;\ \hbar = h/2\pi\right) \quad (1)$$

In a relative recent paper, (2015, [16]), a research team of Science' Institute for Nuclear Research in Debrecen, Hungary, after some experiments for the detection of dark photons, announced that significant deviation from the internal pair creation during the $(e^+ - e^-)$- transition to the ground state of an excited Be8* nucleus was observed at large angles, which indicates that in an intermediate step, was formed a neutral super-light particle with a mass of ~17 MeV/c^2, named X17, (~34 m_e), the excited Be8* state being obtained by proton interaction with a target of Li7, i.e. by a reaction:

$$\text{Li7} + p^+ \rightarrow \text{Be}^* \rightarrow \text{Be8} + b^0;\ b^0 \rightarrow e^+ + e^-,\ m(b^0) \approx 34\ m_e \quad (2)$$

In another paper, [17], a team of American physicists from California concluded that the evidenced new boson- which was reconfirmed experimentally in 2019 [18], could be the evidence for the predicted X-boson of a fifth fundamental force, of lepton- quarks coupling.

The experimentally evidenced particle with 34m_e suggests that- in accordance with CGT, in the decay of Be8* to its ground state, the b^0 –boson was emitted by an excited constituent nucleonic quark in the form of a neutral preon. Its stability can be explained by the conclusion that it is formed as cluster of an even number n = 7x6 = 42 quasielectrons, (integer number of degenerate "gammons", $\gamma^*(e^{*-}e^{*+})$), with mass $m_e^* \approx 34/42 = 0.8095\ m_e$, i.e. reduced to a value corresponding to the charge $e^* = \pm(2/3)e$ by a degeneration of the magnetic moment's quantum vortex $\mathbf{\Gamma_\mu} = \mathbf{\Gamma_A} + \mathbf{\Gamma_B}$, given by 'heavy' etherons of mass $m_s \approx 10^{-60}$ *kg* and 'quantons' of mass $m_h = h \cdot 1/c^2 = 7.37x10^{-51}$ kg. The considered "gammons" were experimentally observed in the form of quanta of "un-matter" plasma, [19].

The m_e^* -value results in CGT by the conclusion that the difference between the masses of neutron and proton: (m_n -$m_p \approx 2.62$ m_e) is given by an incorporate electron with degenerate magnetic moment and a linking 'gammon' $\sigma_e(\gamma^*) = 2m_e^* \approx 1.62$ m_e, forming together a 'weson', $w^- = (\sigma_e(\gamma^*) + e^-)$, which explains the neutron in a dynamide model of Lenard- Radulescu type [11, 12], as being composed by a protonic center and a negatron revolving around it by the $\boldsymbol{\Gamma_\mu}$ -vortex with the speed $v_e^* \ll c$, at a distance $r_e^* \approx 1.36$ fm [13], (close to the value of the nucleon's scalar radius: ~ 1.25 fm- used by the formula of nuclear radius: $R_n \approx r_0 \cdot A^{1/3}$), at which it has a degenerate μ_e^S -magnetic moment and S_e^n -spin.

The conversion: $p_r + e' \rightarrow n_e + \nu_e$, ($e^-$-capture) is explained by the conversion: $\sigma_e(\gamma^*) \rightarrow \nu_e + \in$.

The degenerate value μ_e^P of the magnetic moment of a protonic positron (giving its charge, as in the Anderson's model) or of a quasielectron results in CGT by the decreasing of its Compton radius r_μ^e proportional with the quantum density ρ_n of the protonic N^p-cluster in which is placed the electron's super-dense kernel, (its centroid), from the value:

$r_\mu^e = 3.86x10^{-13}m$, to the value: $r_i = r_\mu^p = 0{,}59\,fm$, (virtual radius of the proton's magnetic moment μ_p) given by the quantum mean density increasing, from the value: $\bar{\rho}_e$ to the value: $\bar{\rho}_n \cong fd \cdot Np \cdot \bar{\rho}_e$, conform to equations:

$$\mu_p = k_p \frac{m_e}{m_p} \mu_e = k_p \frac{\rho_e}{\bar{\rho}_n} \mu_e = k_p \frac{1}{f_d \cdot N^p} \mu_{Bp} = \frac{e.c \cdot r_\mu^p}{2}; k_p = \frac{g_p}{g_e} = 2.79 = \frac{\rho_n(r^+)}{\rho_n^0} = e^{-\frac{r^+}{\eta_d}} \tag{3}$$

in which: k_P-the gyromagnetic ratio; $\bar{\rho}_e; \bar{\rho}_n$ – the mean density of electron and nucleon;

$r_e^+ = 0.96\,fm$ -the position of the protonic positron' centroid in report with the proton center; f_d - coefficient of the quasielectron's mass degeneration, the positron's vortex $\boldsymbol{\Gamma_\mu}$ being diminished by the distribution of its vortical energy to all degenerate electrons of the particle's N^p –cluster of 'gammons', giving the particle's magnetic moment by an un-paired quantonic vortex of radius r_μ^P and of circulation: $\Gamma_\mu^P = 2\pi c \cdot r_\mu^P = g_p(m_e/m_p) \cdot \Gamma_\mu^e$.

The used electron model supposes an exponential variation of its density, (conform to eq. (3)), given by photons of inertial mass m_f, vortically attracted around a dense kernel m_0 and confined in a volume of classic radius a = 1.41 fm, (the e-charge in electron's surface).

The superposition of the (N^p + 1) quantonic vortices, Γ_μ^*, of the protonic quasielectrons, generates inside a volume with radius: $d^a = 2.1$ fm, a total dynamic pressure: $P_n = (1/2)\rho_n(r) \cdot c^2$ which gives an exponential nuclear potential: $V_n(r) = -\upsilon_i P_n$ of eulerian form conform to (3), with: $\eta^* = 0.8\,fm$ *and* $\upsilon_i(0.6$ fm)-the 'impenetrable' volume [11, 12, 20]:

$$V_n(r) = \upsilon_i P_n = V_{n0} \cdot e^{-r/\eta*}; V_{n0} = -\upsilon_i P_{n0}, \tag{4}$$

the nucleon resulting as formed by $N^p \approx 54x42 = 2268$ quasi-electrons which give a proton density in its center of value: $\rho_n^o \approx N^p \cdot \rho_e^o = 5.04x10^{17} kg/m^3$,

($\rho_e^0 = 22.24\ x10^{13}\ kg/m^3$), giving- with $v_i(a_i) = 0.9\,fm^3$: $V_s^0 = 127.5\ MeV$ and: $V_s(d = 2\,fm) \approx 9\ MeV$ – value specific to the mean binding energy per nucleon in the nuclei with the most strongly bound nucleons, (9.14 ÷9.15 MeV/nucleon for ^{56}Fe, ^{58}Fe, ^{60}Ni, ^{62}Ni).

The weak force of beta-emission is explained by the disintegration of a linking "gammon": $\gamma^*(e^{*-}e^{*+})$ whose superdense centroids give the electronic (anti)neutrino, (the photonic shell being transformed into disintegration energy $\in_d$), the couple: $w^{\pm}(e^{\pm}\gamma^*)$ forming a "weson" which added to the u^+-quark (or $\bar{u}^-$) gives the d^--quark, (or $^-d^+$-quark).

The neutron's disintegration is explained in CGT as a reaction of d-quark's transforming:

$$d^- \to u^+ + e^- + \bar{\nu}_e + \in_d; (w^-(e^-\gamma^*) \to e^- + \bar{\nu}_e + \in_d(0.827\ \text{MeV})) \qquad (5)$$

(by: $\sigma_e(\gamma^*) \to \bar{\nu}_e + \in$), the relativist speed $v \to c$ of the beta radiation electron being obtained with the vortex $\mathbf{\Gamma}_\mu = \mathbf{\Gamma}_A + \mathbf{\Gamma}_B$ of the proton's magnetic moment, whose etheronic part $\mathbf{\Gamma}_A$ explains the magnetic potential A and the atomic electrons' perpetual rotation, the quantonic part $\mathbf{\Gamma}_B$ explaining the magnetic induction **B** of the proton's magnetic field.

In this model, the quasi-crystalline structure of the preonic kernels and of quarks is maintained with the electronic centroids m_0 distanced at $d_i \approx 2x10^{-17}$ m, by the equilibrium between the electric and magnetic attraction between paired quasielectrons and a small repulsive field generated by internal photons' destruction (by zero-point vibrations of the electronic centroids) and by heavy 'naked' photons (including also photons corresponding to X-rays and γ-rays- which can be emitted at nucleon's vibration). The current quark's radius results of value: $r_q \approx 0.21 \div 0.3$ fm, conform to some older experiments [21].

The model explains the fact that some nuclear isomers such as Hf-178 (t½-31 y.) decay to a low-energy state by emitting gamma rays in concordance with the fact that the electron's interaction with green laser pulses evidenced γ-rays emission [22, 23].

The number: $n = 42\ m_e^*$ giving the z^0-preon may be explained by considering a "quarcin" $c_0^{*+} = 7x3 = 21\ m_e^*$, (with hexagonal symmetry), resulting that: $z^0 = (c_0^* + \bar{c}_0^*) = 34\ m_e$, ($\bar{c}_0^*$-anti-quarcin), the reaction (2) being explained by the z^0-preon's decay, in the form:

$$z^0(c_0{}^* + \bar{c}_0{}^*) \to c_0{}^+ + \bar{c}_0{}^+ \to e^+ + e^- + \gamma; (c_0{}^+ \to c_0{}^0 + e^{\pm}). \qquad (6)$$

($c_0^+(^2/_3e)$ -'quarcins', identifiable as high energy electrons [13]).

In CGT was deduced a quark model of cold forming quark, with effective (constituent) mass giving the particle's mass by the sum rule, by considering as fundamental stable sub-constituent the basic preon $z^0 = 42\ m_e^* \cong 34\ m_e$ which can form derived "zerons", (preonic neutral bosons: $2z^0$; $z_1(3z^0)$; $z_2(4z^0)$; $z_\mu(6z^0)$, $z_\pi(7z^0)$,), the light and semi-light quarks ($m_qc^2 < 1$ GeV) resulting by only two preonic bosons: $z_2(4z^0) = 136\ m_e$ and: $z_\pi(7z^0) = 238\ m_e$.

The main CGT"s link with the S. M. consists in the fact that- considering the electron's mass confined in its kernel, the z^0-preon can be considered as formed by 3 pairs ($u_c{}^-u_c$) of current u_c -quarks formed by a quasielectron ($e^{*\pm}$) surrounded by 3 gluonic 'gammons', i.e. with mass: $m(u_c) = 7m_e^* \approx 2.9\ MeV/c^2$, the current d-quark resulting with a mass:

$m(d_c) = m(u_c) + w^-(e^- \gamma^*) = (2.9 + 2.62)$ MeV = 5.52 MeV, (in the limits accepted by S.M.).

Also, the degenerate gammons γ^*(0.827 MeV) can be considered a 'gluol' which is the CGT's equivalent of the 'gluon' considered in the S.M. with a mass' upper limit: 1.3 MeV.

The u_c –quarks forming by the gluol's splitting is explainable by the attraction between the resulted 'free' quasielectrons and adjacent 'gluols', with the forming of a $(u_c\overline{u}_c)$-pair.

The cold quarks results as superpositions of preonic bosons $z_2 = 4z^0$ and $z_\pi = 7z^0$ with almost the same symmetry (**Figures 1** and **2**), resulting a constituent quark' mass equation [13, 14]:

$$q = m + k \cdot z_\pi + n \cdot (k-2) \cdot z_2; m = (m_1^+; m_2^-); k = 0 \div i = 3; n < k, (n = 0 \div 2) \quad (7)$$

i.e.: -(k, n = 0) $\Rightarrow q = m_{1,2}$; (k = 1, n = 0) $\Rightarrow q = r^{\pm}$, ("rark"- un-stable quark); (k = 2) $\Rightarrow q = p^+, n^-$; – (k = 3, n = 0) $\Rightarrow q = \lambda^+, \lambda^-$; (k = 3, n = 1) $\Rightarrow q = s^-, s^+$; (k = 3, n = 2) $\Rightarrow q = v^+; v^-$.

From eq. (7), the baryons mass results as combinations (q-q-q) according to equation:

$$M_b = M_q + k \cdot z_\pi + n \cdot (k-6) \cdot z_2; M_q = \sum_{i=1}^{3} m_i; m_i = (m_1^+; m_2^-); k = 6 \div 9; n \leq 2 \quad (8)$$

The particle's mass results by eqs. (7) and (8) in the approximation of the sum rule applied to the particle's cold forming, as consequence of the quantum fields'

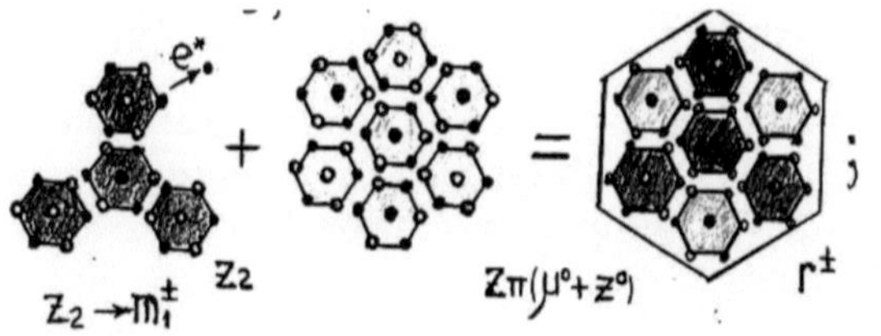

Figure 1.
The m_1^*, z_π^* and r^*- quark pre-clusters forming from z^0-preons.

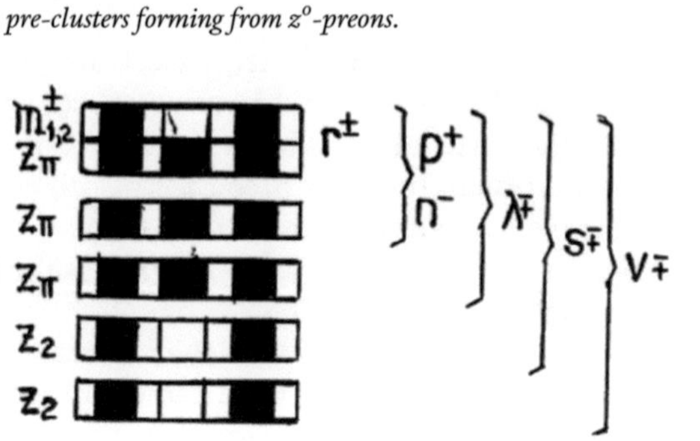

Figure 2.
The cold forming of semi-light quarks by pre-clusters of $m_{1,2}$; z_2 and z_π.

superposition principle applied to the particle's cold forming as sum of degenerate electrons, whose total vortical field Γ_v can explain also the nuclear force $F_n = -\nabla V_n(r)$ conform to eq. (4), [11, 12, 20].

The particles cold forming by clusterizing of preons and quarks forming ('mark', 'park', 'nark', etc.) may result- according to CGT, in a "step-by-step" process [14] consisting in:

- quark pre-cluster forming, as un-collapsed quasi-crystal Bose-Einstein condensate of gammons or of preons, (at $E_k = ½m_\gamma v^2 \rightarrow \mu_c;\mu_c$- attractive potential, of magnetic or also electric interaction between gammonic electrons or between z^0 -preons);

- quark (cluster) forming, as non-destructive (partial) collapsed quark pre-cluster; (at $E_k = ½m_\gamma v^2 < \mu_c$), and:

- elementary particle/dark boson forming as confined cluster of quarks with the current mass in the same baryonic impenetrable quantum volume and completed with a shell of 'naked' photons; (a small impenetrable volume of quasielectrons/preons and a repulsive field of zeroth vibrations impede the destructive collapse).

The known possibility to produce bosons and quarks ($(q^-$-$q^+)$ pairs) by high energy (e^--e^+) interactions is explained by the fact that –at distances $r \rightarrow a = 1.41$ fm, it results that:

$\mu_c \rightarrow e^2/8\pi\varepsilon_0 a = m_e c^2$, so ($e^-$-$e^+$)- pairs can be formed even at $E_k \rightarrow ½m_e c^2$.

The selection rule for the quarks' masses, which explains their relative stability, is given by the fact that the cold preonic pre-clusters of z_{2-} and z_π- bosons have almost the same symmetry (in form of star with three arms an of hexagon-**Figure 1**) and their superposition gives a quasi-stable pre-cluster (of charged pre-quarks or of pseudo-quarks with null charge), which- by cold non-destructive collapsing generated by attraction between electronic/preonic magnetic moments, gives a cold formed quark, with fractional charge.

Even if in CGT is considered that the integer e-charge of a formed particle is given by a single degenerate electron with degenerate magnetic moment, incorporated in the surface of the particle's quantum volume, formally is useful to consider a fractional charge: $(+^2/_3)\cdot e$ for the m_1^+- quark (resulted formally by a z_2-zeron which loosed a degenerate quasielectron- **Figure 1**) and for p^+-quarks, the cold quarks with negative charge $(-^1/_3)\cdot e$: m_2^-, n^-, λ^-, s^-, v^-, resulting as formed by a cold quark with positive charge $(+^2/_3)\cdot e$: m_1^+, p^+ and an attached electron with degenerate magnetic moment ($\rightarrow n^-$) or also a z_π- zeron and one or two z_2– zerons ($\rightarrow \lambda_s^-$, s_s^-, v^-), they having also a positive charged correspondent: λ^+, s^+, v^+, (without attached electron). In CGT the current quarks have a radius of ~0.21 fm and a repulsive shell of ~0.1 given by kinetized light photons.

The resulted structure of the fundamental elementary particles, considered as formed "at cold", in the approximation of the sum rule, by cold quarks with effective mass m_k and fractional electric charge $q^\pm = (+^2/_3 e$ or $-\,^1/_3 e)$, formed as preonic clusters, is given by the sub-structures presented in the **Table 1**, [11, 12]. The disintegration of the negative quark into a positive quark occurs by the emission of the incorporated (degenerate) electron and of an electronic antineutrino resulted as coupled kernels of a destroyed pair of internal degenerate electrons, forming a linking "gammon": $\sigma_e\ (e^{+*} + e^*) \rightarrow \nu\ (\overline{\nu}_e)$, (as in eq. (5)).

Basic quarks: $m_1 = (z_2 - m_e^*) = 135.2\ m_e$,	$m_2^- = m_1^+ + e^- + \sigma_e\ (e^{+*} + e^{-*}) = 137{,}8\ m_e$; $m_2^- \to m_1 + e^- + \overline{\nu}_e$; $\pi^+(m_1^- m_2) \to \mu^+ + \nu(2z^0)$
Derived quarks: $p^+\ (n^-) = m_1\ (m_2) + 2z_{\pi}$; $n = p^+ + w^-$; $n \to p^+ + w'$; $w^- = \sigma_e + e^-$; $\sigma_e \to \overline{\nu}_e + \in$	$(p; n) \approx 611.2 m_e$; $613.8 m_e$; $\lambda^{\pm} = p^+(n^-) + z_{\pi}$; $(\lambda^-) = 851.8\ m_e$; $s^{\pm} = \lambda^{\pm} + z_2$; $(s^-) = 987.8 m_e$; $v^{\pm} = \lambda^{\pm} + 2z_2$; $(v^-) = 1123.8\ m_e$; $[p^+(n^-) + z_2]$.
Mesons: $(q\overline{q})$	**Baryons**: (q-q-q); $(q^+ \equiv q(^2/_3 e)$; $q^- \equiv q(-^1/_3 e)$
$\mu^{\pm} = 2z_1 + e^{\pm} = 205\ m_e$; $/\mu^+\ (206.7\ m_e)$; $(z_1 = 3z^0$; $z_2 = 4z^0$; $z_{\mu} = 6z^0$; $z_{\pi} = 7z^0)$	$-p_r = 2p + n = 1836.2 m_e$; $n_e = 2n + p = 1838.8 m_e$; / exp.: $p_r^+(1836.1\ m_e)$; $n_e\ (1838.7 m_e)$;
$\pi^o = m_1 + \overline{m}_1 = 270.4 m_e$;$/\pi^0(264.2\ m_e)$	$-\Lambda^o = s^- + n + p = 2212.8\ m_e$; / $\Lambda^0\ (2182.7\ m_e)$
$\pi^+ = m_1 + \overline{m}_2 = 273\ m_e$; / $\pi^+(273.2\ m_e)$ $\pi^{\pm} \to \mu^{\pm} + \nu_{\mu}\ (2z_0)$; $\nu_{\mu} = z_{\nu} = 2z_0$	$-\Delta^{(++;+;0;-)} = s^{\pm} + \lambda^{\pm} + p^+(n^-) = 2445.6$; $2453.4\ m_e$; /exp.: $\Delta^{\pm;0} = 2411 \pm 4\ m_e$
$K^+ = m_1 + \overline{\lambda} = 987\ m_e$; / $K^+\ (966.3\ m_e)$	$-\Sigma^+ = v^- + 2p = 2346.2 m_e$; $\Sigma^- = v^- + 2n = 2351.4 m_e$; /exp.: $\Sigma^+(2327\ m_e)$; $\Sigma^-\ (2342.6\ m_e)$
$K^o = m_2 + \overline{\lambda} = 989.6\ m_e$; / $K^o(974.5\ m_e)$	$-\Sigma^o = v^- + n + p = 2348.8\ m_e$ /exp. $\Sigma^0\ (2333\ m_e)$;
$\eta^o = m_2 + \bar{s} = 1125.6\ m_e$; / $\eta^0\ (1073\ m_e)$ $\phi^0 = \lambda + \overline{v} = 1975.6\ m_e$; / $\phi^0\ (1995\ m_e)$ $N^{0^*} = v + s + \lambda^+ = 2960.8\ m_e$ / $N^{0^*}(2972\ m_e)$	$-\Xi^o = 2\ s^- + p = 2586.8\ m_e$; $\Xi^- = 2\ s^- + n = 2589.4\ m_e$; /exp.: Ξ^0 (2572); $\Xi^-\ (2587.7\ m_e)$; $-\Omega^- = 3v^- = 3371.4\ m_e$; / exp.: Ω^- $(3273\ m_e)$; $N_0^{3^*} \approx 3366\ m_e$; predicted: $\Omega^{++} = 3v^+ = 3363.6\ m_e$

Table 1.
Elementary particles (theoretic mass (CGT)/experimentally determined mass).

Also, the charge of the muon is given- in the model, by an electron with degenerate magnetic moment attached to a neutral (collapsed) cluster $z_{\mu} = 6z^0$. According to CGT, the muon's disintegration: $\mu^{\pm} \to e^{\pm} + \nu_{\mu}$ is carried out by the mass conservation, in the form: $\mu^{\pm}(z_{\mu} + e^{\pm}) \to e^{\pm} + \nu_{\mu}(2z^0) + z_2(4z^0)$, which explains the fact that the electron and the muon have the same electric charge even if they have very different values of their masses.

Unlike the S.M., which considers the transforming: $s^- \to u^+ + W^-$; $W^- \to e^- + \overline{\nu}_e$, in CGT we have a weak interaction reactions: $s^- \to p^+ + z_{\pi} + z_2 + w^-$; $w^- \to e^- + \sigma_e \to e^- + \overline{\nu}_e$.

It may be also observed that the mass of the $m_{1,2}$ –quark is approximately equal to: $m_e/\alpha = 137 m_e$, corresponding to Olavi Hellman's relation (1) by $l_0 = a_c/2 = a = 1.41$ fm – classical electron's radius corresponding to its e-charge distributed in its surface $S_e = 4\pi a^2 = k_1 e$.

A compound quark model which used a basic boson with the mass $m(\pi_{1/2}) = m_e/\alpha = 137 m_e$ (α- the fine structure constant) close to the values: $m(z_2) = 136\ m_e$ and $m(m_2^-) = 137.8\ m_e$ used in CGT, was proposed by D.-Yu Chung [24].

The strong force between quarks is explained in CGT by a 'bag' model resulted from the (multi)vortical model of nucleon, of cold genesis, [20].

3. The explaining of the heavy quarks as compound structures

From eq. (7) of the quark's mass, it results in CGT that every quark has and a positive ($q = +^2/_3 e$) and a negative ($q = -1/_3 e$) variant. Also, in report with the S.M. which identified only three quarks with mass < 1GeV/c^2, CGT deduced by eq. (7) other three quasi-stable quarks with mass < 1GeV/c^2: $m_{1,2}$(69.5 MeV)- mesonics and: λ(435 MeV); v(574 MeV)- baryonics.

The masses of some "resonance" particles (*) result also in the variant of "cold" forming:

$\Delta^{o*} = 2v^- + p = 2858.8\ m_e$; $\Delta^{-*} = 2v^- + n = 2861.4\ m_e$; (known mass: 2850 m_e), and: $\Xi^{-*} = 3\ s^- = 2963.4\ m_e$; (known mass: 3004 m_e).

But for explain particles with mass heavier than that of the Ω^-(1672) is necessary to explain the heavy quarks, with mass $m_q > 1\text{GeV}/c^2$, (of second and third generation).

It is believed generally that the baryon "resonances" are identified as excitations of the quarks, but other researches argue the possibility to obtain the energy spectrum of some baryon resonances as interaction between a baryon and vector mesons [25].

Also, M. de Souza [26] deduced the energy levels of baryon resonances with maximal discrepancy of ~5% by a simple formula for the energy of the system of three quarks:

$$E_{n;\,m;\,k} = h\nu_1(n+1) + h\nu_2(m+1) + h\nu_3(k+1) \tag{9}$$

where n; m; k = 0; 1; 2; 3; 4; and $h\nu_1$, $h\nu_2$, $h\nu_3$ are the ground states of the corresponding energy levels of baryonic quarks, being equal to the rest mass' energy of constituent quarks, ($E_q = m_q c^2$) taken as being: $E_q(m_u \approx m_d) = 0.31$ GeV, $E_q(m_s) = 0.5$ GeV, $E_q(m_c) = 1.7$ GeV, $E_q(m_b) = 5$ GeV, $E_q(m_t) = 174$ GeV, (u, d- up, down; s, c- strange, charm; b, t –bottom, top).

A selection rule for the heavy quarks masses, which- with high discrepancy, may be applied also for the values used by de Souza (which are close to but higher than those used by the Standard Model, $m_q^{\bullet}$) was found by R. A. Carrigan Jr. [27] which found an exponential formula for the heavy quark masses, of the form:

$$m_n^{\bullet} \approx m_1^{\bullet} \times 3^{n-1};\ m_1^{\bullet} = m_s^{\bullet} \approx 0.486\text{GeV}/c^2 \tag{10}$$

as an approximation for the quark masses: $m_c^{\bullet} = 1.55\ \text{GeV}/c^2$ and $m_3^{\bullet} = m_b^{\bullet} = 4.73\ \text{GeV}/c^2$, (values used by the Standard Model of the Quantum mechanics). Eq. (10) was explained by the forming of compound quarks with upper mass as tri-quark combinations:

$$q_n = (q\bar{q}\,q)_{n-1} \tag{11}$$

i.e- as triplets formed by quarks with adjacent lower mass, ($m_2^{\bullet} \approx 3\ m_1^{\bullet}$; $m_3^{\bullet} \approx 3\ m_2^{\bullet}$): two paired quarks and an un-paired quark giving the charge and the magnetic moment.

In concordance with G.M. it results that eq. (11) maintain the value p = 1/3 for all quarks.

From **Figures 1** and 2, by a z^0-preon with a ratio: length/diameter → 1, it results in CGT that the cold quark's stability is given by the (quasi)crystalline arrangement of its cluster of preonic kernels (named 'kerneloids') of regular hexagonal polyhedron form, with the ratio: length/diameter = 1÷2, i.e.- with 3÷5 levels of preonic bosons z_π ($7z^0$) and/or $z_2(4z^0)$ and a light $m_{1;2}$ -quark resulted from a z_2-boson by the loosing of a degenerate electron e^*.

By the principle of similitude, it results that a similar (quasi)crystalline arrangement may result by kernels of (semi)light quarks ($m_q < 1\text{GeV}/c^2$) or of heavy quarks (c- or b-quarks), formed as composite quarks, by some (q-$\bar{q}$)-pairs and an un-paired $q^{\pm}$ – quark, (eq. (11)).

We observe also that relation (10) between the values of ($m_s^{\bullet}$, $m_c^{\bullet}$ and $m_b^{\bullet}$) used by the Standard Model and the values (m_s, m_c and m_b) used by Souza (which are slightly higher) gives relative high discrepancy, i.e.: (m_c; $m_c^{\bullet}$) $\neq 3m_s$; (m_b; $m_b^{\bullet}$) $\neq 3$ (m_c; $m_c^{\bullet}$).

It was shown by author [28] that the values m_s, m_c and m_b used by de Souza [26] and those used by the S.M. [1] can be deduced by the CGT's model, with eq. (10).

Using the notations: q^*; m_q^* for the cold quark of CGT and its mass, we observe that:

a. $m_u \approx m_d = 0.31\ GeV/c^2 \approx 612\ m_e \approx m(p;\ n)$ -the mass of the nucleonic quarks of CGT;

b. $m_s = 0.5\ GeV/c^2 = 978.5\ m_e$ ($\approx m_s^* = 987.8\ m_e$, $\sim 0.504 GeV/c^2$) -the mass of s-quark resulting almost as in CGT;

c. $m_c = 1.7\ GeV/c^2 = 3326.8\ m_e$ used by Souza can be obtained by eq. (10), by taking: $m_3 = m_v^+ \approx 1121.2\ m_e \approx 0.574$ GeV (instead of m_s), i.e- the mass of the superior quark named "vark", used in CGT (in its positive ($+^2/_3e$) or negative ($-^1/_3e$) charged variant), which is the adjacent inferior quark mass in report to $m_4 = m_c$ and considering the quark $c(m_c^+)$ as de-excited state of the triplet with mass: $m_4^* = m(c^*) = 3m_v^*(v^+) = 3363.6\ m_e$, ($1.718\ GeV/c^2$), according to the next de-excitation reaction:

$$c^{*\pm}\left[\left(v^{\pm}\overline{v}^{\pm}\right)v^{\pm}\right] \to c^{\pm}\left(\sim 1.702 GeV/c^2\right) + z^0(34m_e) \tag{12}$$

-in concordance with the reaction (2). It results also a new Ω^- -variant: $\Omega^-(3v^-)$: $\Omega^-(1702)$.

d. For m_b, we can obtain similarly the value used by de Souza [26], by eq. (10) in which: $m_4^* = m_c$ and $m_5^* = m(b^{*\pm}) = 3m_c \approx 5.1\ GeV/c^2$, by the next de-excitation reaction:

$$b^{*\pm}\left[\left(c^{\pm}\overline{c}^{\pm}\right)c^{\pm}\right] \to b^{\pm}\left(\sim 5 GeV/c^2\right) + z_3(204\ m_e); z_3 = z_\mu = (2\times 3)z^0 = 2z_1 \tag{13}$$

The considered reaction (13) is concordant with the reaction: $\pi^0 \to z_\mu(6z^0) + \nu_\mu$, and with the known reactions: $\pi^+ \to \mu^+ + \nu_\mu$; $\mu^{\pm} \to e^{\pm} + \nu_\mu(\overline{\nu}_\mu)$, but considered as in CGT, by the mass conservation' rule, i.e. - in the form: $\mu^{\pm} \to z_\mu + e^{\pm} \to e^{\pm} + \nu_\mu(2z^0) + z_2(4z^0)$.

The explanation of the value: $m(b^* - b) = z_\mu = 6z^0$, released at the b^*-cluster's de-excitation, may be obtained by writing the mass of the resulted b-quark by eqs. (12) and (13), in the form:

$$m(b) \approx 3\left(3\left(m_v - z^0/3\right) - 2z^0\right) = 9\left(m_v - z^0\right); \left(v(1123m_e) \to v^{\cdot}(1089m_e) + z^0\right) \tag{14}$$

which shows that –according to the CGT's model and in concordance with eq. (10), at the b-quark' "hot" forming from v-quarks (resulted from nucleonic quarks), each constituent v-quark loose a z^0-preon, as consequence of its meta-stability, the kernel of this z^0-preon being emitted from the v-quark's kernel.

Applying this conclusion to $\Omega^-(3v^-)$ it results that: $m(\Omega^-) = (3371.4 - 3x34) m_e = 3269.4\ m_e$, (compared with the experimental mass: 3273 m_e, resulting a discrepancy of 0.14%).

The conclusion of eq. (14) is applicable and to the current b-quarks contained by the surface of the t-quark's kernel and is specific to high energy interactions.

e. The t-quark results as collapsed cluster: $t^{\pm} = (7x5)m(b^{\pm}) \approx 175GeV$, ($17(b\overline{b}) + b^{\pm}$), the structure of its kernel resulting by similitude with the preonic structure of the s-quark (CGT- **Figure 2**) as formed from a pre-cluster of regular hexagonal polyhedron form with length/diameter→ 1, in acceptable concordance with the experimentally obtained value of its constituent mass ($176.8 \approx 177$ GeV/c^2, [29], p. 135), with upper stability given by the crystalline form of its current mass, formed by kernels of b-quarks, conform to the model.

The form: $(7x3)m(b^{\pm}) = 105$ GeV with un-paired $b^{\pm}$ – quark and the forms: $h^0 = (7x3)m(c) = 35.7$ Mev/c^2 and $(7x5)m(c) = 59.5$ Mev/c^2 results also as possible.

The values of m_c and m_b used by de Souza and of the t-quark result by the eqs. (12) and (13) with discrepancies of 0.31%, 0.08%, and respective- of ~1%.

The resulted compound quark model of heavy quarks suggests also the possibility of a cold genesis variant of these quarks, in the form: $c^*(v\overline{v}v)$ and $b^*(c^*\overline{c}^*c^*)$ and of some heavy baryons of low vibration of their quarks, formed by three heavy 'cold' quarks.

Because the quarks $v^{\pm}$ and $s^{\pm}$ of CGT have close masses, the quarks $c^{\bullet} \approx 1.55GeV$ and $b^{\bullet} \approx 4.73GeV$ of the S.M. [29] result from direct combinations between $s^{\pm}$ and $v^{\pm}$ or as derived from c-, b- quarks, approximately in the form:

$$c^{\cdot} = \left((s \cdot \overline{s} \cdot v) - z^0\right) : c'\left(s^- \overline{s}^- v^{\pm}\right) \rightarrow c^{\cdot\pm} + z^0(34m_e);\ \Rightarrow m(c^{\cdot}) \approx 1.557\ \text{GeV}/c^2 \quad (15)$$

$$b^{\cdot\pm} = \left((c^{\cdot\pm}\overline{c}^{\cdot\pm}c^{\pm}) - z_2\right) : b'\left(c^{\pm}\,\overline{c}^{\pm}c^{\pm}\right) \rightarrow b^{\cdot\pm} + z_2(136m_e);\ \Rightarrow m(b^{\cdot}) \approx 4.75\ \text{GeV}/c^2 \quad (16)$$

the forms: c', b', being "isomeric" (excited, meta-stable) states. The explanation for the value $m(b' - b^{\bullet}) = z_2(4z^0)$ is the fact that the resulted $b^{\bullet}$ -quark's structure is approximately:

$$b^{\cdot} \approx \left[\left(c'(s \cdot \overline{s} \cdot v) - z^0\right)\left(\overline{c}'(s \cdot \overline{s} \cdot \overline{v}) - z^0\right)\left(c^{\wedge}(s \cdot \overline{v} \cdot v) - z^0\right)\right] = [c^{\cdot}c^{\cdot}c] - z_2 = b - 2z_{\pi};$$
$$\left(v \rightarrow s + z_2 \Rightarrow c \rightarrow c^{\cdot} + \pi^0(2z_2)\right). \quad (17)$$

i.e.-formed by three compound quarks: c',$\overline{c}$', $c^{\wedge}$ - de-excited by emission of a z^0-preon whose kernel is released from their quarcic kernel. So, the quarks $s^{\bullet}(486)$ $c^{\bullet}(1550)$; $b^{\bullet}(4730)$ used by the S.M. result from the CGT's quarks: s^-, c^+, b^- by the reactions:

$$c(1700) \rightarrow c^{\cdot}(1561) + \pi^0(2z_2); b(5000) \rightarrow b^{\cdot}(4756) + z_6(2z_{\pi}); s^{\pm}(500) \rightarrow s^{\cdot\pm}(483) + z^0. \quad (18)$$

Eq. (18) indicates that the S.M.'s quarks are quasi-stable 'hot' formed quarks, resulted from excited metastable quark: s^-, c^+, b^-, and that excited quarks v, s loose a z^0-preon.

It is explained why the particles' masses result by the sum rule mainly by the variants of heavy quarks' masses used by de Souza and by CGT, although experimentally were evidenced the mass variants of quarks characteristic to the S.M.

By (18), by: $s^{\pm} \rightarrow s^{\bullet\pm} + z^0$, it results:

$\Delta^{\pm;0}$ = [2445.6; (2453.4) -34] = [2411.6; (2418.4)] m_e;

Λ^o = (2212.8–34) m_e = 2178.8 m_e, (discr. < 1%); $\eta^o = m_2 + \bar{s}^\bullet$ = 1091.6 m_e, (discr. < 1.8%).

It results also that in the formula (9) it may be used also the rest energy of the quarks "v" ('vark') or λ("lark") in the first term, with $h\nu_1 = m_q c^2$ and that is more natural to interpret the numbers (n, m, k) = 0, 1, 2, 3, .., which multiply the baryonic resonance' energy in eq. (9), as being numbers of neutral pseudo-quarks q_i^0 added to the charged $q_i^{\pm}$ – quark which gives the ground state according to a compound quark model, these neutral pseudo-quarks having the same rest mass, i.e.: $q^{\pm}_i = (q\bar{q}q^{\pm})_{i-1}$; $q^0_i = (q^-q^-q^+)_{i-1}$.

This conclusion may explain also some mass differences of baryon resonances, resulted approximately equal to the mass of a boson z^0, z_2, z_π, Σz_2; Σz_π.

For example, the neutral correspondent of the quark $s^{\pm}$ is the scalar meson $f_0(500)^o$ and the neutral correspondent of the quark $c^{\pm}$ is the scalar meson $\rho(1700)^0$.

A next 'cold' heavy quark formed by clusterizing may result, according to the sum rule, by:

$m_4^* = (m_f; m_f^*) \approx (3m_b'; 3m_b^*)$, ($f_q^{\pm}$ – "fark", "frozen" quark, with mass: m $(f_q) \approx m(\Omega_{bbb})$).

An 'exotic' heavy but un-stable quark may be formed by a $q^{\pm}$ – quark surrounded by six q^0-pseudoquarks or three pairs $q - \bar{q}$.

The t-quark' forming from protonic jets: $j(P) + j(\bar{P})$ is explained by the fact that the strong force $F_N = -\nabla V_n$ determines a (quasi)crystalline (re)arrangement of the current quarks, by re-organizing of the internal structure of nucleonic quarks: $m_{u,d} \approx m_{p,n} \approx 612\ m_e$.

If we interpret the energy: $\Delta E_q = m(\delta_k)c^2$, loosen at the de-exciting of the quarcic cluster q^c_{k+1}, as binding energy between quarks, (by similitude with a nucleus), it may be obtained a semi-empiric relation for the mass of the heavy quarks c ≈ 1.7 GeV and b ≈ 5 GeV:

$$m(q_n^{\,c}) = 3^{n-1}\left[m_1 - (z^0/3)\ln\left(3^{n-1}3^{n-2}\right)\right]; n > 1 \tag{19}$$

with: $m_1 = m_v = 0.574\ \mathrm{GeV}/c^2$; $z^0 = 34\ m_e$ and $\ln(3^{n-1}3^{n-2}) = \ln 3^{2n-3}$, which gives:

$n = 2 \rightarrow m(q_2^{\,c})c^2 = 1.703$ GeV; $n = 3 \rightarrow m(q_3^{\,c})c^2 = 4.994$ GeV;

$n = 4 \rightarrow m(q_4^{\,c})c^2 = 14.64$ GeV.

The logarithmic part of the second term of the expression (19) indicates that the considered binding energy per v-quark increases with the number of v-quarks.

A relative similar semi-empiric relation may be found also for the quarks $c^\bullet \approx 1.55$ GeV and $b^\bullet \approx 4.73$ GeV, of the S.M., but in the form:

$$m(q_n^{\cdot}) = 3^{n-1}\left[(m_1^{\cdot} + \delta^{\cdot}) + (z^0/3)\ln 3^{n-2}\right], n > 1; (m_1^{\cdot} + \delta^{\cdot}) = (2m_s + m_v - z^0)/3 \tag{20}$$

with: $m_1^{\bullet} = m_s \approx 500$ MeV/c^2, $\Rightarrow \delta^{\bullet} = 19$ MeV/c^2, (or: $m_1^{\bullet} = m_s^{\bullet} \approx 486$ MeV/c^2, $\Rightarrow \delta^{\bullet} = 33$ MeV/c^2), giving: n = 2 $\rightarrow m(q_2^{\bullet})c^2 = 1.557$ GeV $\approx m(c^{\bullet})$; n = 3 $\rightarrow m(q_3^{\bullet})c^2 = 4.728$ GeV.

The expression (20) is characteristic to mass addition to the tri-quark cluster, (as consequence of the fact that was taken: $m_1^{\bullet} = m_s$ instead of: $(2m_s + m_v)/3$), as in the case of the Sakharov's equation [30], (which adds a term of spin-spin interaction at the total mass of quarks) and shows a link between the masses of s-, $c^{\bullet}$-, $b^{\bullet}$- quarks and $z^0(34\ m_e)$.

4. The explaining of the strong interactions between particles

It is possible to explain similarly – in CGT, also some strong interactions in which preonic z_k- bosons and bosonic pairs $(q - \overline{q})$ of the quantum vacuum, with total mass $M \le Q/c^2$, separated by the interaction energy Q, may generate heavier quarks in combination with the initial quarks, as in the next possible interaction:

$$\begin{aligned}
&(\text{Experimental reaction}) : \pi^-(\overline{m}_1 + m_2) + p_r(2p^+ + n^-) + Q \\
&\quad \rightarrow \Lambda^o(s + n + p) + K^o(m_2 + \overline{\lambda}); \\
&\overline{m}_1 + p^+ + Q \rightarrow \overline{m}_1 + m_1 + z_6 + Q = \pi^o + z_6 + Q \rightarrow (s^- + \overline{s}); \overline{s}^+ \rightarrow \overline{\lambda} + z_2; \quad (21)\\
&s^- + n^- + p^+ \rightarrow \Lambda^o; \overline{\lambda} + m_2 \rightarrow K^o; \\
&- \text{permitted reaction}; \ (z_2 = 136 m_e; z_6 = 2z_\pi = 476\ m_e).
\end{aligned}$$

with (p; n) $\approx$ (u; d). The conservation of the 'generation' number g- in G.M., (or of the 'strangeness' S –in S.M.) corresponds to g(λ; v) = g(s) = 1; g(m_1; m_2) = g(u; d) = g(z_k) = 0 and to the possibility of quarks transforming: $q_1^g \leftrightarrow q_2^{g'}$ only for quarks with the same generation number: g = g', but in CGT this indicates only the low probability of the quark's transforming into another with different generation number: $P(g' \neq g) \ll 1$, because in a weak interaction of type: $q_1^g \rightarrow q_2^{g'} + \Sigma z_k$ the sum ΣS is not mandatory conserved. For example, the reaction: $\pi^-(\overline{m}_1 + m_2) + p_r(2p + n^-) + Q_i \rightarrow \Lambda^o(s + n + p) + \pi^o(m_1 + \overline{m}_1)$- forbidden by the law of S- or g-number's conservation, is theoretic possible in CGT by: $p \rightarrow m_1 + 2z_\pi$; $Q_i \rightarrow z_\pi + z_2$; $m_2 + 3z_\pi + z_2 \rightarrow s^-$, (which not conserve ΣS or Σg), but is low probable.

A (semi)empiric relation for the particles lifetime, sustaining also the particles' cold forming, is obtained in CGT by considering the $\mu^\pm$ – lepton (lifetime: $\tau_0 = 2.2\mathrm{x}10^{-6}$ sec.) as single-particle and taking into account a vibration energy ε_v of the constituent quarks, which- according to CGT, generate a partial destruction of the particle's intrinsic vorticity, with the loosing of a part: Δm_p of internal 'naked' photons, giving: $\tau_k \sim 1/\Delta m_P(T)$.

At an ordinary temperature: T = 300 K of the particles' environment, the lifetime $\tau_k(\varepsilon_v)$ of the baryons (n = 3 quarks and $\tau_B \approx 10^{-10}$ sec.) and of the mesons (n = 2 quarks and $\tau_m \approx 10^{-8}$ sec.) is dependent to its intrinsic temperature T_q conform to equation [11, 12]:

$$\tau_k = \frac{\tau^0}{k_v \cdot 10^{2n}} \approx \frac{\tau_0 \cdot m_p}{\Delta m_P(T)}; \tau^0 \cong 3x10^{-14}\ \text{sec.}; k_v = \frac{\varepsilon_v}{\varepsilon_v^o} = \frac{n \cdot \nu_i}{\nu_c^o} = \frac{n \cdot T}{T_N}; T_N \cong 2x10^{12}\ \text{K} \quad (22)$$

in which: ν_c^{0} represents the critical frequency of the phononic energy ε_v^{0} of quark vibration at which the proton's disintegration take place: $\nu_c^{0} = \nu_c(T_N \approx 2x10^{12}K) \approx 4x10^{22}$ Hz.

Eq. (22) may explain the fact that the heavy baryons with composite heavy quarks can have a longer lifetime at T → 0 K but cannot have a long life at an ordinary temperature, [11, 12].

5. The nucleon's stability and the quasi-crystalline structure of quarks

In CGT, similarly to the S.M.'s constituent quark model, it is considered that the electron's mass is formed by a 'kerneloid' containing the (super)dense kernel m_0 of radius $r_0 \approx 10^{-18}$ m and by a shell of bosons which in the electron's case are 'naked' photons, in concordance with the evidenced possibility to obtain a B-E condensate of photons, [31].

This electronic kerneloid is equivalent to an 'impenetrable' quantum volume (similar to that of the nucleon), having a radius $r_{ie} \approx 10^{-2}$ fm- in accordance to some high-energy scattering experiments reported by Milonni et al. (1994, p.403 [32]).

The possibility to explain reactions of strong interactions between particles by heavier quarks transforming into lighter quark(s) and bosonic preon(s) specific to CGT but also by heavier quarks forming from these subcomponents, as in eq. (21), indicates that these sub-components maintain their higher stability also in strong interactions, by a quasi-crystalline arrangement of the electronic kerneloids k^e of their z^0-preons, the resulted preonic kerneloids forming the quark's kerneloid- which can be considered as being its current mass. The radius of the z^0-preon's kerneloid k^z results of value: $r_z = 3.5x10^{-2}$ fm, [13].

The preonic quasielectrons retain their photonic shell (also at the preon's releasing) by the vortical field of the Γ^e_μ -vortices of the degenerate magnetic moments, maintained by their kernels, in accordance with a classic equation of electron's intrinsic rest energy [11, 12]:

$$m_e c^2 \approx ½ \int \varepsilon_0 E^2 dV(r) \approx ½ \int \mu_0 H^2 dV(r); \left(E = c \cdot B; r = 0 \div r_\mu = \hbar/m_e c\right) \quad (23)$$

which explains the electron' mass m_e as saturation value: $n \cdot m_f$ of magnetically (vortically) confined 'naked' photons. These Γ_μ^{e} -vortices are maintained by the negentropy of the quantum vacuum given by etherono-quantonic winds (fluxes) which explain also the constancy of the magnetic moment of the free charged particles, in CGT [11, 12].

Eq. (23) explains the maintaining of the constituent mass also to quarks changed in strong interaction between interacting particles conform to the sum rule, as that of eq. (21).

The quasi-crystalline arrangement of preonic kerneloids of quarks formed by clusterizing is 'inherited' from the quarcic non-collapsed quasi-crystalline pre-cluster formed by pre-clusters of z_{2} - and z_{π} - preonic bosons, (**Figures 1** and **2**), the quarks confining force resulting in CGT by magneto-electric interaction between quasielectrons and by a pressure of kinetized photons giving a repulsive shell of radius 0.6 fm in accordance with a "bag" model of strong interaction with a bag' radius $r_i^{*} = a_i \approx 0.6$ fm [20], (as in the model of Toki & Hosaka).

The proton's high stability- compared with the stability of heavier baryons and with that of non-leptonic mesons, is explained conform to eq. (22) by a stronger coupling of its quarks by magneto-electric interactions between pairs of quasielectrons also to the radial direction, (**Figure 3**), as consequence of the same size of the constituent quarks, which reduces the quark's vibrations. Also, to the axial direction, the magnetic moments of the quasielectrons of a z^0-preon are axially coupled, the central, un-paired Γ_μ^* -vortex-tube giving the preon's magnetic moment; similarly, the axially coupled preons give the magnetic moment of the quarcic cluster of preonic bosons (z_2, z_π).

From **Figure 4** representing a preonic z_π -layer of a quarcic kerneloid it results that the calculated radius value: $r_z = 3.5 x 10^{-2}$ fm [13] of the preon's kerneloid, ensures a mean distance: $d_i \approx (^2/_3) \cdot r_z \approx 2.3 x 10^{-2}$ fm between the electronic centroids m_0 on the radial direction, which gave a value: $r_{ie} = 1.15 x 10^{-2}$ fm for the electron's mechanical radius, (as in Ref. [32]), the minimal value of the preon's length resulting of value: $l_z = 6xd_i \approx 0.14$ fm.

Because the quasi-crystalline structure of (u, d)- quark's kerneloid have three layers- in CGT, ($m_{1;2}$; z_π; z_π -**Figure 2**), with (4; 7; 7) z^0-preons, it results a minimal length of the (u; d)- quark' kerneloid: $l_q = 3l_z \approx 0.42$ fm, (at $T \to 0$ K).

The minimal radius of the quark's kerneloid (specific to its cold state) results of value:

$r_q^0 \approx 3xr_z = 0.105$ fm, value which- compared with the experimental radius value of the quark's current mass: $r_q \approx 0.21 \div 0.3$ fm [21, 33], indicates a small vibration

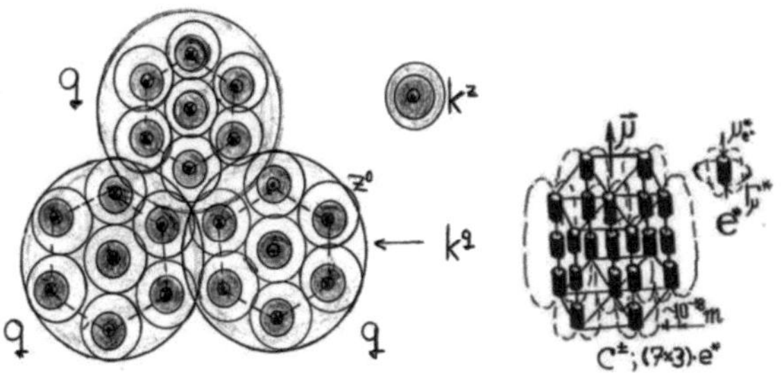

Figure 3.
Baryonic and preonic kerneloid.

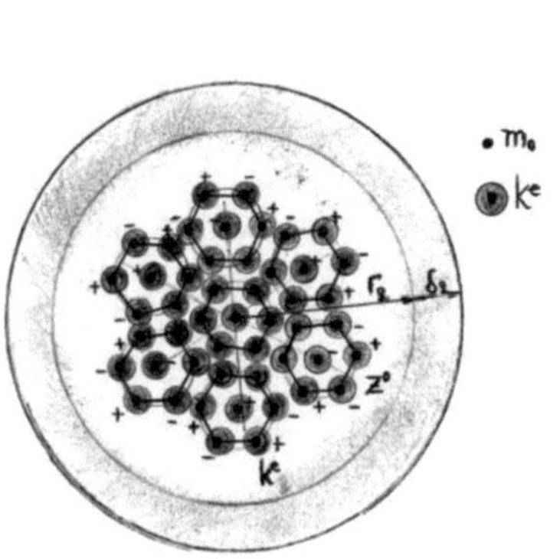

Figure 4.
Preonic z_π -layer of quarcic kerneloid.

liberty ($l_v \approx 0.1 \div 0.2$ fm) of the z^0-preos inside the quark's kerneloid, equivalent to a repulsive shell δ_q.

The mechanical radius of the nucleonic impenetrable volume υ_i is given by three coupled quarcic kerneloids (**Figure 3**) and it results of value: $r_i^n \approx 2r_q + \delta_q \approx 0.45 \div 0.5$ fm –close to the value of l_q, $(r_i^n/l_q \to 1)$ –in good accordance with the experiments of electrons and protons scattering to nucleons, [21], ($\sim 0.45 \div 0.5$ fm) and with the "bag" model of strong interaction resulted in CGT [20].

The last experimentally determined value for the quark's radius: $\sim 4\text{x}10^{-19}$ m [6] corresponds in this case to the radius of the super-dense electronic centroid, [11, 12, 20], being close to the upper limit determined by X- rays scattering on electron [4, 5].

The quarcic cluster's rotation, giving the quantum number l of orbital angular momentum, is generated by the magnetic moment's vortex Γ_μ, according to CGT [11, 12].

6. The existence of new heavy quarks' variants

By re-obtaining the constituent quark $c^{\bullet} \approx 1.55$ GeV of the S.M. by CGT, in the form:

$c^{\bullet} = [(s\bar{s})v) - z^0]$, it results that other composed heavy quarks of type: $q_c = (q_1 \bar{q}_1 q_2)$ may be formed with the semi-heavy quarks: $s^{\pm}$, $v^{\pm}$, and by a de-excitation reaction.

From the de-excitation reactions (12)–(16) we observe also that the emitted preonic boson z_k at the 'hot' forming of compound quarks q_c is smaller for lighter composed quarks and bigger for heavier ones. With the notations:

$\delta_1 = (z_0)$; $\delta_2 = [z_1(3z^0); z_2(4z^0)]$; $\delta_2' = (z_3 = z_\mu = 6z^0)$; $\delta_3 = (z_4 = z_\pi = 7z^0)$; $\delta_3' = (z_5 = 2z_2 = \pi^0)$,

$q_1 = (s; v)$; $q_2 = [c^{\bullet}(1.55\text{GeV}); c(1.7\text{GeV}); c^{\wedge}]$; $(q_3; q_3'; q_3^*; q_3^{*\prime}) = (b^{\bullet}; b; b^*; b^c)$; $q_4 = f_q(b; b^{\bullet})$; $q_t = t(b; b^{\bullet})$, it results as possible the variants ("flavors") of composed quarks presented in **Table 2**.

q^c (compound)	q_2(CGT)	q_3 (CGT)	q_3^* (CGT)	q_3^{**} (CGT)	q_4 (CGT)	$q_t = (t;h)$
$q_1 = (s; v)$	$c^{\bullet}(s\bar{s}v - z_0)$	$b^{\bullet}(c^{\bullet}\bar{c}^{\bullet}c - z_2)$	$b^{\bullet}(c^{\bullet}\bar{c}^{\bullet}c)$	$b^c(c^*\bar{c}^*c^*)$	$f_q(b\bar{b}b)$	t(7x5)b
$q_2' = q^s$	$c(v\bar{v}v - z_0)^s$	$b^{\wedge}(c^{\bullet}\bar{c}c - z_2)$	$b^{\wedge}(c^{\bullet}\bar{c}c)$	$b^{c'}(c\bar{c}^*c^*)$	$f_q(b^{\bullet}\bar{b}^{\bullet}b^{\bullet})$	t(7x5)$b^{\bullet}$
^ - new	$c^{\wedge}(v\bar{v}s - z_0)$	$b^{\wedge}(c^{\bullet}\bar{c}^{\bullet}c^{\bullet} - z_2)$	$b^{\wedge}(c^{\bullet}\bar{c}^{\bullet}c^{\bullet})$	$b^c(\bar{c}^*c^*c^{\wedge})$	$f_q(b^{\bullet}\bar{b}b)$	t(7x5)b^*
	$c^{\wedge}(s\bar{s}s - z_0)$	$b(c\bar{c}c - z_3)$	$b(c\bar{c}c)$	$b^{c'}(\bar{c}cc^*)$	$f_q(b^{\bullet}\bar{b}^{\bullet}b)$	h'(7x5)c
$m(q^c_k)$ GeV($/c^2$)	1.557 (SM) $(1.70)^{s+}$ $(1.704)^{s-}$1.631 1.483	4.744 (SM) 4.887 4.601 $(5.008)^- = b^{s-}$ $(4.996)^+ = b^{s+}$ ($c \approx 1.7$) ($c^{\bullet} = 1.557$)	4.814 4.957 4.671 $(5.1)^+$ $(5.112)^-$	5.154;5.16 5.144;5.13 5.084;; 5.014;	15.000 14.232 14.744 14.488 15-z_5= =14.861	175 $(166)^{\bullet}$ $(180.4)^{*-}$ $(178)^{*+}$ h'(59.5) t'($h'\bar{h}'h'$) =178.5
$\delta(z_k)$ (emitted)	$\delta_1 = z_0$	δ_2 (z_2; z_3) ($z_3 = z_\mu$)			δ_3(z_4; z_5)	?

Table 2.
Possible variants of compound heavy quarks (theoretical masses).

The used notations allow a general formula for the compound heavy quark forming, in the form of a weak de-excitation reaction (leptonic or non-leptonic):

$$q^c_{k+1} = q^0_{k+1}(q_k \overline{q}_k q_k)' - \delta_k; k = 1; 2; 3; m(\delta_k) \le m(\pi^0); m(\delta_k)/m(q^0_{k+1}) \le 2\% \quad (24)$$

which do not depends on the quarks' spins, in concordance with the experimentally resulted conclusion that the quarks' spins have a low contribution to the particle's spin, (4 ÷ 24%, [34]). Eq. (24) differs from the Sakharov's equation [30] which- at the total mass of quarks, adds a term of spin–spin interaction of the quarks.

The quantity δ_k can be experimentally observed as gamma-quantum with energy $E_\gamma \ge 17$ MeV corresponding to preonic bosons: z^0, $2z^0$, z_1, z_2, or –respective, as μ^0-, π^0-mesons, the value of δ_k being proportional to the interaction energy, ($\delta_k \sim E_i$), the reaction (24) corresponding to a weak interaction of leptonic or non-leptonic type.

Also, the similitude with the semi-light quarks forming by bosonic preons (**Figure 2**), which explains the t-quark's mass, indicates as possible also meta-stable states:

q_{xc} = (7xn)m($q^\pm$) with n = 3÷6 kerneloids of $q^\pm$ – quarks with m($q^\pm$) - effective mass, the forms of type: q_{hc} = (7x3)m($q^\pm$) or of type: q_{hc} = (7x5)m($q^\pm$) = [17($q\overline{q}$) + $q^\pm$] being compound quarks and the forms with n = 4, 6, being pseudo-quarks, (neutral clusters).

Because- in the case of a t-quark, the de-excitation of at least the centrally contained b_t –quarks' kernels is impeded by the others, we can take a value m (b_t) ≈ ($c\overline{c}$ c –z_2) ≈ 5.03 GeV which gives: m_t ≈ (7x5)m(b_t) = 176 GeV/c^2, (closer to the known value: 177 GeV/c^2).

Heavier quarks (q_4, q_t) can also result according to eq. (24) or with the kernel formed as cluster of (7x5) kerneloids of s-, v-, c- or f_q-quarks, (ex.: h'('hark') =17($c\overline{c}$)+ $c^\pm$, m(h') = 59.5 GeV), or in the form: [3($q_1 \overline{q}_1$)+ $q_2^\pm$] with m(q_2) < m(q_1), (ex.: 3 ($b\overline{b}$)+ $c^\pm$ = 31.7 GeV).

A variant of t-quark results also in the form: t' = ($h'\overline{h}'h'$) = 178.5 GeV.

In **Table 2** are presented the mass variants of compound heavy quarks resulting by eq. (24). The notations q^*, q (q^s) and $q^\bullet$ indicate cold quark, 'hot' meta-stable quark and respective-de-excited stable quark, (of S.M. type, resulted from a meta-stable quark, of Souza type).

7. The explaining of the heavy particles' forming

Eq. (24) results as specific also to heavy particles' 'hot' forming from de- excited heavy quarks, such as the in the case of the heavy pseudo-scalar or vector mesons (**Tables 3** and **4** - Annex A and B) and of the baryons with J^P $^3/_2$,, (**Table 6**), by a generalization of the form:

$$M_b{}^d = M_b{}^0\left(q_i q_j q_k\right)' - m(\delta_Q); m(\delta_Q) \le m(\pi^0), m(\delta_Q)/M_b{}^0 \le 2.7\%. \quad (25)$$

m(δ_Q) = Σm(z_k) being given as sum of z_k -bosons emitted by the most excited q - quarks.

7.1 The heavy baryons

The most part of the heavy baryons with $J^{P½}$ (**Table 5**, Annex C) results with discrepancy of less than 1% as ground states ()' of (q_i-q_j-q_k)' combinations of de-excited quarks, without de-excitation reaction of the quarcic cluster, excepting some excited baryons with $J^{P\frac{1}{2}}$ as those having u- or d- quark(s) in their structure (considered in the p-; n- variant of CGT), for which the de-excited state $()^d$ is obtained by emission of at most (1÷2) z^0-preons, with the exception of the double charmed Xi, whose de-excited state is obtained by emission of a z_2 bosonic preon, (the case of Ξ_{cc}^{++}(3621), resulted as [(pcc) - z_2]).

This fact indicates that by their magnetic attraction, the quarks with anti-parallel spins give an upper stability to the formed particle. The possibility to obtain the mass of the majority of baryons with $J^P = \frac{1}{2}$ as simple sum of masses of component de-excited quarks indicates that the lost quantity δ_k is explained rather by eq. (24), as released preonic boson(s), than by eq. (19)- as matter corresponding to the binding energy.

The heavy baryons with $m_q c^2 > 1$ GeV and $J^{P3/2}$ (**Table 6**, Annex D), result in the model as de-excited states $(q\text{-}q'\text{-}q'')^d$ obtained by the emission of z^0- or z_k- "zerons" (k = 1÷4), $\Sigma m(z_k)$ representing less than 2.7% of its mass, (excepting Ξ^{*0}(1532), resulted as: [(λ^+vv) –z_1], for which $m(z_1) \approx 3.4\%$), being explained similarly also the recently discovered particles:

Σ_b^-(6.097); X_b^-(6.100) [37], as: $(6.096)^d[(vvb)'\text{-}z_1]$; $(6.113)^d[(vvb)'\text{-}2z^0]$ and: $Z_c(3900)$; $X_c(3872)$ as: $[(v^-c^+c^+)'\text{–}z_2]$; $[(v^-c^+c^+)'\text{ -}z_3](3870)^d$.

The higher instability of the non-de-excited baryons with $J^{P3/2}$ may be explained by the conclusion that the rotation of the quarcic cluster –giving the orbital angular momentum l, determines a lower magnetic attraction and a higher vibration of the current quarks.

The most de-excited states of heavy baryons with $J^{P3/2}$ are obtained by emission of a "zeron" $z_1 = 3z^0$, i.e.- of a z^o- preon per each q-quark. A good argument for the proposed model is also the fact that- by eq. (24) applied for heavy $J^{3/2}$- baryons, it can be explained also the particles: $Z_c(3900)$; $X_c(3872)$, (considered as tetraquarks) and: $B^*_j(5732)$ (considered as meson [35, 36]), as forms: $Z_c'(v^-\ c^+\ c^+)'$– $z_2 = (ccs)^d = 3904.5$ MeV/c^2; $X_c'(v^-\ c^+\ c^+)'$ –z_3 = 3869.7 MeV/c^2, and: B^*_j = 5748(b λ n) –z^0 = 5731 MeV/c^2 (discrepancy < 0.2%). Small differences, under 17 MeV, between baryon resonances of the same family, such as between Δ(1900), Δ(1905), Δ(1910), Δ(1920), Δ(1930), may be explained by partial destroying of z^0- preons at relative high interaction energies, with the loosing of some "gammonic" pairs of electronic kerneloids from its structure. An argument can be the fact that- at the c-quark's 'hot' transforming, are emitted more electrons than muons, [38].

Also, heavy neutral bosons, quasi-stable at T → 0 K, can result in form of clusters with mass: $M_Z = \sum_K(n_1.z_\pi + n_2.z_2)$; $(n_{1;2} \leq 5)$, conform to CGT.

7.2 The heavy mesons

The theoretically resulted masses of heavy mesons are presented in **Tables 3** and **4**, which show that these mesons may be explained as pairs $(q_k^-q_l)$ by considering the both variants of the charm and of the bottom quarks: c^s = c and $c^\bullet$; b^s = b and $b^\bullet$, (the Souza's and the S.M.'s mass variants), the pair of heavier quarks being de-excited by the emission of a preonic 'zeron' z_k, in accordance with eq. (25) but considered for $M^d(q_k\,\bar{q}_l)$.

For example, differences between vector mesons, such as: ω^0(0.7826 GeV) and ϕ^0(1.0195GeV), (difference: ~ 464 $m_e \approx 2\ z_\pi = 476\ m_e$) can be explained by the structure: $\omega^0((d\bar{s}) - z^0) = 0.795$ GeV and $\phi^0((s\bar{v}) - z_1(3z^0)) = 1.022$ GeV.

The discovered particle X(5568), with un-clarified structure (supposed to be a tetra-quark [36]) results by the model as charged tetra-quark $X^{\pm}[b\overline{\lambda}^{\pm} m^{\pm}\overline{m}^{\pm})]$ or as heavy meson: $X_b(v\overline{b})$ = 5574 MeV, which decays by a reaction of type: X^{+} $[b\,\overline{\lambda}(m_1\,\overline{m}_2)] \rightarrow B_s^{0'}(b\overline{\lambda}) + \pi^{+}$ or: X^{-} $[b\,\overline{\lambda}\,(m_2\,\overline{m}_1)] \rightarrow B_s^{0'}(b\overline{\lambda}) + \pi^{-}$, respective – by a reaction: $X_b^{\pm}(v\overline{b}) \rightarrow B_s^{0'}(\lambda\overline{b}) + \pi^{0}(2z_2)$, resulted in CGT by the v-quark's transforming: $v^{\pm} \rightarrow \lambda^{\pm} + 2z_2$, (**Table 1**).

It results also that neutral W-, Z-bosons can be explained as neutral clusters, in the form: $W^0 = (2x2^3)b \approx 80$ GeV; $Z^0 = 2x3^2b \approx 90$ GeV (doublets of cubic or squared clusters of b-quarks), the cubic form: $3^3b = 27b = 135$ GeV resulting as possible heavy quark.

Also, a neutral variant of the recently obtained new value for the t-quark's mass: 199 GeV/c^2 [39] may be explained as a pseudo-quark with regular hexagonal polyhedron form of its kerneloidic cluster, in the form: $m(t^{\bullet}) = (7x6)m(b^{\bullet})$, ($b^{\bullet} \approx 4.75$ GeV).

The resulted meson $(h'\overline{h}')$ = 119GeV has a mass close to those of the boson H^0 (125.7GeV).

The heavy τ- particle (∼1.77GeV), considered as heavy lepton in S.M., with mass resulted from the empirical Koide's formula, is obtained as exotic meson: $\tau(m_{1,2}\overline{c}^{\pm})$ = (1.7695 GeV).

7.3 The multi-quark particles

Some particles evidenced with a multi-quark type structure may be also explained similarly, by de-excitation reactions, specific to a reaction of weak type, (Annex E, **Table 7**).

$\Theta^{+}(1540)$ (considered pentaquark $(uudd\overline{s})$) results as: $\Theta^{+}[(p\overline{n}\overline{s}^{\bullet}\overline{\lambda}]$ by: $s \rightarrow s^{\bullet} + z^0$.

The determined masses of relative recently discovered 'charmed' particles (july 2022, [40]) of penta-quark and tetra-quark type: -P_{Ψ}^{Λ}(4338 MeV/c^2) /$(sucd\overline{c})$ and -$T^{a}_{cs}(2900)^{++}/(\overline{s}c\overline{u}d)$ with a neutral variant: $T^{a}_{cs}(2900)^{0}/(\overline{s}c\overline{d}u)$, whose constituent quarks considered by S.M. give by the sum' rule: 4262 MeV/c^2, respective: 2712 MeV/c^2, (discrepancy higher than 1.7%, respective 6%), can be explained by the sum' rule with the mass/structure: $P_{\Psi}^{\Lambda}(ducd\overline{c})$ = (2x313 + 312 + 2x1700) = 4338 MeV/c^2 and:

$T^{a}_{cs}(\overline{v}c\overline{d}u)$ = (574 + 1700 + 312 + 313) = 2899 MeV/c^2, (with discrepancy ≈ 0, in CGT).

Also, it result as possible octo-quark particles in cubic form, as: $\frac{1}{2}W^0 = (2^3)b^{\pm} \approx 40$ GeV.

8. Concordance with experiments

The main experiments that sustain indirectly the used explanatory model of CGT, are:

i. the transforming of gamma quantum of 1 MeV into a pair ($e^{+}e^{-}$) by an electrostatic energy $E_e \geq 2m_e c^2$ and the experimentally obtaining of "gammons" in the form of quanta of "un-matter" plasma, [19], (which argue the model of gamma quantum of 1 MeV as pair of degenerate electrons and the possibility of particles' forming from "gammons");

ii. the experimentally obtaining of a Bose-Einstein condensate of photons, (a "super-photon"), by a German team (2010, [31]), indirectly proving the existence of the rest mass of photons (considered in CGT), as in the case of the 'dark photon' theorized in the Q. M.);

iii. the "stopped light" experiment [41], that evidenced a ultra-slow passing of a photon through a Bose-Einstein condensate, and the field-like nature of the "dark" energy evidenced in astrophysics [42] (which sustains the vortical photon/electron model of CGT);

iv. the decay: $t \rightarrow W + j(b^- b)$, [43], (evidencing that b-quarks are components of t-quark);

v. the experimentally obtaining of W, Z-bosons and $q - \overline{q}$ pairs by (e^- -e^+) or (e^--p^+)-interactions and the charm-quark decaying into muons and electrons [36], (which sustain the possibility of particles' forming as clusters of pairs of degenerate electrons).

vi. the same size order for the experimentally determined radius' values of the scattering center(s) inside the electrons and inside the nucleons, (electronic dense centroids, in CGT).

9. Conclusions

- The possibility to explain the mass spectrum of a large number of particles by preonic compound quarks, by the approximation of the sum rule, with the heavy quarks: c, b, formed as triplets: $q_n = (q_k \overline{q}_k q_k)_{n-1}$ or $q_n = (q_k \overline{q}_k q_l)_{n-1}$ of quarks q_k, q_l with adjacent lower mass: $v^{\pm}$(573 MeV) or/and $s^{\pm}$($\sim$500 MeV) resulted in CGT (as clusters of z^0–preons) and with participation of the (semi)light quark $\lambda^{\pm}$ (435 MeV) of CGT, sustains indirectly the particles' model of CGT and the possibility of a cold genesis scenario in conditions of B-E condensate' pre-clusters forming, ($T \rightarrow 0$ K; $E_k = p^2/2m_q \rightarrow \mu_q$), and argues the conclusion that the entire mass of the particles comes from constituent quarks. Also, the resulting model indicates the cold forming of a quasi-equal number of negatrons and positrons in the Proto-universe's period (before Big-Bang), by chiral quantum fluctuations, (vortices).

  In this case, in CGT, the genesis of z^0-preon (X17-boson) from degenerate electrons results as being fundamental in the particles' natural explaining, (as the Higgs boson in the S.M).

  Also, by the fact that the cold quarks result as superpositions of preonic bosons $z_2 = 4z^0$ and $z_\pi = 7z^0$ with almost the same symmetry, the resulting model explains naturally not only the fact that quarks can be transferred in strong interaction from a particle to its interaction partner with the same constituent mass (carrying only its bosonic shell), but also the possibility of paired quarks forming from relativist jets of negatrons and positrons, in the sense that the explanation of the S.M. for this phenomenon, supposing the possibility of a quarks pair forming from a pair of relativist electrons with increased mass via an intermediary virtual photon is identified in CGT as a formal explanation, based on a relation that depends on the reference frame of the observer and not to a microphysical phenomenon, the Einsteinian speed-depending mass increasing resulting as apparent in CGT, (being explained as apparent effect generated by a real decreasing of the values of the electric field, $E_L \sim \gamma_c^{-3}$; $E_T \sim \gamma_c^{-1}$, ($\gamma_c = 1/\sqrt{(1 - v^2/c^2)}$), and of the magnetic field: $B \sim \gamma_c^{-1}$, [44]).

- The fact that the quarks: 'sark', 'chark','bark', ('strange', 'charm', 'bottom'- in S. M.) were experimentally evidenced in the mass variant of S.M. can be explained by the conclusion that they result from the Souza's/CGT's variant by loosing of z^0-, π^0-, $z_6(2z_\pi)$- boson.

- The CGT's model do not exclude the possibility of small additions (< 2%) of gammons and z^0-preons to the mass of quarks also at some excited heavy particles as in the case of $Z_b(10650)$ [38], ($Z^*[((u\bar{u}\bar{b}^- b^+)^+ + 2z^0] = (10624)^d + 34 = 10{,}658$ MeV) -relative similar to the m_1-quark or to the proton's case- which acquiring a negatron and a linking 'gammon' σ_e (γ^*) becomes neutron, in CGT, (the limit 2% being given by: $(m_2^- - m_1^+)/m_2 \approx 2.6/137$, explained by eq. (5), by the mass of the "weson" $w^- = \sigma_e(\gamma^*) + e^- = 2.6\ m_e$).

- It results also that - if the z^0-, z_k- preonic neutral bosons or/and pseudo-quarks q^0 (theoretically considered in CGT) are in a cold state ($T \to 0$ K), being neutral particles they interact weakly with photons of electromagnetic radiation and they can explain in this case a part of the Universe's dark matter.

- The main CGT's link with the Standard Model consists in the fact that- considering the electron's mass confined in its kernel, the z^0-preon (identified as being the experimentally evidenced X17-boson) can be considered as formed by 3 pairs ($u_c\,\bar{u}_c$) of current u_c -quarks formed by a quasielectron ($e^{*\pm}$) surrounded by 3 gluonic 'gammons', ('gluols'- in CGT), i.e. with mass: $m(u_c) = 7m_e^* \approx 2.9$ MeV/c^2, the current d-quark resulting with a mass:

  $m(d_c) = m(u_c) + w^-(e^-\gamma^*) = (2.9 + 2.62)$ MeV $= 5.52$ MeV, (in the limits accepted by S.M.), with the difference that in CGT the radius $r_0 \leq 10^{-18}$ m is the radius of the electron's kernel.

  Also, the degenerate 'gammon' $\gamma^*(0.827$ MeV) can be considered a 'gluol' which is the CGT's equivalent of the 'gluon' considered in the S.M. with a mass' upper limit: 1.3 MeV, but in the CGT's model of quark the cluster of 'gluols' generates a saturated state, as a "glasma" in the S.M.

  The u_c –quarks forming by the gluol's splitting is explainable by the attraction between the resulted 'free' quasielectrons and adjacent 'gluols', with the forming of a ($u_c\,\bar{u}_c$)-pair.

  Another similitude consists in the fact that the 'weson' $w^\pm$ of CGT and the $W^\pm$-boson of SM, even if they have different nature, they transform similarly, ($w^\pm \to e^\pm + \nu_e(\bar{\nu}_e) + \in$), the second generation variant, ($\to \mu^\pm + \nu_\mu(\bar{\nu}_\mu)$), being specific to the transforming of the couple: ($m_{1,2}z_2$) or ($^-m_{1,2}z_2$) incorporated in the preonic cluster of a semi-light quark, conform CGT.

- Regarding the generation model (G.M.), the previous observation explains the difference of generation number (g) between (e, ν_e)- and (μ + ν_μ)- leptons and between (u, d)- and (s, c)-quarks, as consequence of the fact that the first leptonic doublet results by the transforming of a current quark (u_c, d_c), (i.e.- leptonic) and the second leptonic doublet results by the transforming of a constituent quark, (i.e. -non-leptonic). But in CGT the conserving of Σg indicates only the low probability: $P(g' \neq g) \ll 1$ of quark's transforming of type: $q_1^{g} \leftrightarrow q_2^{g'} + \Sigma z_k$ with $g' \neq g$, because- in these weak interaction reactions, the strangeness' conserving is not mandatory, they being allowed by the conservation of Σp.

  Also, the sum rule used in CGT corresponds to the conservation of the particle number, p, (without the use of the rishons model), which explains also the form (21) of the strong interaction process by the conclusion that the neutral bosons γ^*, z_k, (*gammons*, '*zerons*') have p = 0.

  So, it results that the G.M. is compatible with CGT's model as specific formalism, the natural restrictions of the CGT's model being based on the total charge and mass-energy conservation rule, (with mass→energy conversion but without the relativist speed-depending mass increasing), which explains also the selection rule for the quarks' masses.

  It also results the possibility to extend naturally the list of quarks, with the advantage of the explaining of the particles' mass spectrum by the sum rule, by de-excitation reactions.

A. Annex

Heavy mesons (MeV/c²) -experimental mass- [35]	**Theoretic mass, (de-excited meson, CGT)^d, MeV/c²; (p; n) ≈ 312; 313**	**Observations / Predictions, (MeV/c²)**
η' (957.6); f^0(980) $^{1/}{}_3((u\bar{u}) + (d\bar{d}) + (s\bar{s}))$	$(965)^d = \eta''(1000)' - 2z^0$; ( )' -excited $(983)^d = \eta''(1000)' - z^0$;	$\eta''((s+\bar{s})' = 1000$ $(s+\bar{\lambda}) = (935)'$
η_c(2980.3) $(c\bar{c})$; D^*_{sJ}(2860) η_b (9300) $(b\bar{b})$	2992.5: $(c^\bullet\bar{c}^\bullet)^d = (c^\bullet\bar{c}^\bullet)' - z_\pi(7z^0)$ 9338.4: $\left(b^\bullet\overline{b^\bullet}\right)^d = \left(b^\bullet\overline{b^\bullet}\right)' - z_\pi$	$[(c\bar{c})' - z_\pi] \approx 3278$; $(c \to c^\bullet)$ $\left(b^\bullet\overline{b^\bullet}\right) = 9460$; $(b \to b^\bullet + z_6)$
D^+;D^0(~1865); $(c\bar{d})$; $(c\bar{u})$	$(1863)^d = (c^\bullet\bar{n})^d = (c\bar{n})' - \pi^0$; $(c \to c^\bullet)$	$[c(v\bar{v}v^\pm)\bar{n}]^* \approx 2031$
X(5568) (?)^x [36]	5574 $(v\bar{b})^d \to B_s(\lambda\bar{b})' + \pi^0$; $(v \to \lambda + \pi^0)$	(x-un-clear structure)
D_s^+(1968.4) $(c\bar{s})$	1968 $(c^\bullet\bar{\lambda})^d = (c^\bullet\bar{\lambda})' - z^0$	$[c(v\bar{v}v^\pm)\bar{s}]^* \approx 2218$

Heavy mesons (MeV/c²) -experimental mass- [35]	Theoretic mass, (de-excited meson, CGT)^d, MeV/c²; (p; n) ≈ 312; 313	Observations / Predictions, (MeV/c²)
$B^+;B^0(\sim 5279)$ $(u\bar{b})$; $(d\bar{b})$	$5278\ (p\bar{b})^d = (p\bar{b})' - z_1(3z^0)$	$(p\bar{b}^{\bullet})$; $(n\bar{b}^{\bullet}) \approx 5042$
$B_s{}^0(5366.3)$, $(s\bar{b})$ $B_s{}^{0'}(5372)$; $(\Lambda_b(5425)^r$	$(5365.5)^d = (\lambda\bar{b})' - z_2(4z^0)$ $B_s(\lambda\bar{b})' = 5435$; $(5382)^d = B_s(\lambda\bar{b})' - z_1$	$(s\bar{b})' - z_2 = 5430.5$ (Λ_b) $(s\bar{b}^{\bullet})' - z_2 = 5244$
$B_c{}^+(6276\pm 4)$ $(c\bar{b})$	$(6297)^d = \left(c^{\bullet}\bar{b}^{\bullet}\right) = (c\bar{b}^{\bullet})' - \pi^0(2z_2)$	$(c\bar{b}^{\bullet})' = 6440$; $(c\bar{b}) = 6700$

Table 3.
Theoretic masses of heavy pseudo-scalar mesons, (CGT).

B. Annex

Heavy mesons (MeV/c²) -experimental mass- [35]	Theoretic mass, (de-excited meson, CGT)^d, MeV/c²;	Observations + Predictions (MeV/c²)
$\rho^+(775)$; ρ^0 (775.26); $\rho^-(775)$; $(u\bar{d})$; $(d\bar{d})$; $(d\bar{u})$	$(777.3)^d = (p\bar{s})' - 2z^0$ $(778.3)^d = (n\bar{s})' - 2z^0$	$z^0 = 17.374$ MeV/c² $z_2 = 69.5$; $z_3 = z_\mu = 6z^0$
$\omega(782.65)$; $(u\bar{u} + d\bar{d})/\sqrt{2}$	$(795.6)^d = (n\bar{s})' - z^0$	$813 = (n\bar{s})$
ϕ (1019.46); $(n\bar{s})$	$(1022)^d = (s\bar{v})' - z_1(3z^0)$	$(1009.5) = (s\bar{s})$
J/ψ (3096.9); $\Theta^0{}_c(3.099)$; $(c\bar{c})$	$(3096.7)^d = (c^{\bullet}\bar{c}^{\bullet})' - z^0$	$m(c^{\bullet}) = 1557$ (CGT)
$\Upsilon(1S)$ (9460.3); $(b\bar{b})$	$(9462.6)^d = (b^{\bullet}\bar{b}^{\bullet})' - z^0$	$m(b^{\bullet}) = 4744$ (CGT)
$K^{*+}(891.6)$ $(u\bar{s})$; $K^{*0}(895.8)(d\bar{s})$	$(\sim 900)^d = (s\bar{\lambda}^{\pm})' - 2z^0$	870; 871, $(\lambda\bar{\lambda}^{\pm})'$
$D^{*+}(2010.3)(c\bar{d})$ $D^{*0}(2007)(c\bar{u})$;	$2012(c\bar{n})'$; $2011(c\bar{p})'$	( )'-excited state
$D^{*+}{}_s$ (2112.1) $(c\bar{s})$	$(2117.6)^d = (c\bar{\lambda})' - z^0$	$m(c) \approx 1700$
B^{*+}; $B^{*0}(5325.2)$ $(u\bar{b})$; $(d\bar{b})$	$(\sim 5331)^d = (\lambda^{\pm}\bar{b})' - z_3$	$5312(p\bar{b})$; $5313(n\bar{b})$
$B^{*0}{}_s(5415.4)$ $(s\bar{b})$	$(5417.6)^d = (\lambda\bar{b})' - z^0$	$m(b) \approx 5000$
$B^{*+}{}_c$ (unknown); $(c\bar{b})$	$6700\ (c\bar{b})'$; $6557(c^{\bullet}\bar{b})'$	$(c^{\bullet}\bar{b}^{\bullet})' = 6300$

Table 4.
Theoretic masses of heavy vector mesons, (CGT).

C. Annex

Baryons, J^P ½ experimental mass (GeV/c^2), [29, 36, 37], CDF collab.2009*	**Theoretic mass, (cold baryon, CGT)* p^*; n^*(~0.312;); λ^{-*}(0.435); s^{-*}(0.504); v^{-*}(0.574); c^{*+}(1.718); b^{*-}(5.166); b^{*+}(5.154)**	**Ground st./De-excited baryon ()-ground st.; $()^d$ -de-exc.(theor.) s^-(0.504); $c^{\pm}$(~1.7); $b^{\pm}$(~5.0) -predicted baryons $()^p$; (GeV)**
N (0.938÷0.939); (udd)	~0.939; (ppn); (pnn),	$(^r)$ = recent dis.; ()'-excited st.
-$\Delta^{(++;+;0;-)}$ (1.232)	~1.25; ($s^{\pm} + \lambda^{\pm} + p^+(n^-)$)*	(1.25)' – z^0 = $(1.233)^d$; ($s \rightarrow s^*$)
-Λ^0 (1.116); Λ^{0*}(~1.6) (uds)	~1.13 (n + p + s)*; 1.583 (v + 2 s)*	(1.13)'-z^0 = 1.113; 1.583(v + 2 s)'
-Σ^+; Σ^-; Σ^0 (1.189; 1.192; 1.197) (uus; uds; dds)	~1.199; ~1.2; (v + 2p)*; (v + p + n)*; (v + 2n)*	(1.199)'; (1.2) -discrepancy (CGT): 0.25%
- K^{0*}(1.522); Ξ^{-*} (1.535) (dss)	1.514 (v + s + λ)*; (s s s)* = 1.51	- (1.51);
-Λ_c^+ (2.286) (udc) -D_{s0} $(2317)^r$ [36]	2.343 (pnc)* = (pnc*); (m_c^* = $3m_v^*$)	-(2.325)' (≈ m(D_{s0}); (pnc)' - z^0 = $(2.308)^d$; d.0.4%
Λ_c (2627)	2.653 (λsc)*	(λsc) = 2.635; –discr. 0.3%
-Λ_b^+(5.619)(udb); $\Lambda_b^0(6.071)^r$ [37]	5.791(pnb)*;(nnb)*; 6.228 (svb$^+$)*	(5.625); –dis. 0.1%; (6.074)
-Σ_c^{++}(2.454)(uuc); Σ_c^+(2.453) (udc)	2.465 (pλ^+c)*; 2.466 (pλ^-c)*	~(2.448); –discrep. 0.3%
-Σ_c^0(2.4537) (ddc)	2.467 (nλ^-c)*;	(2.449); –discrep. 0.3%
-Σ_b^+(5.811) (uub)	5.79 (ppb)*; 5.913 (pλ^+b)*	(ps^+b) ≈ 5.812; –dis. 0.05%
-Σ_b^0(5813.5) (udb)	5.791(pnb)*; 5.914 (pλ^-b)*; (ps^-b)*	(ps^-b) =5.813; $(pnb)^d$ = 5.62
-Σ_b^-(5.815) (ddb)	5.79(nnb)*;5.91(nλ^-b)*; 5.96 (nsb)*	(ns^-b) =5.814; –dis. 0.12%
-Ξ_c^+(2.467) (usc) - Ξ_c^0 (~2.469) (dsc)	2.526 (psc)* 2.527 (nsc)*	(nsc)' = (2.513)'; (nsc)'–2z^0 = $(2.479)^d$;(dis.0.5%)
-$\Xi_c^{'+}$ (2.575); (usc); (') = "prime"	2.604 (pvc)*	(2.586); –discrep. 0.4%
$\Xi_c^{'0}$ (2.578); (dsc); Λ_c(2593)	2.605 (nvc)*	(2.587); – discrep. 0.35%
-Ξ_{cc}^{++} (3.621); (3.627); (ucc); Λ(3621); ψ'(3686)	3.748 (pcc)*; ($s^{\pm}c^*c^*$)* = 3.648 (pcc)' – z^0 = $(3694,6)^d$	(3.712)'; (3.618)($s^{\pm}c^*c^*$) (vc^*c^*)'–z_2 = $(3.618)^d$; (dis. ≈ 0)
Ξ_{cc}^+ (unknown) (dcc)	3.749 (ncc)*	(3.713)';(ncc)'– z_2 = $(3.643)^d$
Ξ_b^0 (5.788) (usb)	(psb)* = 5.978;	(5.812)' –z^0; (pvb) = 5.886
Ξ_b^0 $(5.792)^0$; Ξ_b^- $(5.796)^-$ (dsb)	(n$sb^{\pm}$)* ≈ 5.979;	$\Xi_b(5.796)^d$ ≈ Ξ_b^*(5.813)'–z^0
$\Xi_b^{0'}$ (usb); $\Xi_b^{-'}$(dsb); (unknown)	[(p;n)vb]* ≈ 6.053; 5.914[(p; n)λb]*	[(p;n)vb) ≈ (5.887)'; (5.748)'
$\Xi_b^-$$(5.935)^r$(ssb);$\Lambda_b'$$(5.92)^r$(usb) [37]	(λsb)* = 6.1	(5.935)' (λsb)'; –discr. → 0
Ξ_{bb}^0; Ξ_{bb}^- (unknown); (ubb), (dbb)	10.644 (pbb)*; 10.645 (nbb)*	(10.312)'
Ξ_{cb}^+; Ξ_{cb}^0 (unknown);(ucb), (dcb)	7.196 (pcb)*; 7.197 (ncb)*	(7.012)'; (7.013)'

Baryons, J^P ½ experimental mass (GeV/c^2), [29, 36, 37], CDF collab.2009*	**Theoretic mass, (cold baryon, CGT)* p^*; n^*(~0.312;); λ^{-*}(0.435); s^{-*}(0.504); v^{-*}(0.574); c^{*+}(1.718); b^{*-}(5.166); b^{*+}(5.154)**	**Ground st./De-excited baryon ()-ground st.; $()^d$ -de-exc.(theor.) s^-(0.504); $c^{\pm}$(~1.7); $b^{\pm}$(~5.0) -predicted baryons $()^P$; (GeV)**
$\Xi_{cb}^{+'}$; $\Xi_{cb}^{0'}$ (unknown); (ucb), (dcb)	7.317 $(\lambda^+cb)^*$; 7.319 $(\lambda^-cb)^*$	(7.135)'; (7.136)'
-Ω_c^0 (2.695); (2.698) (ssc^+)	2.718 $(ssc)^*$;	(2.7)(ssc); (λλc) =2.57
Ω_b^{*-}(6.054),(ssb); $\Lambda_b^{0**}(6.072)^r$, (*)	$(svb)^*$ = 6.240; $(ssb)^*$ = 6.166;	(6.074)'(sv^-b^-)'- z^0 = $(6.056)^d$
-Ω_{cc}^+(scc); Z_c(3900); $\psi_2(3823)^r$	3.936 $(scc)^*$; $(\lambda cc)^*$ = 3.871	Z_c(scc)' = 3.9; (λcc)' = 3.835
-Ω_{cb}^0 (unknown) (scb)	7.384 $(scb)^*$; 7.458 $(vcb)^*$	(7.2)'; (7.274)'
-Ω_{bb}^- (unknown) (sbb)	10.832 $(sbb)^*$; $(vbb)^*$ = 10.906	(10.5)'; (vbb)' = (10.574)'
-Ω_{ccb}^+; $-\Omega_{cbb}^0$,(ccb); (cbb), (unkn.)	8.602 $(ccb)^*$; 12.046 $(cbb)^*$	(8.4)'; (11.7)'
-$\Lambda_b'(6.146)^r$; $(6.152)^r$ (ssb)	6.302 $(vvb)^*$;	(vvb)' = (6.148)';
$D_3^{*0}(2.760)^r$;$\Omega_c(2.768)^r$; B^*_j(5.732)	2.796 $(v^- s^- c^+)$*; 5.902(b^* λ n)	(2.774); (5.748)'- z^0 = $(5.730)^d$
-A_c^0; A_c^- (unknown - predicted)^P	$(\lambda vc^+)^*$ = 2.727; $(\lambda vc^-)^*$ = 2.73;	(λvc)' ≈ (2.7)'; $(2.71)^P$
-$\Lambda_c(2.86)^*$); (Ω_c^*)?; (Ω_{cc}^*)?	$(vvc)^*$ = 2.866; $(vcc)^*$ = 4.01	$(2.85)^P$(vvc)'; $(3.974)^P$(vcc)'
-$(\Omega_b^*)?^-$ (unknown- predicted)	6.036 $(\lambda\lambda b)^*$; 6.175 (λvb)*	$(5.87)^P$(λλb)'; $(6.01)^P$(λvb)'
? R_{hcc}; R_{hcb} (unknown-predicted)	$(h'c\ c)^*$ = 62.936; $(h'c\ b)^*$ = 66.384	$(62.9)^P$; $(66.2)^P$
? R_{hhc}; R_{hhb} (unknown-predicted)	$(h'h'c)^*$ = 120.7; $(h'h'b)^*$ = 124.16	(h'h'c)' = $(120.7)^P$; $(124.)^P$

Table 5.
Theoretic masses of de-excited ("hot" formed) baryons, J^P ½, (CGT).

D. Annex

Baryons experimental mass (MeV), [29] J^P $^3/_2$	**Theor. mass,, GeV (de-excited quarks)': λ^-(0.435) u; d = p; n (0.312); s(0.5); v^-(0.574); c (1.7); b (~5.0)**	**Theoretic mass, (GeV) of de-excited baryon $()^d$ + predicted baryons- $z_1(3z^0)$; $z_2(4z^0)$; $z_3(6z^0)$; ()'-excited state**
-N^{0*}; Λ^{0*} (1520); udd; sdu	(v + s + λ)' ≈ $(s\ s\ s)^*$ = 1.514	(1.514)
-Λ_b^0 (5912); $-\Lambda_b^0$ (5920); (?);	(λ s b)' = 5.935	(λ s b)'- z^0 ≈ 5.917
Σ_s^{*-}(1385) dds; Σ_c^{*++}(2518) uuc	(v s p)' = 1.390; $(ps^+c)'$ = 2.511	1.390; 2.511
Σ^{*+}_c(2517.5); Σ^{*0}_c(2518.8)	$(ps^-c)'$ = 2.512; $(ns^-c)'$ = 2.513	2.512; 2.513
Σ^{*+}_b (5832.1), uub; Σ^{*0}_b, udb	(pvb)' = 5.886; $(nvb^+)'$ = 5.887	$((p;n)\ v\ b^+)' - z_1$ ≈ $(5.835)^d$
Σ^{*-}_b (5835.1) ddb	$(n\ v\ b^-)'$ = 5.899	$(n\ v\ b^-)' - z_1$ = $(5.847)^d$
Ξ^{*0} (1531.8) uus	$(\lambda^+\ v\ v)'$ = 1.58; $(s^+s\ v)$ = 1.57	$(\lambda^+\ v\ v)' - z_1$ = $(1.531)^d$

Baryons experimental mass (MeV), [29] $J^{P}\ {}^{3}/_{2}$	**Theor. mass,, GeV (de-excited quarks)': $\lambda^{-}(0.435)$ u; d = p; n (0.312); s(0.5); $v^{-}(0.574)$; c (1.7); b (~5.0)**	**Theoretic mass, (GeV) of de-excited baryon $()^{d}$ + predicted baryons- $z_1(3z^0)$; $z_2(4z^0)$; $z_3(6z^0)$; ()'-excited state**
Ξ^{*-} (1535) uds	$(\lambda^{-}\ v\ v)' = 1.584$	$(\lambda^{-}\ v\ v)' - z_1 = (1.532)^{d}$
$\Xi^{*+}_{c}(2645.9)$; $\Xi^{*0}_{c}(2645.9)$,(u/d)sc	$(s^{\pm} + s^{-} + c)' = 2.7$	$(s^{\pm}s^{-}\ c^{+})'-z_1 \approx (2.647; 2.648)^{d}$
$D_j^{*}(2760)$; $D_j(2740)$	$(s^{\pm} + v^{-} + c^{*})' = 2.774$	$(s^{\pm}\ v^{-}c^{*})'-z^{0};2z^{0} \approx 2.76; 2.74$
$\Xi_{cc}(3519)$; $X_{c1}(3510)$ ucc;	$(\lambda^{\pm}\ c^{\bullet}\ c^{\bullet})' = 3.549 \approx X_{c2}(3556)$	$((3.549)' - 2z^{0} = (3.515)^{d}$
Ξ^{*+}_{cc}; Ξ^{*++}_{cc} (unknown), (d/u) cc; ψ (3770) ucc; X(3842.7)	$(\lambda^{\pm} + c + c)' = 3.836$ $(s^{\pm}c\ c)' = 3.900$;	$(\lambda^{\pm}c\ c)' - z_1 = (3.783)^{d}$ $(s^{\pm}\ c\ c)' - z_1 = (3.848)^{d}$
Ξ^{*0}_{b} (5945.5); $\Xi^{*-}_{b}(5955)^{r}$ (u/d)sb	$(s^{-} + s^{-} + b^{\pm})' = 5.996; 6.01$	$(2\ s^{-} + b^{\pm})'-z_1 \approx (5.94; 5.95)^{d}$
Ξ^{*0}_{bb}; Ξ^{*-}_{bb} (unknown) (u/d)bb	$(\lambda^{\pm} + b^{-} + b^{-})' = 10.459$	$(s^{\pm} + 2b^{+})' - z_2 = (10.431)^{d}$
Ξ^{*+}_{cb} (unknown) ucb	$(\lambda^{-} + c^{+} + b^{+})' = 7.135$	$(s^{-}c^{+}b^{+})' - z_1 = (7.147)^{d}$
Ξ^{*0}_{cb} (unknown) dcb	$(\lambda^{-} + c^{-} + b^{+})' = 7.139$	$(s^{-}c^{-}b^{+})' - z_1 = (7.151)^{d}$
$\Omega^{-}(1672.45)$ sss	$(v^{-}\ v^{-}\ v^{-})' = 1.722$	$(v^{-}\ v^{-}\ v^{-})' - z_1 = (1.670)^{d}$
Ω^{*0}_{c} (2766) ssc	$(v^{-} + s^{-} + c)' = 2.774$	discr.0.3% (isomeric.)
Ω^{*-}_{b} (unknown) ssb	$(v^{-} + s^{-} + b^{-})' = 6.074$	$(v^{-}\ s^{-}\ b^{-})'-z_3 = (5.9697)^{d}$
Ω^{*+}_{cc} (unknown) scc; $Z_{cs}(3985)$ $X_{c2}(3930)$; $Z_c(3900)$; $X_c(3872)$ [36]	$(v^{-} + c^{+} + c^{+})' = 3.974\ (Z_{cs})$ $(3.974)' - 2z^{0} = X_{c2}(3.939)^{d}$	$(v^{-}c^{+}c^{+})'-z_1 = (3.922)^{d}$ $(v^{-}\ c^{+}\ c^{+})' - z_2 = (3.9045)^{d}$ $(v^{-}\ c^{+}\ c^{+})' - z_3 = (3.8697)^{d}$
Ω^{*0}_{cb} (unknown) scb	$(v^{-} + c^{+} + b^{-})' = 7.274$	$(v^{-}c^{+}\ b^{-})' - z_3 = (7.1697)^{d}$
Ω^{*-}_{bb} (unknown) sbb	$(v^{-} + b^{-} + b^{-})' = 10.574$	$(v^{-}b^{-}\ b^{-})' - z_3 = (10.470)^{d}$
Ω^{++}_{ccc} (unknown) ccc	$(c^{+} + c^{+} + c^{+})' = 5.1$ $(c^{+\bullet} + c^{+\bullet} + c^{+\bullet})' = 4.67$	$(c^{+}\ c^{+}\ c^{+})' - z_3 = (5.0)^{d}$ $(c^{+\bullet}c^{+\bullet}c^{+\bullet})' - z_2 = (4.6)^{d}$
Ω^{*+}_{ccb} (unknown) ccb	$(c^{+} + c^{+} + b^{-})' = 8.4$ $(c^{+\bullet} + c^{+\bullet} + b^{-\bullet})' = 7.858$	$(c^{+}\ c^{+}\ b^{-})' - z_3 = (8.296)^{d}$ $(c^{+\bullet}\ c^{+\bullet}b^{-\bullet})' - z_3 = (7.753)^{d}$
Ω^{*0}_{cbb} (unknown) cbb	$(c^{+} + b^{-} + b^{-})' = 11.7$	$(c^{+}\ b^{-}\ b^{-)'} - z_3 = (11.595)^{d}$
Ω^{*0}_{bbb} (unknown) bbb	$(b^{+} + b^{-} + b^{-})' \approx 15.0$	$m(\Omega^{*0}_{bbb})' \approx m\ (f^{\pm})$
$\Sigma_b^{-}(6.097)^{r}$; $X_b^{-'}(6.100)^{r}$ [37] $\Xi_b^{-'}(6.100)^{r}$; $()^{r}$-recently discover.	$(s^{-} + v^{-} + b^{-})' = 6.074$; $(v^{-} + v^{-} + b^{-})' = 6.148$	$(6.096)^{d} = (vvb)' - z_1 = \Xi_b^{*}\pi^{0}\ (vvb)' - 2z^{0} = (6.113)^{d}$

Table 6.
Theoretic masses of de-excited ("hot" formed) baryons, $J^{P\,3}/_{2}$, (CGT).

E. Annex

Multi-quark particle (CGT) (q-q-q ... q) or $(q-\bar{q}-q-\bar{q})$ $(c = c^{+}(^{2/}{}_{3}e))$; $b = b^{-}(-^{1/}{}_{3}e))$ $(t = t^{+}(^{2/}{}_{3})e$; $t \approx (7x5)b$	**Theoretic mass, GeV/c^2, (de-excited q. $()^{d}$./excited bar.()' u;d; $(p;n)^{d}$ $(\sim 0.312)^{d}$; $\lambda^{-}(0.435)^{d}$; $s^{-}(\sim 0.504)^{d}$; $v^{-}(0.574)^{d}$; $c^{+}(1.7)^{d}$; $b^{-}(\sim 5.0)^{d}$; m(t) = (7x5)m(b) = 175**	**Closest experimental value $()^{e}$, GeV; [36, 37] (particle/mass value) -observations-**
$[p\,\bar{d}\,s\bar{\lambda}]$; $[p\,\bar{d}\,s^{\bullet}\bar{\lambda}]$	(1.56)'; $[(1.56)' - z^{0}] = (1.543)^{d}$;	$K^{+}(1.555)$; $\Theta^{+}(1.54)$; $(s \rightarrow s^{\bullet})$
$[p\,\bar{p}\,c\,\bar{v}^{-}]^{+}$; $[p\,\bar{p}\,b\,\bar{v}^{+}]^{-}$ $[p\,\bar{n}\,c\,\bar{v}^{-}]^{++}$; $[n\,\bar{p}\,c\,\bar{v}^{-}]^{0}$	(2.898); $[(6.198)'^{-} - z_1] = (6.146)^{d}$ $(2.899) - 2z^{0} = (2.864)^{d}$	$X_c(2.9)^{e}$; $\Lambda_b^{0}(6.146)^{d}$ [37] $T^{a}_{cs}\ (2.9)^{e}$ [40] ; $D^{*}_{sJ}(2860)^{-}$
$[p\,\bar{s}\,s\,\bar{v}^{-}]^{+}$; $[s\,\bar{p}\,c\,\bar{v}^{+}]^{-}$;	(1.886)' ; $[(1.886) - z^{0}] = (1.868.6)^{d}$	$\Phi^{--}(1860)$

Multi-quark particle (CGT) **(q-q-q…q) or (q-$^-$q-q -$^-$q)** **($c = c^+(^{2/}{}_3e)$) ; $b = b^-(-^{1/}{}_3e)$)** **($t = t^+(^{2/}{}_3)e$; t ≈ (7x5)b**	**Theoretic mass, GeV/c^2,** **(de-excited q. ()^d./excited bar.()$'$** **u;d; (p;n)^d (∼0.312)^d; λ^-(0.435)^d;** **s^- (∼0.504)^d; v^-(0.574)^d ; c^+(1.7)^d ;** **b^-(∼5.0)^d; m(t) = (7x5)m(b) = 175**	**Closest experimental** **value ( )^e , GeV; [36, 37]** **(particle/mass value)** **-observations-**
$[\overline{p}\,p\,\overline{c}\,b^-]^-$; $[\overline{p}\,p\,b^+\,\overline{b}^-]^+$	(7.325) ; [(10.624)$'_{bb}$ ± (z^0;2z^0)]	$Z_b^{\pm}$(10.610); (10.65) [36]
$[c\,\overline{c}\,s^-v^-b^-]$;$[b\,\overline{b}\,s^-v^-c^+]^0$	(9.474)$'$; (12.81)$'$	Υ^-(1S)(9.46)^e
$[s\,\overline{s}\,c\,\overline{v}^-]^+$; $[s\,\overline{s}\,b\,\overline{v}^+]^-$ $[s\,\overline{s}\,c\,\overline{\lambda}^-]^+$; $[\lambda\,\overline{\lambda}\,c\,\overline{s}^-]^+$	(3.274); (6.574); (3.135)’; (3.135)$'$ – z^0 = (3.118)^d (3.135)’–2z^0/3z^0 ≈ (3.100)/(3.083)^d (3.070)’ – z^0 = (3.053)^d	Ω_c^0(3.119)^e; $\Theta^0{}_c$(3.099); J/ψ(1S)(3.097)^e; Ω_c^0(3.090)^e ; Ω_c^0(3.066)^e; Ξ_c(3.055)^e; Ω_c(3.05)^e Ξ_c(3.07)’→Ξ_c(2965)^e + μ;
$[s\,\overline{s}\,c\,\overline{b}^-]^+$; $[v\,\overline{v}\,c\,\overline{s}^-]^+$	(7.7)$'$; (3.348)$'$; (2.948)’ ; (2.948)’ – z^0 = (2.931)^d	Ξ_c^0(2939)^r ; Ξ_c^0(2923)^r
$[\overline{v}\,v\,\overline{b}\,s^+]^+$; $[v\,\overline{v}\,c\,\overline{b}^-]^+$	(6.648)$'$; (7.848)$'$	
$[\overline{c}\,c\,\overline{v}\,s^+]^+$; $[\overline{c}\,c\,\overline{v}\,v^-]^0$	(4.474)’ – z^0 = (4.456)’; (4.548)^d (4.474)’- z_μ = (4.37)^d; (4.548)’-z_1	P_c(4.457)^e -z^0 = P_c(4.44)^e Z_c(4.380)^e; X(4.500)^e [37]
$[\overline{c}\,c\,\overline{c}\,c^-]^0$; $[\overline{s}\,b\,\overline{\lambda}\,\lambda^-]^0$	(6.8)$'$; (6.370)’ – z^0 = (6.352)^d	X(6.900)^e; Ω_b(6.35)^e; (6.34)^e
$[\overline{c}\,c\,\overline{d}\,p^+]^+$; $[\overline{b}\,b\,\overline{d}\,v^-]^0$	(4.025)$'$; (10.887)$'$ (6.248)$'$; (6.230)^d ≈ (6.248)’ - z^0	Z_c(4.020)^e ; Y_b(10.890)^e [36]; Ξ_b(6.227)^r
$[\overline{c}\,c\,\overline{b}\,s^-]^0$; $[\overline{c}\,c\,\overline{b}\,v^-]^0$	(8.9)$'$; (8.974)$'$	
$[\overline{c}\lambda\,\overline{\lambda}\,\lambda^-]^-$; $[c\overline{c}\lambda\,\lambda\,\lambda]^-$ $[c\overline{c}\,v\,\lambda\,n]^{-*}$; $[c\overline{c}\,n\,p\,n]^0$	(3.005)$'$; (4.705)$'$; (4.148)$'$(c^-c d $^-\lambda$) ; (4.338)^0(d u c d^-c)^d	D_j(3.0)^e; X_c(4.700)^e, [37] X(4.140)^e-2z^0 ≈ Z_c(4.100)^e $P_\Psi{}^\Lambda$(4.338)^0 [40]
$[b\,\overline{\lambda}\,\overline{m}_2m_1]^+$; $[\overline{b}\,b\,\overline{v}\,s^-]^0$	(5.574)$'$; (11.074)$'$; (11.074)’– z_1(3z^0) = (11.022)^d	X(5.568)^e → $B_s{}^0\pi^{\pm}$ Υ^-(11.020)^e
$[v\,\overline{v}\,c\,\overline{c}\,\overline{m}_2m_1]^+$; $[c\overline{c}\,n\,\lambda\,s]^-$	(4.687)$'$; 4.687’–z^0 = (4.670)^d 4.687’–z_1(3z^0) =(4.635)^d ; (4.648)$'$;$'$ (4.648)$'$ – z^0 = (4.631)^d	X(4.685)^e ; Y(4.660)^e X(4.630)^e [37]
$[\overline{c}\,c\,\overline{n}\,p^+]^+$; $[\overline{c}\,c\,\overline{s}\,s]^0$	(4.026)$'$– z^0 = (4.009)^d ; (4.400)$'$	Z_{cs}(4.003)$^+$; ψ(4.415)^e
$[\overline{c}\,c\,\overline{s}\,v]^0$; $[\overline{c}\,c\,\overline{v}\,v]^0$	(4.474)$'$– z^0/2z^0 = (4.457)^d; (4.474)$'$– z_1 = (4.422)^d; (4.548)$'$	P_c(4.450); P_c(4.440)^e Z_{cc}(4.430)^e ;
$[\overline{c}\,c\,\overline{\lambda}\,p]^+$; $[\overline{c}\,c\overline{\lambda}\,\lambda]^0$; $[\overline{c}\,c\,\overline{\lambda}\,s]^0$	(4.147)$'$; (4.270)$'$; (4.335)$'$ (4.270)$'$ – z_1 = (4.218)^d	ψ(4.160); X(4.274); *Zcs*(4.220)$^+$
$[s\,\overline{s}\,c^\bullet\,\overline{s}]$; $[s^\bullet\,\overline{s}^\bullet c^\bullet\,\overline{s}^\bullet]$; $[\overline{c}\,s\,\overline{s}\,n]^-$	(3.057)’; (3.057)’ – z_1 = (3.005)^d (3.012)’; (3.012)’ – z_1 = (2.995)^d	Ω_c(3.050)^e, Ω_c(3.000)^e, (3.012)’ – z_2 = (.2977)^d ; Ξ_c(2.965)^e
$[c^\bullet\,v\,v\,v\,\overline{v}]^0$; $[c^+\,v\,v\,v\,\overline{v}]^0$	(3.996)’;	Z_{cs}(3.985)$^+$;
$[c\,\overline{v}\,v\,\overline{v}\,v\,\overline{n}]^+$;$[c\,\overline{v}\,s\,\overline{v}\,v\,\overline{p}]^0$	(4.309)’; (4.234)’- z^0 = (4.217)^d	P_c(4.312)$^+$; *Zcs*(4.220)$^+$
$[c\,\overline{\lambda}\,\lambda\,\overline{s}\,s\,\overline{s}]^+$; $[c\,\lambda\,\lambda\,s\,s\,s]^-$	(4.070)$'$– z^0/2z^0 = (4.053)/(4.035)^d	ψ(4.040);
$[\overline{b}\,b\,\overline{s}\,c]^+$; $[\overline{b}\,b\,\overline{v}\,c]^+$	(12.218)$'$; (12.292)$'$	
$[s\,\overline{s}\,v\,c\,b]^0$; $[v\,\overline{v}\,s\,c\,b]^0$	$[b\,\overline{b}\,p\,n\,\lambda\,v\,v\,c]^0$	(8.292)$'$; (8.366)$'$; (13.833)$'$
$[p\,\overline{p}\,v\,c\,b]^0$; $[p\,\overline{p}\,v\,\overline{s}\,c\,\overline{b}]^+$	(7.9) ; (8.4)	

Table 7.
Theoretic masses of non-excited and de-excited multi-quark particles, (explained or predicted by CGT).

Author details

Marius Arghirescu
State Office for Inventions and Trademarks, GCI-DLMFS, Romanian Academy, Bucharest, Romania

*Address all correspondence to: arghirescu.marius@osim.ro

References

[1] Patrignani C et al. (particle data group), "quarks". Chinese Physics C. 2016;**40**:100001

[2] Andersson B, Gustafson G, Ingelman G, Sjöstrand T. Parton fragmentation and string dynamics. Physics Reports. 1983;**97**(2–3):31-145

[3] Yndurain F. Limits on the mass of the gluon. Physics Letters B. 1995;**345**(4):524

[4] Chipman RD, Jennings LD. Physics Review. 1995;**132**:728

[5] CERNCOURIER. Precision Pins down the electron's Magnetism. 4 Oct 2006.

[6] ZEUS Collaboration. Limits on the effective quark radius from inclusive e-p scattering at HERA. Physics Letters B. 2016;**757**:468-472

[7] Georgi H, Jarlskog C. A new lepton-quark mass relation in a unified theory. Physics Letters B. 1979;**86**(3–4): 297-300

[8] Barut AO. Stable Particles as Building Blocks of Matter. Surveys High Energy Physics. Turkey; 1980;**1**:113

[9] Robson B. The generation model of particle physics". Book: Particle Physics, Ed. by E. Kennedy, London, UK: Intechopen.com. 2012. p. 1-28

[10] Marshall R. The pair production of quarks. Physikalische Blatter. 1997;**53**(7/ 8):685-688

[11] Arghirescu M. The cold genesis of matter and fields. Ed. Science PG. 2015. pp. 83-105. viXra:1104.0043

[12] Arghirescu M. The Material Structures Genesis and Fields Effects. Bucharest, RO: Ed. MatrixRom. 2006

[13] Arghirescu M. A preonic quasi-crystal quark model based on a cold genesis theory and on the experimentally evidenced neutral boson of 34 me. Global Journal of of Physics. 2016;**5**(1): 496-504

[14] Arghirescu M. A model of particles cold forming as collapsed Bose–Einstein condensate of gammons. PAIJ. 2018; **2**(4):260-267

[15] Olavi H. A Proposed Elementary Particle Theory Based on Interacting Classical Fields. Turku: Turun Yliopisto; 1961 preprint

[16] Krasznahorkay AJ et al. Observation of anomalous internal pair creation in 8Be: A possible signature of a light, neutral boson. Physical Review Letters. 29 Jan 2016;**116**(4):042501. arXiv: 1504.01527v1, [nucl-ex]. 2015

[17] Feng JL et al. Evidence for a Protophobic fifth force from 8Be nuclear transitions. Physical Review Letters. 12 Aug 2016;**117**(7):071803. arXiv: 1604.07411v1, [hep-ph]. 2016

[18] Krasznahorkay AJ et al. On the X(17) light-particle candidate observed in nuclear transitions. Acta Physica Polonica B. 2019;**50**(3):675 [arXiv: 1910.10459v1]

[19] Sarri G, Poder K, Cole J, et al. Generation of neutral and high-density electron– Positron pair plasmas in the laboratory. Nature Communications. 2015;**6**:6747

[20] Arghirescu M. The nuclear force explaining by a bag model resulted from a vortexial, cold genesis model of nucleon. Physics & Astronomy International Journal. 2018;**2**(4):349-358

[21] Yan Y, Tegen R. N-N scattering and nucleon quark core. Science Asia. 2001; **27**:251

[22] Burke DL et al. Positron production in multiphoton light -by light scattering. Physical Review Letters. 1997;**79**: 1626-1629

[23] Levi BG. Kondo physics seen in quantum dot. Physics Today. 1998;**51**: 17-18

[24] Chung D-Y. The masses of elementary particles and hadrons. Speculations in Science and Technology. 1997;**20**:259–268. arXiv:hep-ph/ 0003237v4. 2001

[25] Oset E, Sarkar S, et al. Baryon resonances. Nuclear Physics A. 2010;**835** (1–4):271-278. arXiv:0911.2580v1. 2009

[26] Souza ME. Calculation of almost all energy levels of baryons. Papers in Physics. 2011;**3**:art. 030003

[27] Carrigan RAJ. Compound Quarks as an Explanation for the Apparent Quark Mass Spectrum. Batavia, Illinois: Fermi Nat. Accelerator Lab; 1978. p. 60510

[28] Arghirescu M. An explanatory model of heavy quarks and particles generating resulted by a cold genesis theory. Theoretical Physics Letters. 2021;**9**(11): 190-217

[29] Griffiths DJ. Introduction to Elementary Particles. 2nd revised ed. WI LEY-VCH Verlag GmbH & Co. KGaA. 2008. pp. 181-188

[30] Sakharov AD. Mass formula for mesons and baryons. Zhurnal Eksperimental'noi i Teoreticheskoi Fiziki. 1980;**78**:2112-2115

[31] Dung D, Kurtscheid C, et al. Variable potentials for thermalized light and coupled condensates. Nature Photonics. 2017;**11**:565-569

[32] Milonni PW. The Quantum Vacuum-an Introduction to Quantum Electrodynamics. N. Y: Ed. Academic; 1994

[33] Islam MM, Luddy RJ. High energy elastic scattering in condensate enclosed chiral bag model and TOTEM elastic measurements at LHC at 7 TeV. Finland: The EDS Blois. 9-13 Sep 2013. pp. 1-4. arXiv:1310.5602

[34] Ashman J. EMC collaboration "a measurement of the spin asymmetry and determination of the structure function g1 in deep inelastic muon-proton scattering". Physics Letters B. 1988; **206**(2):364

[35] Oliveet KA, al. (particle data group), "meson summary table". Chinese Physics. 2014;**C38**:090001

[36] Wang W, Zhuyz R. Can X (5568) be a tetraquark state? Chinese Physics C. 2016;**40**(9):093101. arXiv:1602.08806v2 [hep-ph]. 2016

[37] CERN Courrier: Physics News. LHCb observes four new tetraquarks. 3 Mar 2021; Fascinating quantum mechanics. 12 Apr 2021

[38] Harry C et al. Evidence of brand new physics at CERN? The Conversation. 23 Mar 2021

[39] Abachi S et al. (D0 Collaboration) "Observation of the top quark" FERMILAB-Cord-95/103-E. La Thuile, Italy, March 5-11,1995. arXiv: hep-ex/ 9503003.1995

[40] SCI-NEWS. CERN Physicists Discover Three New Exotic Particles. 5 Jul 2022

[41] Harris SE, Hau LV. Nonlinear optics at low light levels. Physical Review Letters. 1999;**82**:4611

[42] Gong-Bo Z, Raveri M, Pogosian L, et al. Dynamical dark energy in light of the latest observations. Nature Astronomy. 2017;**1**:627-632

[43] Vejcik S. The CDF Collab., FERMILAB CONF-97-249-E. Proceedings. The CDF Collab., FERMILAB CONF-97-249-E. Proceedings. La Thuile, Italy. Mar 1997. p. 3

[44] Arghirescu M. Observations concerning the mass variation in a Galilean – Type relativity. IJHEP. 2018; **5**(1):44-54

Chapter 2

Phasic Structure of the Standard Model

Rafael Cañete Mesa

Abstract

We present a physical relationship able to justify by itself the whole spectrum of fundamental particles of matter (fermions), that is, the hierarchical structure, the value of their masses, the basis of those values and the analytical and positional relation of the same. A hierarchy supported on an element of periodicity or sequence of differentiated states (phases) that makes it resemble a periodic table of elementary particles, which allows us to define material states not well defined until now and others unregistered, with which it is precisely evidenced that there has to exist a particle $\mathfrak{M} = 171.87$ eV, or an equivalent energy entity, which represents a common zero generation to (anti)quarks and charged (anti)leptons, as well as a mediator particle in the connection of these with (anti)neutrinos, for which, by applying this methodology, we obtain a massive base ν_1, ν_2, ν_3, consistent with the theoretical and experimental requirements for neutrino oscillation, which allows us to explain the CP violation. Previously, we obtained, by means of the quantum wave formulation, a symmetrized wave packet (SWP) and an associated energy transmutation equation, equivalent, although more general and precise, to the one obtained by means of a corpuscular treatment, this one that we will use and accounts for all the transitions between particles.

Keywords: neutrino oscillation, massive base, dark matter, CP violation, SM

1. Introduction

The description of particles in the quantum mechanics environment is good for particles in a certain state of confinement, but not for free particles. In this case, by means of the wave functions, either of travelling waves or of waves groups, only a partial or not free correspondence has been achieved, and always by means of the use of wave quantum mechanics, since the matrix mechanics, despite the mathematical equivalence between one and the other, carried out by Von Neumann [1], is not applicable to these continuous states of energy or, said in another way, to the case of the free particle not subject to potentials, which does not lend itself to the application of the mathematical methods of Matrix Mechanics since its very essence is based on Fourier expansions valid for periodic systems that necessarily require the particle to be confined to a bound system [2]. It is precisely this non-applicability of the matrix quantum to the problem and its predominance, however, for many other issues as a consequence of its versatility, which has caused in a certain sense that research routes aimed at overcoming the problem are not sought, i.e. aimed at providing a true theoretical support to the wave-corpuscle duality, in accordance with the experimental evidence and the relevance of the question. To some extent, it has been decided to explore to satiety that which allows itself to be treated by matrix formalism, and what is not has been left aside, in spite of the fact that it is a particular area (unapproachable by other methods) capable of providing information analogous to reality, since the wave function is a physical entity and not a mathematical representation like that constructed with functions in matrix mechanics. Consequently, we know that particles are created by means of a creation operator, but we do not know what that creation act consists of because we know nothing about the created object.

It is for all these reasons that, going back a century, we are going to return to the path from the beginning, and give it another focus, another push, from the conviction that this reality, that of duality, is a superior (more primordial) reality by far to other realities or work schemes of physics that after many years of work know how to say many phenomenological things, but very few of the matter itself or its organization.

The theoretical pretensions of this exposition will be divided here into two objectives or developments: one in which by means of a wave treatment, we will find the same equations that govern corpuscular dynamics in all its energetic forms, but already related (as terms of a single equation) and with additional information of a wave character; and another development in which we will treat part of this additional information, referring mainly to the term of kinetic energy and the crucial role it plays in the energy conversion and in the creation-organization of the material particles of our universe or physical reality, while the study of mass (formed by means of wave constituents) and the rest of the energy forms associated with the other terms of the equation will be the object of another study. More specifically, we will demonstrate with this treatment and the aforementioned equation that all matter and all energy have a wave character, which is present both in their creation (which is precisely the process carried out from a wave packet in the first development) and in all processes or interactions, as a consequence of being precisely the resident wave component, which we call phase factor (the wave is not extinguished with the formation of matter), which allows them, or, to be more exact, because every energetic interaction and every possibility of constituting itself in particle is precisely an interaction carried out through that perpetual wave component.

In accordance with these purposes, we will see that in our itinerary we will arrive, indeed, as a first-order objective to the mentioned equation [Eq. (32)], but that it will lose protagonism (it will be developed term by term in another work) because the protagonism is taken by the phase factor of a phase ϕ [defined from the first one through Eq. (37)], as the motor of all energetic transformation, which, however, also loses the protagonism (it will be developed for all the families of particles in detail in another work) because this importance is taken by the relation between phases χ defined in Eq. (39) from the previous one (being reduced to a formula without system constants), for being, according to the title, our main objective and of greater repercussion, or of a more contrastable impact due to the analytical character of what we want to postulate as true.

Indeed, the relationship between phases will allow us to establish by means of a simple formula between the masses of a triad (such as that of the three generations) a scheme or organization of all the material particles, that is, their masses, calculated analytically, and the distance between them in phase distances within that scheme, which will allow us to determine which points of it are likely to contain a mass (located or locatable).

In this context, besides verifying all the existing fermions, and bringing the adjustment of the masses of the massive neutrinos to experimental order with an unprecedented degree of theoretical precision, to the point of being able to determine their hierarchical ordering (which Katrin's experimental precision has not resolved), we will find a particle or energy entity at a central or convergence point, necessary in this scheme, which curiously occupies the gap or energy jump of five orders of magnitude existing between the first two flavours of neutrinos. A central particle that will turn out to be the generating or intermediate element, in such a way that all the others are reached by it to a certain phase, as it happens with the set of three (locatable) particles or genuinely primordial masses that constitute a natural neutrino base with which we can compose the base of massive neutrinos already referred to (and a transformation matrix), in the same way that this is later transformed into the different flavours by another one.

A particle and a methodology provide answers to other diverse questions. Questions such as the seesaw mechanism or the CP violation, on which it is not intended to reach categorical conclusions within the scheme of work or to address it exhaustively, only to show its connection and potential, and to leave a note, advance or point of interest.

2. Symmetrized wave packet (SWP)

A free particle can be represented by a one-dimensional or elemental wave, which we call the plane wave, $\psi = Ae^{i(kx-wt)}$, which describes the possibility of the particle to be located in space in the interval $[-\infty, +\infty]$, as occurs, for example, with its real part $\psi = ACos(kx - wt)$, called also travelling wave, which we can easily represent (see **Figure 1**), but which at the same time evidences the impossibility by means of it of determining in what position of space the particle is, as a consequence of the impossibility of normalizing the wave function in the mentioned $[-\infty, +\infty]$ interval,

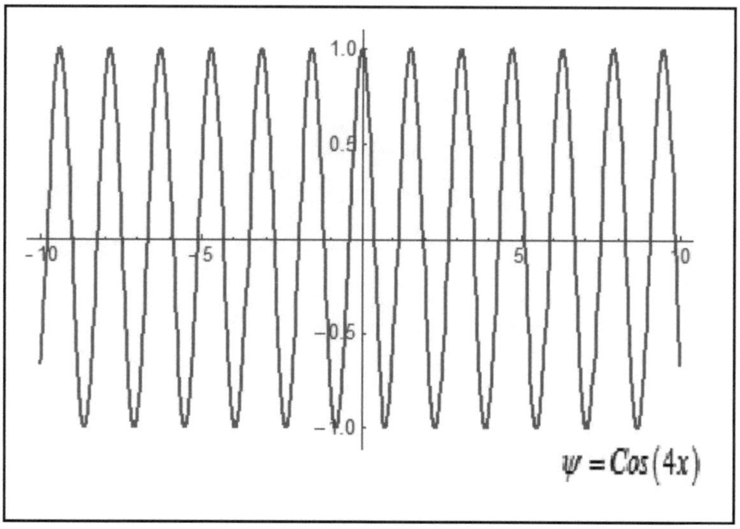

Figure 1.
Traveling wave.

that is, of integrating in the interval and finding a defined value for A, of which results an insufficient description. We could say, in accordance with the principle of indeterminacy, that if the quantity of motion (p) is fixed exactly, the position (x) of the particle is completely undefined. And we can say that it is this impossibility of being normalized that makes (beyond the Fourier reference) matrix mechanics or Neumann unification (isomorphism of mathematical spaces L_2 and ℓ_2) not applicable for plane waves associated with free particles, since it requires that the functions be square integrable, and the failed normalization process is nothing but the solution of an integral or sum of this type [3]. A question, by the way, that makes that, paradoxically, the vectors of that mathematical space (of quantum mechanics) do not belong to the space itself (see Section A-3 of Chapter 2 in [4]).

This is somehow solved by the wave packets of different typology or distribution (Gaussians, Lorentzians) that can be obtained, since they are equivalent (see Chapter 20 of [5]), by constructing light pulses (wave train of finite length) or by superimposing many waves of approximate wave vector, that is, of different values p, and consequently of an undetermined value p, which restricts the indeterminacy of x to an interval Δx, accounting for a defined location and a certain sense of occupation (the wave of matter of de Blogie).

These distributions are functional arrangements that make the original wave extinguish beyond a few boundaries on either side of a central point, so that it is a discrete spatial arrangement that can even evolve over time [6]. Explicitly, by applying the function $\theta = \exp(-x^2/20)$ (see **Figure 2**), to the previous function ψ, we get (see **Figure 3**):

This is what we also know by wave modulation or modulated wave used for the transmission of information associated precisely with a certain energy distribution. It is this way of distributing energy in the group that is intended to be used to identify the particle, as an evident form of energy information, in it.

The attempts, however, to associate this wave model with the real constitution of the elementary particle have failed, given that, contrary to what happened with the particles they try to represent, they are extinguished in time (see p. 64, complement G_I of [4]), which is unavoidable in the current paradigm.

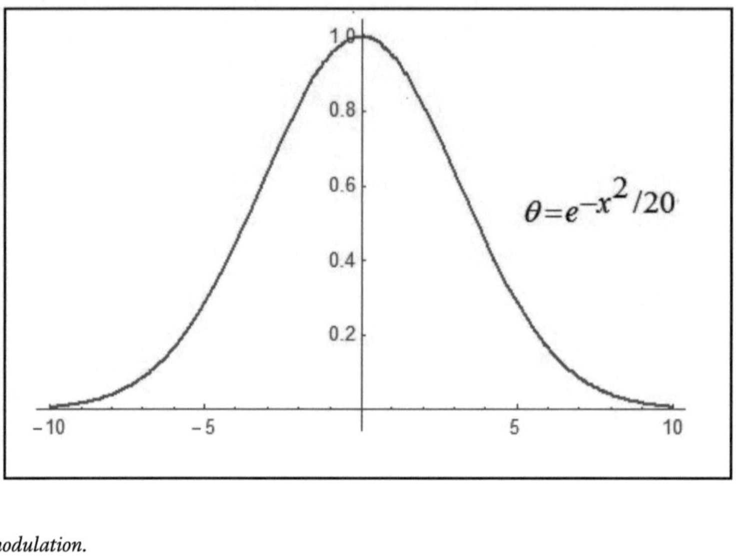

Figure 2.
Amplitude modulation.

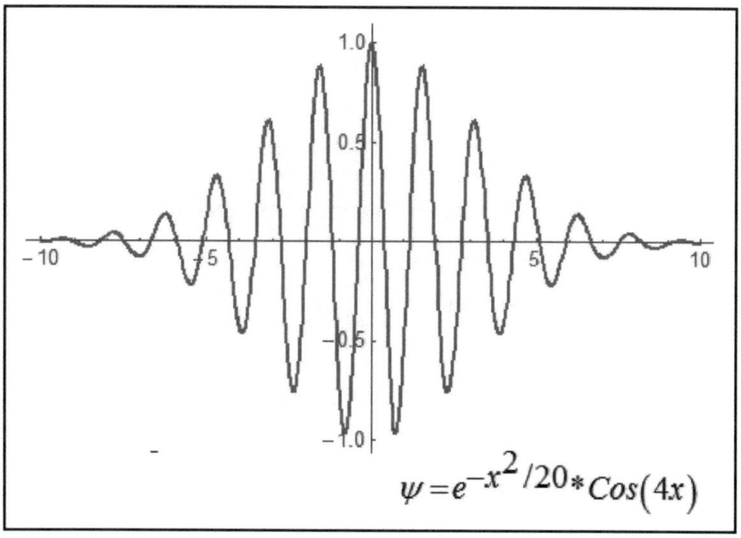

Figure 3.
Wave packet.

In the first place, we are not going to try to overcome in our analysis the inescapable in the current paradigm, but to characterize corpuscular behaviour by using some kind of distribution. From there on, the results will be those that force us to interpret things in a certain way, among which there may be some paradigmatic. For this, and in particular way, we are going to use the superimposition of a set of plane waves of variable frequency and constant amplitude, such as:

$$\Psi_k(x,t) = \int_{-\infty}^{+\infty} e^{i(w_k t - kx)} dk, \tag{1}$$

which for the cosine of the complex exponential, restricted to the interval $k = k_0 \pm \Delta k/2$, gives rise to a modulated monochromatic wave of frequency w_0,

$$\Psi_k(x,t) = B \frac{\sin[\Delta k/2(vt - x)]}{vt - x} e^{i(w_0 t - k_0 x)}, \tag{2}$$

that is, to a plane wave and a modulating or envelope factor accompanied by a normalization constant B, which includes the factor $(\times 2)$ that should accompany the sine function. The function (1) has in reality the form (Taylor series):

$$\Psi_k(x,t) = e^{i(w_0 t - k_0 x)} \int_{k_0 - \Delta k/2}^{k_0 + \Delta k/2} e^{i[(k - k_0)/2(vt - x)]} dk, \tag{3}$$

whose complete and mathematically most manageable solution is (see Chapter 12 of [5]):

$$\Psi_k(x,t) = B(i)\frac{e^{-i[\Delta k/2(vt-x)]}}{vt-x}e^{i(w_0t-k_0x)}, \tag{4}$$

to which we can later demand its real character, and we have added the factor (i) precisely so that in this case, the real part is analogous to that presented in (2).

On this basis, we can build two functions that keep a certain symmetry. The first can be one similar to (4) but reversing the propagation of the travelling wave, that is:

$$\begin{aligned} \Psi_1(x,t) &= B(i)\frac{e^{-i[\Delta k/2(vt-x)]}}{vt-x}e^{-i(w_0t-k_0x)} \\ &= B(i)\frac{e^{-i(M/2)}}{vt-x}e^{-i(N)} = B(i)\frac{e^{-i(M/2)}}{vt-x}\ \phi^{-1}(\mathrm{x},\mathrm{t}), \end{aligned} \tag{5}$$

a transformation that is analogous to having originally considered a function conjugated to the original (1) with the integration limits transposed:

$$\Psi_1(x,t) = \overset{\leftrightarrow}{\Psi}{}_k^{*}(x,t) = \int_{+\infty}^{-\infty} e^{-i(w_kt-kx)}dk. \tag{6}$$

The second, with respect to the original one (4), would be a wave function that preserves the propagation of the travelling wave but inverts the envelope:

$$\begin{aligned} \Psi_2(x,t) &= B(-i)\frac{e^{i[\Delta k/2(vt-x)]}}{vt-x}e^{i(w_0t-k_0x)} \\ &= B(-i)\frac{e^{i(M/2)}}{vt-x}e^{i(N)} = B(-i)\frac{e^{i(M/2)}}{vt-x}\ \phi(\mathrm{x},\mathrm{t}), \end{aligned} \tag{7}$$

action that is equivalent (according to the corresponding Taylor series) to exchanging the limits of integration in the original function, that is:

$$\Psi_2(x,t) = \overset{\leftrightarrow}{\Psi}_k(x,t) = \int_{+\infty}^{-\infty} e^{i(w_kt-kx)}dk, \tag{8}$$

The result will be two wave functions, (5) and (7), on which we will be able to carry out a combination, which will be nothing more than a process of symmetrization, that is to say, an adaptation similar to that carried out on certain quantum observables to ensure their classical validity. In this case, we will have two plane waves propagating in opposite directions that intersect and interfere into the mutual space of the envelopes, in a constructive way for the real component and destructive in what refers to the imaginary one, reaching its only real character naturally and not, as in (4), by means of physical criteria or impositions. A situation that we can associate to a real composition of the waves, that is, to the way of structuring themselves physically (that is, our thesis, in fact) and not only to a well-disposed operational model, which is also to participate in the operational advantages of the forming functions, already mentioned. Therefore,

$$
\begin{aligned}
\Psi(x,t) &= \frac{(\Psi_1(x,t)+\Psi_2(x,t))}{\sqrt{2}} = B\frac{(i)e^{-i(M/2)}e^{-i(N)}+(-i)e^{i(M/2)}e^{i(N)}}{\sqrt{2}(vt-x)} \\
&= (i)B\frac{e^{-i(M/2)}e^{-i(N)}-e^{i(M/2)}e^{i(N)}}{\sqrt{2}(vt-x)} = (i)B\frac{e^{-i(M/2)}\phi^{-1}(x,t)-e^{i(M/2)}\phi(x,t)}{\sqrt{2}(vt-x)} \\
&= (i)B\frac{\left[\vartheta_1\phi^{-1}(x,t)-\vartheta_2\phi(x,t)\right]}{\sqrt{2}} = (i)B\frac{\left[\vartheta_1\phi_1-\vartheta_2\phi_2\right]}{\sqrt{2}},
\end{aligned} \tag{9}
$$

where $\phi_2 \equiv \phi, \phi_1 \equiv \phi^{-1}$, the value $\sqrt{2}$ is a new normalization factor that leaves the initial B factor intact, which we can calculate on the initial function (4), expressed in a more generic way (without the factor i, which was intentionally introduced), for $t = 0$ and for $b \equiv 2/\Delta k$, as:

$$
\Psi_1(x) = B\frac{e^{i(x/b)}e^{i(k_0 x)}}{x}, \tag{10}
$$

which we will be able to normalize,

$$
\begin{aligned}
\int_{-\infty}^{+\infty} f dx &= B^2\int_{-\infty}^{+\infty}\Psi_1{}^*(x,0)\Psi_1(x,0)dx = B^2\int_{-\infty}^{+\infty}|\Psi_1(x)|^2 dx \\
&= B^2\int_{-\infty}^{+\infty}\frac{e^{i2(x/b)}}{x^2}\phi^*(x,t)\phi(x,t)dx = B^2\int_{-\infty}^{+\infty}\frac{e^{i2(x/b)}}{x^2}dx = 1,
\end{aligned} \tag{11}
$$

and solve by means of the integral calculus[1]:

$$
B^2\oint\frac{e^{i2(z/b)}}{z^2}dz = \int_{-\infty}^{+\infty} f dx + \oint f = \int_{-\infty}^{+\infty} f dx = B^2(-\pi i)\frac{2i}{b} = B^2\frac{2\pi}{b} = 1. \tag{12}
$$

From which it results:

$$
B = \left(\frac{b}{2\pi}\right)^{1/2} = \left(\frac{1}{\Delta k\pi}\right)^{1/2} \tag{13}
$$

3. Energy form of the SWP

As the aim of our discussion is to obtain the corpuscular energy from the wave function, we can state that the target is to differentiate from the value of the total energy those values that are clearly undulatory, and incontrovertible, given the default undulatory character of our starting physical system, and to distinguish them from the remaining values, that is to enunciate, to separate the value of the energy in at least two energetic terms, the strictly undulatory and the one that is not. For this, let

[1] By calculating the residual, we solve the integral value for the second-order pole in z = 0 that passes through the contour, since we are calculating x in the real line, and whose contribution is consequently half $(-i\pi)$ in the same (main value of Cauchy) and zero in the rest of the contour.

us again take the function of the symmetric wave packet (SWP) given in (9) and calculate with it the expectation value of the energy, which will be given by:

$$\overline{E} = \int \xi P dx = \int \Psi^*(x,t) i\hbar\partial_t \Psi(x,t) dx, \tag{14}$$

and which must coincide with the energy of the system described by a free particle such as the electron, including that derived from its purely undulatory character.

The expected value is an application of the operator energy ξ over a probability density P, which must be a positive real value. In fact, the normalization process (11) is intended by the conjugation of the function to achieve a positive real value coinciding with this density value. This is:

$$B^2 \int_{-\infty}^{+\infty} \Psi^*(x,0)\Psi(x,0)dx = B^2 \int_{-\infty}^{+\infty} |\Psi(x)|^2 dx = \int_{-\infty}^{+\infty} P dx = 1 \tag{15}$$

This directs us to reconsider the normalization process and observe that it is done through conjugation to achieve an integrable square function, which brings forth a positive real value. That square integrable function is, as seen in (11), the one that results from annulling the imaginary part of the original function through the relation $\phi^*(x,t)\phi(x,t) = 1$, being usually that resulting function is a real function since the imaginary factor has been weeded out by this method.

But neither does the resulting function necessarily have to be real, as in fact is the case in (11), nor does it necessarily have to be achieved by the conjugation of the complete original function as it also happens in our case. That is, the conjugation is a method (the most usual) but not an inescapable condition or an end in itself, the latter being to reach a positive real value for the probability density and the consequent normalization constant. For this very reason, and because of the possibility of solving the value of the complex integral as it is (because it has a solution), we have only needed to conjugate the plane wave and not the envelope. Such a resolution is a consequence of the extrapolation of the complex behaviour to the real one, or, in other words, of the possibility of carrying it out because the contribution of the imaginary part is null, as shown in $\oint f$. That is, taking $\Psi = \vartheta \times \phi$, generically, we have not made $\Psi^* = \vartheta^* \times \phi^*$ to obtain the real value, since this would have led us to $\Psi^* \times \Psi = 1$, but $\Psi^* = \vartheta \times \phi^*$, being the idiosyncrasy of the integral itself that is responsible for accounting for the value (main Cauchy's) where it has, that is, on the real line.

This real character that is derived from the properties of the integral in a natural way, and that we originally adopted in (5) and (7) by means of physical criteria, we will reach it in what follows revalidating $\Psi^* = \vartheta \times \phi^*$ and leaving (without imposing criteria) that the operative itself is the one in charge of rescuing the real value of the function where it corresponds without adulterating it, that is, without suppressing the imaginary part (forcing it to be real) nor conjugating it to obtain a real value in a forced way. All this can be summarized by saying that the probability density has the same form for any use of it.

With this in mind and taking into consideration what is already established for $M \equiv [\Delta k(vt - x)]$, $N \equiv (w_0 t - k_0 x)$, $\phi(x,t) \equiv e^{i(N)} = \phi_2 = \phi_1^{-1}$ and $\Psi_i = \vartheta_i \phi_i$, we can calculate Eq. (14). An equation what, according to the result obtained in Eq. (9), we can separate into its two components:

$$\overline{E} = \int \Psi^*(x,t) i\hbar\partial_t \Psi(x,t)dx = \left(\overline{E}_1 + \overline{E}_2\right)$$
$$= (i)^2 B^2 \frac{\left[\int \zeta_1(x,t)dx - \int \zeta_2(x,t)dx\right]}{\left(\sqrt{2}\right)^2} \tag{16}$$
$$= B^2 \frac{\int (\zeta_2(x,t) - \zeta_1(x,t))dx}{2} = B^2 \int \zeta(x,t)dx,$$

where it meets:

$$\zeta_2(x,t) = \Psi_2{}^*(x,t) i\hbar\partial_t \Psi_2(x,t) = \vartheta_2 \phi_2{}^* i\hbar\partial_t(\vartheta_2\phi_2)$$
$$= \vartheta_2 \phi^{-1}(x,t) i\hbar\left[\vartheta_2 \partial_t \phi(x,t) + \phi(x,t)\partial_t \vartheta_2\right]$$

$$= \frac{e^{i(M/2)}e^{-i(N)}}{(vt-x)} i\hbar \left[\frac{e^{i(M/2)}e^{i(N)}}{(vt-x)}\left(iw_0 + \frac{iv\Delta k}{2}\right) - v\frac{e^{i(M/2)}e^{i(N)}}{(vt-x)^2}\right]$$
$$= i\hbar e^{i(M/2)}\left[\frac{e^{i(M/2)}}{(vt-x)^2}\left(iw_0 + \frac{iv\Delta k}{2}\right) + \frac{e^{i(M/2)}}{(vt-x)^3}(-v)\right] \tag{17}$$
$$= \hbar\left(\frac{e^{i(vt-x)\Delta k}}{(vt-x)^2}\left(-w_0 - \frac{iv\Delta k}{2}\right) + \frac{e^{i(vt-x)\Delta k}}{(vt-x)^3}(-iv)\right)\Bigg],$$

and

$$-\zeta_1(x,\mathrm{t}) = -\Psi_1{}^*(x,t) i\hbar\partial_t \Psi_1(x,t) = -\vartheta_1\phi_1{}^* i\hbar\partial_t(\vartheta_1\phi_1)$$
$$= \vartheta_1 \phi(x,t) i\hbar\left[\vartheta_1 \partial_t \phi^{-1}(x,t) + \phi^{-1}(x,t)\partial_t\vartheta_1\right]$$
$$= -\frac{e^{-i(M/2)}e^{i(N)}}{(vt-x)} i\hbar\left[\frac{e^{-i(M/2)}e^{-i(N)}}{(vt-x)}\left(-iw_0 - \frac{iv\Delta k}{2}\right) - v\frac{e^{-i(M/2)}e^{-i(N)}}{(vt-x)^2}\right]$$
$$= i\hbar e^{-i(M/2)}\left[\frac{e^{-i(M/2)}}{(vt-x)^2}\left(iw_0 + \frac{iv\Delta k}{2}\right) - \frac{e^{-i(M/2)}}{(vt-x)^3}(-v)\right]$$
$$= \hbar\left(\frac{e^{-i(vt-x)\Delta k}}{(vt-x)^2}\left(-w_0 - \frac{iv\Delta k}{2}\right) - \frac{e^{-i(vt-x)\Delta k}}{(vt-x)^3}(-iv)\right)\Bigg] \tag{18}$$

From which results:

$$\zeta(x,t) = \frac{\zeta_2(x,t) - \zeta_1(x,t)}{2}$$
$$= \left[\frac{\hbar v \sin(\Delta k(vt-x))}{(vt-x)^3} - \frac{\hbar v \Delta k \cos(\Delta k(vt-x))}{2(vt-x)^2}\right] \tag{19}$$
$$+ \left[-\frac{\hbar w_0 \cos(\Delta k(vt-x))}{(vt-x)^2}\right] = \zeta^m(x,t) + \zeta^{\varpi}(x,t),$$

where we have taken into account the following relations:

$$2i\sin M = e^{iM} - e^{-iM} \text{ and } 2\cos M = e^{iM} + e^{-iM}, \tag{20}$$

That is, it results an expression without imaginary components per se (which if we had not used SWP would have appeared and we would have had to discard them a posteriori as is usually done in physics), on which, by regrouping terms, the clearly undulatory part of the corpuscular one is clearly differentiated, as we have already mentioned, characterized by the parameters w_0 and v respectively. Indeed, in the second term of the integrand, $\zeta^{\varpi}(x,t)$, the light characteristics of the integrand are revealed, and therefore, of the function as a whole, a matter that, given its nature, is normal. It will be, therefore, the term $\zeta^{m}(x,t)$ of the integrand that will be able to show, from its integral value, the double wave-corpuscle nature, despite having a light origin.

4. Energy treatment of the SWP

We have divided the proposed issue into several parts or actions. On the one hand, (which are actually two) to obtain a function with a particular distribution or configuration from which a plausible energetic form is derived, and on the other, which we are now addressing, to relate that energetic form to that of a particle. It is the moment, therefore, to face the task of obtaining a result, a corpuscular energetic expression from the previous one of undulatory origin. Put more precisely, it is time to obtain a recognizable corpuscular energy value, which is the only way to ensure that the wave function does indeed describe a particle.

The wave function is expressed on the variables x and t, which is very good for the wave part in which we want to know its evolution in time and space on which its energy value depends. This does not happen for a free particle, which has an energy value, regardless of its position (unless there is a potential) or of time if it is at rest, and it is always at rest for its frame of reference. In addition to this, the calculation of energy as a function of the variable involves an error derived from the double consideration of the variable x, as a space for evolution and a space for occupation.

Although the question seems a priori unaddressable, we can nevertheless restrict to some extent this double consideration or ambiguity if we take into account that the wave function of the wave packet is practically cancelled out, that is, the probability that outside this interval a particle is represented (or, that the particle is outside this interval) and that outside this interval there are significant values of energy that contribute to the mean value of the particle is almost zero. Consequently, we can restrict this double consideration if we calculate the measured value to the interval of the group. This is:

$$\overline{E} = B^2 \int_{-\infty}^{+\infty} \zeta(x,t)dx \cong B^2 \int_{x_0-\Delta x/2}^{x_0+\Delta x/2} [\zeta^{m}(x,t) + \zeta^{\varpi}(x,t)]dx, \tag{21}$$

However, we see that despite the approach we have not managed to overcome the duality of x given that despite not existing outside the limits of integration (size) is still needed to assess the interior, also happening to the energy value differentiate positions internal that in no way can be differentiated into a particle, except for processes of formation of the same. In addition to this we see that, even assuming that in this way we were able to achieve a real energy value of the particle, at best it would only serve to evaluate the energy value in a state of rest (we have restricted the variable x to size) but not for the dynamic states that we know as a function of the relative speed, which here is undefined, which forces us to go further in our way of reasoning the problem.

It is clear that in the case presented, the expected value is the energy value of the particle at rest ($t = 0$), and that size Δx is coincident with the size of the particle in that situation. The question is whether, based on this, we might be able to find out the expected value in a general no-rest situation, that is, for another frame of reference. The answer is yes, and that it is so regardless of all the delimitations made above, if we consider the magnitude of that size Δx as the relativistic proper length of it, because, once the proper length $a = \Delta x$ is defined, it is defined for all inertial systems by the Lorentz transform, in the consideration that any value $vt - x$ will be a fixed value for some speed v, and for any instant (being so, in particular, for $t = 0$), that is, with

$$vt - x = a\left(1 - v^2/c^2\right)^{1/2} = a\gamma^{-1}, \tag{22}$$

we naturally move from a position space to a momentum space, transforming one nonspecific variable into another, v, of clear significance and better treatment.

Summarizing the approach made, it will only be possible to calculate the energy of a group of waves (that reside in their envelope) as a whole, if we consider it at a geometric level as a whole, that is, as an individual entity capable of representing or supporting the different states. In this case, and returning to the starting point of the original Eq. (14), known by integrating as a function of speed and the new limits of integration relative to the original $\pm\infty$, we only have to define the new integral for the expectation value of energy and calculate.

$$\begin{aligned}\overline{E} &= \int_0^c \Psi^*(v) i\hbar\partial_t \Psi(v) t dv = \int_0^c B_c^2 \zeta(v) t dv \\ &= B_c^2[T] \int_0^c [\zeta^m(v) + \zeta^\varpi(v)] dv = A^2 \int_0^c [\zeta^m(v) + \zeta^\varpi(v)] dv\end{aligned} \tag{23}$$

where the normalization constant has been modified in two steps as a result of the change to the new variable and the integration interval assigned to it, comparable with the real half line, which, according to (13), meets:

$$B_c^2 = 2B^2 = \frac{b}{\pi} = \frac{2}{\Delta k \pi} \quad \text{and} \quad B_c^2[T] = A^2 \tag{24}$$

and where, considering (22), the integrands expressed in (19) take the form:

$$\zeta^\varpi(v) = -\frac{\hbar w \cos(\Delta k(vt - x))}{(vt - x)^2} = -\frac{\hbar w \cos(\Delta k(a\gamma^{-1}))}{a^2\gamma^{-2}} \tag{25a}$$

and

$$\begin{aligned}\zeta^m(v) &= \frac{\hbar v \sin(\Delta k(vt-x))}{(vt-x)^3} - \frac{\hbar v \Delta k \cos(\Delta k(vt-x))}{2(vt-x)^2} \\ &= \frac{\hbar v \sin(\Delta k(a\gamma^{-1}))}{a^3\gamma^{-3}} - \frac{\hbar v \Delta k \cos(\Delta k(a\gamma^{-1}))}{2a^2\gamma^{-2}},\end{aligned} \tag{25b}$$

that we can still put, according to (19), in a more compact way if we make: $\Phi \equiv [\Delta k(a\gamma^{-1})]$

$$\zeta(v) = \left[\frac{\hbar v \sin(\Phi)}{a^3\gamma^{-3}} - \frac{\hbar v \Delta k \cos(\Phi)}{2a^2\gamma^{-2}}\right] + \left[-\frac{\hbar w \cos(\Phi)}{a^2\gamma^{-2}}\right] \tag{26}$$

We can observe that after the change of variable, we have integrated time with the normalization constant ($A^2 = B_c^2[T]$), which means that t has no functionality as a variable either within or outside the integral, that is, that energy is not a function of time, and that the only functionality it has, which is assumed by the constant (A^2), is dimensional, that is, that of ensuring that energy remains dimensionally homogeneous. Time dimension $[T]$ which, according to (24), ultimately (for the moment) assumes b through A^2. Resulting in this case:

$$A^2 = \frac{b[T]}{\pi} = \frac{2 \times [T]}{\Delta k \pi} \equiv \frac{\hbar}{\pi} \tag{27}$$

The reason is that although we are obliged to obtain the value of energy in a certain way, given that by definition $\xi\Psi(x) = i\hbar\partial_t\Psi(x)$, any change of variable only needs to be dimensionally homogeneous, while the space in which it develops is arbitrary (not necessarily an extrapolation of the first), on the condition of having well-defined limits for normalization. In our case, the new space of study is not $x = x(v)$ but v with its corresponding limits, which we could have set without the need to carry out all this transit. To realize this, we only have to remember that:

$$\begin{aligned}\overline{E} &= \int_1^2 \Psi^*(x) i\hbar\partial_t\Psi(x)dx = \int_1^2 \frac{\Psi^*(x) i\hbar\partial_t\Psi(x)dx}{\Psi^*(x)\Psi(x)dx} \\ &= \int_1^2 \frac{\Psi^*(x) i\hbar\partial_t\Psi(x)tdv}{\Psi^*(x)\Psi(x)tdv} = \int_a^b \frac{\Psi^*(v) i\hbar\partial_t\Psi(v)dv}{\Psi^*(v)\Psi(v)dv},\end{aligned} \tag{28}$$

that is to say, that in the expectation or most probable value of a variable, elements that accompany it can be dispensed with if they are also eliminated in the denominator, that is, in the probability function (hidden by default). This can be done if in all cases it is normalized and complies with:

$$P = \int_1^2 \Psi^*(v)\Psi(v)tdv = \int_a^b \Psi^*(v)\Psi(v)dv = 1 \tag{29}$$

In any case, and in a physical sense, we do not want to calculate $\overline{E}$ in the probabilistic space of the positions, but of the velocities, and this is not achieved by making a copy of the first, but by defining it correctly. Solving the terms of the integral in Eq. (23), expressed in Eqs. (25), as we do in Appendix A, we are left with a sum of integrals as follows:

$$\begin{aligned}
\overline{E} &= A^2\int \zeta(v)dv = A^2\int \zeta^m(v)dv + \zeta^{\varpi}(v)dv \\
&= A^2\int\Big(\zeta_k(v) + \zeta_f(v) + \zeta_{\varpi}(v)\Big)dv \\
&= A^2\left(\frac{\hbar}{a^3}\right)\sin\left[\Delta k\left(a\gamma^{-1}\right)\right]_v \int_{v_o}^{v_f}\frac{v}{\gamma^{-3}}dv \\
&\quad -A^2\left(\frac{\hbar\Delta k}{2a^2}\right)c^2\int\frac{\cos[\Delta k(a\gamma^{-1})]}{\gamma^{-1}}d\left(\gamma^{-1}\right) \\
&\quad +A^2\left(\frac{\hbar w}{a^2}\right)c\int\frac{\cos[\Delta k(a\gamma^{-1})]}{(1-\gamma^{-2})^{1/2}\gamma^{-1}}d\left(\gamma^{-1}\right),
\end{aligned} \tag{30}$$

in which the mass term of the integrand has been separated into two from another species, one kinetic and the other corpuscular or formation.

Expression that, taking into account the normalization factor (24), clarifying the notation of the variable by using $\nu \equiv \gamma^{-1}$ in the last two terms and distinguishing the arguments of the trigonometric functions in them through variable, with respect to that defined for (26), despite being mathematically equal in Eq. (30),

$$\Phi \equiv \left[\Delta k\left(a\gamma^{-1}\right)\right] = [\Delta k(a\nu)] \equiv \Phi_\nu, \tag{31}$$

to note that such arguments are affected by the integral through that variable, we can write as:

$$\begin{aligned}
\overline{E} &= A^2\int \zeta^m(v)dv + \zeta^{\varpi}(v)dv \\
&= A^2\int\Big(\zeta_k(v) + \zeta_f(v) + \zeta_{\varpi}(v)\Big)dv = \overline{E}_k + \overline{E}_f + \overline{E}_{\varpi} \\
&= \left(\frac{\hbar \text{ꝁ}}{\pi a^3}\right)\sin\left[\Phi\right]_v\int_{v_o}^{v_f}\frac{v}{\gamma^{-3}}dv \qquad (a) \\
&\quad -\left(\frac{\hbar\Delta k\text{ꝁ}}{2\pi a^2}\right)c^2\int\frac{\cos[\Phi_\nu]}{\nu}d\nu \qquad (b) \\
&\quad +\left(\frac{\hbar w\text{ꝁ}}{\pi a^2}\right)c\int\frac{\cos[\Phi_\nu]}{(1-\nu^2)^{1/2}\nu}d\nu, \qquad (c)
\end{aligned} \tag{32}$$

that we can assimilate to the pro-corpuscular energy in its three forms of expression, kinetic, material and electromagnetic and to the (more general and precise) energy transmutation equation (ETE).

We will not go into the analysis of all the terms here, which will be the subject of another work. But we are going to put in evidence for what matters to us in this one that they are already beginning to take on, in effect, a corpuscular appearance, a question that we will appreciate more notably in the first of them if, dispensing for now with the factor $\mathrm{sen}[\Phi]_v$, we develop it between the limit $v_o = 0$ and any final v:

$$\begin{aligned}\overline{E}_k &= \left(\frac{\hbar \&}{\pi a^3}\right)\int_0^v \frac{v}{\gamma^{-3}}dv = \left(\frac{\hbar \&}{\pi a^3}\right)\int_0^v \frac{v}{(1-v^2/c^2)^{3/2}}dv \\ &= \left(\frac{\hbar \&}{\pi a^3}\right)c^2\left(\frac{1}{(1-v^2/c^2)^{1/2}}-1\right) = m_r c^2\left(\frac{1}{(1-v^2/c^2)^{1/2}}-1\right) \\ &= m_r c^2(\gamma-1) = m_r\gamma c^2 - m_r c^2 = E_t - E_r = E_k,\end{aligned} \tag{33}$$

since it corresponds to the expression describing the relativistic kinetic energy of a particle of mass m_r, which in this case, taking into consideration (27), is:

$$m_r = A^2\frac{\hbar}{a^3} = \frac{\hbar \&}{\pi a^3} = \frac{\hbar b[T]}{\pi a^3} = \frac{2\hbar \times [T]}{(\Delta k)\pi a^3} = \frac{2\,\hbar}{(\Delta k)\pi a^3}, \tag{34}$$

that is, a mass expressed in wave terms or, more accurately, expressed by its wave constituents, which as a whole validates the treatment carried out for $[T]$. Result (33) that implicitly establishes that the other two terms of Eq. (32) correspond to the energetic part of the process that is not kinetic, and that is necessarily related to the corpuscular formation or energetic transit between the initial and final objects of the (dis) encapsulated material, which will require an appropriate and differentiated definition of the limits of integration, but what for the moment we can associate with E_r in Eq. (33) itself and to a pure electromagnetic form or pulse (32b) and (32c) respectively.

In this context, the term (32a) represents an extension or generalization of the relativistic dynamic expression (33), through the presence of an undulatory term $\text{sen}[\Phi]_v$ that is there, but is not perceived because its participation in phenomenology is null except for those training processes that culminate precisely, as we will see, with (because of) its annulment or cancellation.

Having achieved with this, at the expense of being properly characterized, what we initially intended, that is, a recognizable corpuscular energy expression from the wave expression, and even more so if we realize that (32) is of the form:

$$-E_{em} = -hv = mc^2 = m_0c^2 + E_k. \tag{35}$$

which brings us face to face with the mechanism of energy transmutation and, more precisely, with the need for this additional element to achieve it, as the only one capable of making it possible for kinetic energy to dissipate suddenly from a certain value or, in other words, to lose the possibility of expressing itself in this way and be forced to take (reconfiguring itself in the other terms) another more essential (wave) or more optimized (mass) energy form, which is what gives rise, as we will study, to the whole spectrum of elementary particles.

An additional element by which it is evidenced that matter is not inert, since the process of formation of the corpuscle or material encapsulation represented in (32) does not annihilate the wave outside the limits of the particle or does not assimilate all of it in the process of integration (mathematically carried out in the last two terms) but leaves the "suplus" oscillating part as a vestige, and recovery mechanism, of its wave nature in the kinetic term, or in the mass itself that serves as support.

An additional element or factor that will allow us to reach, consequently, many of the fundamentals of Eq. (32), without the need to treat it, through the knowledge or the correct interpretation of the transition processes for each one of the arguments Φ, that takes us to the concept of phase, which we can put for m_r by means of (34) in (31).

$$
\begin{aligned}
[\Phi] &= \left[\Delta k\left(a\gamma^{-1}\right)\right] = \left[2b^{-1}\left(a\gamma^{-1}\right)\right] = \left[2b^{-1}\left(b^{1/3}\left(\frac{\hbar}{\pi m}\right)^{1/3}\gamma^{-1}\right)\right] \\
&= \left[2b^{-2/3}\left(\left(\frac{\hbar}{\pi m}\right)^{1/3}\gamma^{-1}\right)\right] = \left[\beta_i\left(\left(\frac{1}{m}\right)^{1/3}\gamma^{-1}\right)\right]
\end{aligned} \tag{36}
$$

from which will be derived, in agreement with certain conditions all phenomenology of the matter, this is, its creation and transformation, as we will show, in correspondence with the phase changes associated with said processes of transition.

To be more precise, it is the more general Eq. (37) that will allow us to account for all these processes,

$$
\phi = [\Phi]^{-q} = \left[\beta_i\left(\left(\frac{1}{m}\right)^{1/3}\gamma^{-1}\right)\right]^{-q}, \tag{37}
$$

from which we conclude that all our previous development, with which we arrive at Eq. (36), is valid for the particles of charge $q = -1$, being through a generalization of that development (which here we obviate and we place to another work in order not to deviate from the pretensions of this one) how we would reach the last expression.

In a particular way we can find a value Φ_1 for one transition (γ_1) and a value Φ_2 for another transition (γ_2), being γ_i the same as that which would circumstantially appear in the root Eq. (32a), and the relationship (ratio) between both transitions the relationship that really interests us for this purpose, by allowing us to achieve, through the simplification of common parameters, a scheme of all possible transitions, i.e. phases and particles.

It will be as a consequence of reaching this recognizable phase structure through a single condition

$$
\mathrm{sen}[\Phi]_v = 0, \tag{38}
$$

in the phase factor, which entails various conditions or states in the phase Φ, that we can certify what role this factor plays (already theoretically argued) in all the transitions and, properly speaking, its existence, that is to say, to assure the validity of the term Eq. (32a), as promoter, and that of the Eq. (32) of starting point (as well as the pertinence of the estimations carried out to obtain it), insofar as the equation certifies the non-casual character of said phase structure or the physical nature (reality) of the same, sustained in the energetic transit, and in the capacity to reach discrete material states by means of it.

In other words, the equation establishes simple phase relations that are only satisfied by the masses of the elementary particles that exist in our universe, which thus serve as an experimental verification.

5. Sole and simple relation for all particles in the standard model

Consequently, and based on the above expression, we will indeed look at a simple relationship among all particles of the Standard Model that has to do with an even simpler relation in a certain phase or state diagram, whereby the different generations of particles are in different phases from $\phi \equiv (\phi_0, \phi_1, \phi_2, \ldots)$ each other (which gives an idea of periodicity), and consequently phasic distances among them, which are determined by the limit value of the speed they can develop in the same, which in turn is determined by the inertial mass of the particle in question by means of the Lorentz factor γ. The above represents the outline of a theoretical framework derived from the one we have just seen and from the potential of the relationship itself. Here we will start from the relationship, taken as a formula, i.e. treated as if it were empirical, without going into theoretical questions beyond what is strictly necessary to the understanding of the factors that make it up, because by means of this relationship we are establishing a structured and hierarchical relation of the particles or organization of all matter with an argument as free of analysis or manipulation, simple and unequivocal, as that provided by the equation of a straight line. This approach, on the other hand, will allow us to carry out a summary treatment and achieve an overall view of the entire particle system.

Consequently, and now explicitly, we will develop what has already been expressed according to the following **statement**:

All fundamental particles of matter (fermions) in the SM satisfy the relationship of phases expressed as follows:

$$\chi = \frac{\phi_2}{\phi_1} = \frac{\left[\beta_2\left(\frac{1}{m_2}\right)^{1/3}\gamma_2^{-1}\right]^{-q}}{\left[\beta_1\left(\frac{1}{m_1}\right)^{1/3}\gamma_1^{-1}\right]^{-q}} = \left[\frac{\beta_2}{\beta_1}\left(\frac{(m_2^2/m_3^3)}{(m_1^2/m_2^3)}\right)^{1/3}\right]^{-q}$$

$$= \left[\theta_0^2\left(\frac{m_2^5}{m_1^2 m_3^3}\right)^{1/3}\right]^{-q} = \left[\alpha\left(\frac{m_2^5}{m_1^2 m_3^3}\right)^{1/3}\right]^{-q} = \alpha^{-q} \times \mathfrak{R}_m{}^{-q}, \tag{39}$$

for phase $\phi_2 = p\pi$ and $\phi_1 = q\pi \quad : \; p, q \in \mathbb{N}$,

with $\gamma_1 = m_2/m_1$ and $\gamma_2 = m_3/m_2$,

with m_1, m_2 and m_3 being the triad or natural sequence of particles of a given class, i.e. the generations of that class, of electric charge q, susceptible of being related by means of Lorentz factors γ_1 and γ_2 according to Eq. (37), for the numerator and denominator.

That is, Eq. (39) establishes that, just as there is a phase through (37) for the first two particles of a triad, and there is a phase through (37) for the second two (the central one participates in both), there is a simple ratio χ between both relationships (37). A simple ratio that will not only allow us to verify the relationship between the known particles but also to propose a third one (as we will see later), either in ascending or descending sequence, knowing two others.

The relationship is simpler, if possible, with respect to the one presented if we take into account that $\beta_2 \approx \beta_1$, and that, consequently, $\alpha \cong 1$. That is, according to (36) β_i depends on b, associated with the normalization constant B in (13), which from (24) evolved as the constant A, so that it has a single value for all particles of the same value q, except for the fact that the integral (23) from which the constant A derives is not established for each particle in the interval $[0, c]$ defined there but for the one defined by the final v, according to γ_1 and γ_2, which satisfies (38), which may introduce a small deviation. It can be verified with the cases solved in the following sections that this deviation is only significant in one of them (that of the charged leptons, with $1/\alpha^{-q} \cong 1.04$), being defined, in any case, through $\alpha = \theta_0^2$ as the established ratio between the angle subtended by the speed v_2, associated with γ_2, and the one subtended by c, that is, $\theta_0^2 \equiv \theta_2/\theta_0 \equiv \theta(v_2)/\theta(c)$. The latter will be well understood if we pay attention to the fact that β_1 is characterized by the transit defined by γ_1, and β_2 by the one defined by γ_2 on the first, so that $\beta\theta_0^1 = \beta_1$ and $\beta\theta_0^1\theta_0^2 = \beta_1\theta_0^2 = \beta_2$, and consequently $\alpha = \beta_2/\beta_1 = \theta_0^2$.

On the other hand, and as a consequence of the previous simplification, we will be able to apply Eq. (39) without needing to know the phases. That is, we can apply it and find the value χ (valid for different numerator and denominator phase values) that relates to three given particles of a triad. In fact, the reason and difference of introducing and working with Eq. (39) are the suppression of the internal constants or variables (undulatory) and the possibility of making an external treatment to the particles, unlike the internal one carried out for each one of them through (37) or, being more exact, of (36) for being the one that supports all those variables.

6. The charged particles

The requirement presented can be made explicitly for each of the types of charged particles and assigned to each of them according to Eq. (39) a simple theoretical χ value compatible with the value of its masses:

6.1 Leptons

$$\chi = \frac{\phi_2}{\phi_1} = \left[\alpha\left(\frac{m_2^5}{m_1^2 m_3^3}\right)^{1/3}\right] = \left[\alpha\left(\frac{m_\mu^5}{m_e^2 m_\tau^3}\right)^{1/3}\right] = 2 \quad \left(\propto\frac{2\pi}{1\pi}\right), \tag{40a}$$

Value, in this case, is achieved by $\alpha^{-q} = \alpha = 0.962$ and $\mathfrak{R}_m^{-q} = \mathfrak{R}_m$ for the values m_1, m_2, m_3 known. In effect, with:

$$\gamma_2 = m_3/m_2 = 16.82 \quad (\Rightarrow v_2/c = 0.99823 \equiv \sin\theta_2),$$

$$\text{and} \quad \alpha = \theta_0^2 = \frac{\arcsin(\sin\theta_2)}{\theta_0} = \frac{86,59^o}{90^o} = 0.962,$$

$$\text{we have} \quad \chi = \alpha \times \mathfrak{R}_m = 0.962 \times 2.079 = 2.$$

Treatment with which, as we have already mentioned, we do not obtain the phases but a simplified ratio χ between them, on which we can establish different equivalent proportionality relationships, which we can only solve or make unique with the knowledge of β_i or assuming that the simplest relationship is the correct one, which will allow us to calculate β_i for each of the two phases applied to Eq. (37).

6.2 Quarks (2/3 q)

$$\chi = \frac{\phi_2}{\phi_1} = \left[\alpha\left(\frac{m_2^5}{m_1^2 m_3^3}\right)^{1/3}\right]^{-2/3} = \left[\alpha\left(\frac{m_c^5}{m_u^2 m_t^3}\right)^{1/3}\right]^{-2/3} = 1.5 \quad \left(\propto \frac{3\pi}{2\pi}\right), \tag{40b}$$
$$\left(\text{with} \quad \alpha^{-2/3} = 1.005 \cong 1\right)$$

Analogous to the comment made above, for $\beta_2 = \beta_1 = \beta$ in this case, with which it could be shown, inversely and known β, that by means of Eq. (37) we can reach phase $\phi_1 = 2\pi$ when the rest mass $m_1 = m_u$ reaches the value $m_2 = m_c$ as inertial mass, and phase $\phi_2 = 3\pi$ when the rest mass $m_2 = m_c$ reaches the value $m_3 = m_t$ as inertial mass.

6.3 Quarks (−1/3 q)

$$\chi = \frac{\phi_2}{\phi_1} = \left[\alpha\left(\frac{m_2^5}{m_1^2 m_3^3}\right)^{1/3}\right]^{1/3} = \left[\alpha\left(\frac{m_s^5}{m_d^2 m_b^3}\right)^{1/3}\right]^{1/3} = 0.5 \quad \left(\propto \frac{1\pi}{2\pi} = \frac{2\pi}{4\pi}\right), \tag{40c}$$
$$\left(\text{with} \quad \alpha^{1/3} = 1.004 \cong 1\right)$$

Analogous comment to the previous case, for their respective phases and masses, since the treatment (which we will exemplify in the following section) is universal and valid for all particle triads.

6.4 Summary

The above results are shown in **Table 1**, where it is more clearly evidenced that we have been able to adjust the χ values with certain values, chosen (except for the leptons, which are known and well defined) from among the possible experimental ones, for the masses of the different particles (see Appendix B) for reasons that we

Triad	Particle	Mass (MeV)	$\chi(\phi_2/\phi_1)$
m_0	e	0.511	
m_1	μ	105.658	2
m_2	τ	1776.840	
m_0	u	2.368	
m_1	c	1361.698	1.5
m_2	t	172,950	
m_0	d	5.395	
m_1	s	83.883	0.5
m_2	b	4180.425	

Table 1.
Value χ, according to Eqs. (40), for each triad of values m_1, m_2, m_3, corresponding to the three generations of particles, for each of the three classes of charged particles.

will show in the following section, derived from the ability to propose the masses of the particles already mentioned.

Results that are equally applicable and valid for the antiparticles, which, with the same values for the masses and charge of the opposite sign, will give as result in Eq. (39) a phase inversion with respect to the particles, fulfilling that:

$$\chi_i \overline{\chi}_i = 1. \tag{41}$$

7. The $\mathfrak{M}$ particle

Although in a first evaluation, we did not say that expression (39) determined the value of the masses, but that the value of the masses is regulated by (meets) the Eq. (39), in a second evaluation, we will be able to discriminate the values by means of it, that is, to overcome the indetermination of these experimental data or to correct them. And the latter as a consequence of the fact that the relationship (39) participates in another relational element, by which it can be stated that for all types of charged particles we can find a particle of previous hierarchy m_0 which equally fulfils the expression (39),

$$\chi_0^L = \frac{\phi_1}{\phi_0} = \left[\alpha\left(\frac{m_1^5}{m_0^2 m_2^3}\right)^{1/3}\right]^{-q} = \left[\left(\frac{m_e^5}{m_0^2 m_\mu^3}\right)^{1/3}\right] = 1, \tag{42a}$$

$$\chi_0^{q+} = \frac{\phi_1}{\phi_0} = \left[\alpha\left(\frac{m_1^5}{m_0^2 m_2^3}\right)^{1/3}\right]^{-q} = \left[\left(\frac{m_u^5}{m_0^2 m_c^3}\right)^{1/3}\right]^{-2/3} = 1, \tag{42b}$$

$$\chi_0^{q-} = \frac{\phi_1}{\phi_0} = \left[\alpha\left(\frac{m_1^5}{m_0^2 m_2^3}\right)^{1/3}\right]^{-q} = \left[\left(\frac{m_d^5}{m_0^2 m_s^3}\right)^{1/3}\right]^{1/3} = 4. \tag{42c}$$

with the particularity that such a particle is only one for all of them, or of equal mass, which we will now identify as $\mathfrak{M} = 171.87\text{eV}$, for the different masses m_1, m_2 previously defined, with which it forms a triplet of new triads, being by means of the first, Eq. (42a), as we have calculated the exact value of $\mathfrak{M}$ through the exact value of e and μ, which gives rise to well-defined $\chi_0^{q\pm}$ values for the other expressions (42).

Triad, in which, according to the position of the particles and the equations, the ϕ_1 value for χ of **Table 1** takes the position of the numerator, while the new denominator is ϕ_0, as we present in **Table 2**.

In conclusion, taking the ϕ phases as distances, what we are saying is not only that the different charged fermions are found one from the other at defined distances, quantized or marked by the natural numbers, but that following the quantization process we find the same particle defined in the vicinity of each of the three differentiated electric charges, even equidistant ($\phi_0 = \phi_1$) for two of the three classes of particles. From this point on, being particles of the same value, we can think, as we have done initially, that they are the same and common particle $\mathfrak{M}$, or an equivalent energetic entity, which is at a level prior to the first generation for the three classes of elementary charged particles, which, consequently, can be taken as a zero generation or precursor particle and as evidence of a shared origin.

Triad	Particle	Mass (MeV)	$\chi(\phi_1/\phi_0)$
m_0	$\mathfrak{M}$	0.00017187	
m_1	e	0.511	1/1
m_2	μ	105.658	
m_0	$\mathfrak{M}$	0.00017187	
m_1	u	2.368	2/2
m_2	c	1361.698	
m_0	$\mathfrak{M}$	0.00017187	
m_1	d	5.395	4/1
m_2	s	83.883	

Table 2.
Phase, according to Eq. (42), established between two first generations of particles m_1, m_2 for each of the three families with charged, and a particle of a zero generation shared, m_0, named as $\mathfrak{M}$.

A common particle m_0 of lower hierarchical order for the establishment of a new triad, in accordance with Eq. (39) and for $\alpha \cong 1$, that, on the other hand and by virtue of the exact definition of its value, conditions the values of the particles m_1, m_2, i.e. their choice from the whole experimental spectrum, as we have already mentioned (Section 6) in connection with the choice of these values in the previous triad.

Expressed in a more exhaustive form, we see that m_0 imposes an additional condition on (39), so that it sets m_1^5/m_2^3 by χ at Eqs. (42), according to the experimental values (m_1, m_2), which in turn determines m_2^5/m_1^2, which in turn determines exactly m_3 at Eqs. (40), and vice versa. In this case, if we alter the m_1, m_2 values while keeping the m_1^5/m_2^3 ratio at Eqs. (42), the m_2^5/m_1^2 ratio is changed to Eqs. (40) and consequently m_3. It follows that only one pair of values fulfils the double condition, and that this pair cannot fluctuate in their values more than the fluctuation in the value of m_3, that is, its experimental error.

We can see this in another way, if we notice that associated to the expression (39) are the separate conditions about ϕ, that is, ϕ_1 and ϕ_2, conditions that, particularized to expressions (40) and (42) take for each type of charge the following forms, which are simply the corresponding Eq. (37) with the γ_i developed:

$$\begin{aligned} \phi_2 &= \left[\beta\left(m_2^2/m_3^3\right)^{1/3}\right]^{-q}, \quad \phi_1 = \left[\beta\left(m_1^2/m_2^3\right)^{1/3}\right]^{-q}, \quad \text{(a)} \\ \phi_1 &= \left[\beta\left(m_1^2/m_2^3\right)^{1/3}\right]^{-q}, \quad \phi_0 = \left[\beta\left(m_0^2/m_1^3\right)^{1/3}\right]^{-q}, \quad \text{(b)} \end{aligned} \qquad (43)$$

Consequently, we have a system of three equations (ϕ_0, ϕ_1, ϕ_2) with three unknown values (m_1, m_2, m_3), which has a single solution (a triad of values), and a free parameter β (for the more general case $\alpha = 1 \Rightarrow \beta = \beta_1 = \beta_2$), which adjusts this solution within the experimental values or its neighbourhood, in a solidary way (by the dependence reached through the equations) which in fact supposes the restriction of the experimental spectrum or exact definition of its values. Specifically, from m_0 and fixed a value β we have the whole itinerary until m_3, we only have, therefore, to choose the value β that reaches our value $m_3 \pm \varepsilon_3$, and we will have $m_1 \pm \varepsilon_1$ and $m_2 \pm \varepsilon_2$, which we can particularize for a negatively charged quark which satisfies its transitions for a fixed value $\beta = 5411406.96$, characteristic of the family of particles $(-1/3)$-quarks, in this case, for which it is not difficult to calculate the speeds for which such transitions occur (see **Figure 4**), according with $\gamma_i = m_{i+1}/m_i$, just like we did for the leptons in Eq. (40a). We will be able to see the Itinerary in detail in Appendix C.

We can understand all the transitions described by Eqs. (43), and the original Eq. (37), like the equation of a curved line of curvature β which passes over a point B, and with origin in A. In this case, the origin A is m_0, and the point B on which it passes is m_1, which is the origin of the following equation, as we have represented in **Figure 4**, where it is shown that the phase is, in fact, the evolution of a particle as a function of speed up to the limit, for a determined speed, in which it stops being that particle and is another one (phase change), what, taken to an extreme, could be understood as a process of energetic recombination, that is, of real conversion of kinetic energy into formation energy, demonstrating that both energies are the same thing, nothing extraordinary if we take into account that these phases $\phi \equiv (\phi_0, \phi_1, \phi_2, \ldots) = n\pi$, or maximum value of v, correspond (see Eq. (38)) with the zeros ($\text{sen}[\phi]_v = 0$) of the kinetic energy for the same values γ_i developed in Eq. (43) for ϕ_i, while the expression (39) itself already determines that the size of the phases are defined through the velocities at which the respective Lorentz factors make the inertial mass of a particle coincide with the real mass of the hierarchically subsequent one, evidencing that this inertial mass can only reach this maximum value, and that another value higher than this belongs to another phase, whether the interface is

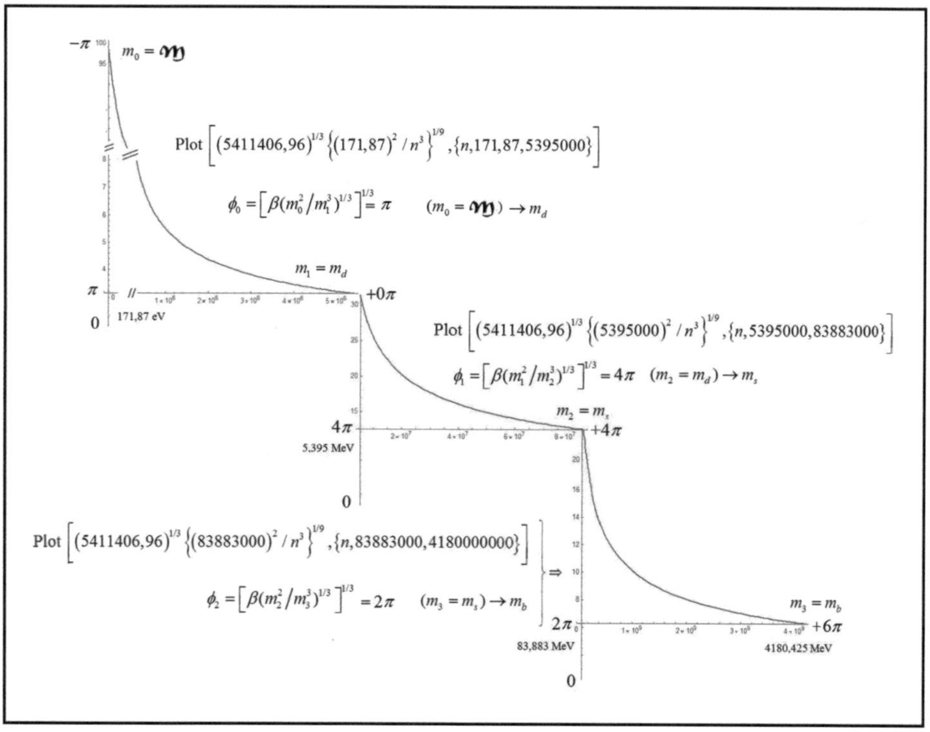

Figure 4.
Phase transitions for negatively charged quarks (−1/3). Evolution of phase as a function of inertial mass according to Eqs. (5) and concatenation of these equations by converting the inertial mass (kinetic energy) into matter (formation energy) of a higher generational order particle at the point of equivalence. Where we have taken m_1 as the reference and, consequently, ϕ_0, negative.

presented as a place of transit or as a differentiation and separation of states (Appendix D).

In this Appendix D where the physical reality underlying the introduction of the particle $\mathfrak{M}$ is evaluated, that is to say, on the one hand, it is emphasized that the phase structure is governed by γ_i, in such a way that whenever we talk about a particle we talk about the one reached, for a phase, as inertial mass of another, fulfilling (38), and with this that γ_i is the only way to relate the candidates to particles, and the phases ϕ_i the way to measure their energy distance between them, which introduces a new limitation (discretization) of the values. This allows to establish or filter the possible values (a discrete set) from a whole universe of values. On the other hand, we study the degree of convergence and, consequently, the impossibility that, according to the phasic structure, a supposed and unknown particle $\mathfrak{M}$ (which underlies the others) is not there: it must be that one, in case there is one, and at those phasic distances.

We say in case there is one because it can be a particle that has fulfilled a function, and no longer exists, or an energetic entity that conforms, given its characteristics, the dark matter and all the genesis of the other (we start from the SWP process). From there, we cannot assure its existence beyond the fact that it is through it, through the phase reached, and repeating the process, that we also reach particles of which we do have knowledge, as we will study in the next section.

8. Neutrinos

We see that all first-generation particles converge on the particle m_0 i.e. the values increasingly diverge from it, differentiated by charge (e, u, d), for different phase values. Reciprocally, it is possible to consider the existence of a divergent and decreasing previous generation from m_0 that posed in the line of progress of charged leptons and/or those not charged involve, for different phases, values lower-order particles that we could identify with neutrinos.

In that consideration, we find in the progress line of charged leptons, also used in Eq. (42a), a number of candidates m_ν in the role of m_{-1}, that is, those that would be reached by varying the values $\phi_\nu = (1, 2, 3, 4, 5\ldots) \times \pi$ with respect to ϕ_0 in relationship (39) for $m_0 = \mathfrak{M}$ and from which we will have to choose those that could truly represent our neutrinos. That is, in this case it is not relevant to find the points (possible neutrino masses) by means of Eq. (39), but to verify that some of them, quantized or marked by means of the natural numbers by ϕ_ν in (39) in the neighbourhood of m_0, truly correspond to those neutrino masses, a verification that can only be done by means of physical criteria.

This verification is achieved in a first order as a consequence of the fact that the number of points reached by the line of progress of charged leptons is restricted (as we will see) by the line of progress of uncharged leptons, from which it results that ν_a, ν_b, and ν_c (by $\phi_\nu = 3\pi, 6\pi, 9\pi$), represent a set of previous (and necessary) generations for charged leptons, that is (with $\alpha \cong 1$ and $\nu_i \equiv$ mass of ν_i):

$$\chi_a^e = \left[\alpha\left(\frac{m_0^5}{\nu_a^2 m_1^3}\right)^{1/3}\right] = \left[\left(\frac{\mathfrak{M}^5}{\nu_a^2 m_e^3}\right)^{1/3}\right] = 1/3, \tag{44a}$$

$$\chi_b^e = \left[\alpha\left(\frac{m_0^5}{\nu_b^2 m_1^3}\right)^{1/3}\right] = \left[\left(\frac{\mathfrak{M}^5}{\nu_b^2 m_e^3}\right)^{1/3}\right] = 1/6, \tag{44b}$$

$$\chi_c^e = \left[\alpha\left(\frac{m_0^5}{\nu_c^2 m_1^3}\right)^{1/3}\right] = \left[\left(\frac{\mathfrak{M}^5}{\nu_c^2 m_e^3}\right)^{1/3}\right] = 1/9. \tag{44c}$$

for $m_a = 0.00551$ eV, $m_b = 0.01559$ eV, $m_c = 0.02863$ eV.

While $\nu_d = \nu_{12} = 0.0448$ eV for $\phi_\nu = 12\pi$ would be plausible but not functionally necessary (it is not, as we shall see in next section, to construct a basis), and the rest of the subscripts, not plausible or invalidated by the experimental reality derived from neutrino transitions in decays or, as we said, of the progress line of uncharged leptons. Assertions that require to demonstrate that the same treatment can be carried out with neutrinos as that already carried out with charged particles, and that, taking as a reference the values supplied by the disintegration processes, which we will assume to be reliable, we will demonstrate, even though they are not accompanied by a lower limit:

$$\nu_e \cong 0 \text{ eV} \quad \nu_\mu \leq 0.17 \text{ MeV} \quad \text{and } \nu_\tau \leq 17 \text{ MeV}, \tag{45}$$

and although it is evident that our main working relationship is not fulfilled for these values, given the energetic distance between them, and mainly the five orders of magnitude between the first two.

It is precisely because of this disproportionate distance that we contemplate the possibility that there is an intermediate particle intercalated between ν_e and ν_μ that allows the transit and verification of Eq. (39). It does indeed exist, and not only that, but it is also, surprisingly, the particle m_0, by means of which not only one solution for (39) is reached but, as we have already anticipated, three ($\phi_\nu = 1, 2, 3$), satisfying some equations equivalent to Eq. (44) for the same (ν_a, ν_b y ν_c), as candidates for ν_e, and the particle ν_μ (with $\alpha \cong 1$ and $\nu_\mu \equiv$ mass of ν_μ):

$$\chi_a^\nu = \left[\alpha\left(\frac{m_0^5}{\nu_a^2 m_1^3}\right)^{1/3}\right] = \left[\left(\frac{\mathfrak{M}^5}{\nu_a^2 \nu_\mu^3}\right)^{1/3}\right] = \frac{\phi_\mu}{\phi_\nu} = 3/3\ (\propto 1/1), \tag{46a}$$

$$\chi_b^\nu = \left[\alpha\left(\frac{m_0^5}{\nu_a^2 m_1^3}\right)^{1/3}\right] = \left[\left(\frac{\mathfrak{M}^5}{\nu_b^2 \nu_\mu^3}\right)^{1/3}\right] = \frac{\phi_\mu}{\phi_\nu} = 3/6\ (\propto 1/2), \tag{46b}$$

$$\chi_c^\nu = \left[\alpha\left(\frac{m_0^5}{\nu_a^2 m_1^3}\right)^{1/3}\right] = \left[\left(\frac{\mathfrak{M}^5}{\nu_c^2 \nu_\mu^3}\right)^{1/3}\right] = \frac{\phi_\mu}{\phi_\nu} = 3/9\ (\propto 1/3). \tag{46c}$$

which correlate with a subsequent triad, for which m_0 is also a solution, i.e. which also satisfies Eq. (39) for the same in another hierarchical situation, thus demonstrating that it is truly the concatenation element of the sequence:

$$\chi_0^\nu = \left[\alpha\left(\frac{m_1^5}{m_0^2 m_2^3}\right)^{1/3}\right] = \left[\left(\frac{\nu_\mu^5}{\mathfrak{M}^2 \nu_\tau^3}\right)^{1/3}\right] = \frac{\phi_\tau}{\phi_\mu} = 3/3\ (\propto 1/1), \tag{47}$$

and in which the last equivalence (member) guarantees in Eq. (46) that the values ϕ_ν reached (which we know what they are because they are associated with the masses ν_a, ν_b y ν_c) correspond to those obtained in the Eqs. (44), what, on the other hand, determines the real value of phase ϕ_μ of connection with m_0 for ν_μ in (46), with respect to phase ϕ_0 for e in Eqs. (44) for these same values $\phi_\nu = 3, 6, 9$. Results that we can put in respective **Tables 3** and **4**:

where the values reached for ν_a, ν_b y ν_c can be considered as theoretical and exact in **Table 3**, since the other two factors of the triad are exact, while those obtained in **Table 4** (and the triad (47) for $\nu_\tau = 16636417$ eV) can be considered as experimental or subject to the imprecise experimental limits reflected in Eq. (45). A difference of values that we will later contextualize and revalidate in another way (Section 9).

We see, in conclusion, that the infinite generations or possibilities of neutrinos are restricted to three in the line or chain of charged leptons in correspondence with the first three generations ($\phi_\nu = 1, 2, 3$), for $\phi_\mu = 1$, in the chain of uncharged leptons. Likewise, we see that, just as there is the sequence m_0, m_1, m_2, m_3, for charged particles, there is for uncharged particles in the form $\nu_e, m_0, \nu_\mu, \nu_\tau$ with its two corresponding triads. One of the triads is Eq. (47) and the other, the resultant of the Eqs. (46), in which ν_e does not appear, but ν_a, ν_b, ν_c, which we can understand as a natural base of neutrinos (supplied

Triad	Particle	Mass (eV)	$\chi(\phi_0/\phi_\nu)$
$m_{\nu 1}$	ν_a	0.00551	
m_0	$\mathfrak{M}$	171.87	1/3
m_1	e	511,000	
$m_{\nu 1}$	ν_b	0.01558	
m_0	$\mathfrak{M}$	171.87	1/6
m_1	e	511,000	
$m_{\nu 1}$	ν_c	0.02863	
m_0	$\mathfrak{M}$	171.87	1/9
m_1	e	511,000	

Table 3.
Phases and values of corresponding to those phases, according to Eq. (44), for the sequence ν_1, m_i, e^-, corresponding to the first generation of each type of lepton and an intermediate particle m_i represented by $\mathfrak{M}$.

Triad	Particle	Mass (eV)	$\chi(\phi_\mu/\phi_\nu)$
$m_{\nu 1}$	ν_a	0.0056	
m_0	$\mathfrak{M}$	171.87	3/3 = 1/1
$m_{\nu 2}$	ν_μ	168,564	
$m_{\nu 1}$	ν_b	0.0158	
m_0	$\mathfrak{M}$	171.87	3/6 = 1/2
$m_{\nu 2}$	ν_μ	168,564	
$m_{\nu 1}$	ν_c	0.0291	
m_0	$\mathfrak{M}$	171.87	3/9 = 1/3
$m_{\nu 2}$	ν_μ	168,564	

Table 4.
Phases and values of corresponding to those phases, according to (46), for the sequence ν_1, m_i, ν_2, corresponding to the first two generations of neutrinos and to an intermediate particle represented by $\mathfrak{M}$.

by the chain of charged leptons) with which we can build, as we will see later, a massive base and, where appropriate, a flavour (ν_e). That is, the natural neutrino base is created, associated to e, through m_0 by the charged lepton chain (m_ν, m_0, m_e), and then progresses (occasionally with the generational change of e itself) with certain configurations (flavours) through m_0 by its chain of distribution (ν_μ, ν_τ), which involves a modification with respect to the massive base (oscillations).

We also see that the above creation corresponds to a seesaw mechanism [7] that does not require the mass of a charged fermion (electron) under a large mass (GeV-TeV) M_R ($m_e^2 = \nu_i M_R$), but an intermediate mass m_0 ($\mathfrak{M}$), which had a reserved energy space (between ν and e), under the medium value of the electron mass, for the potential function developed in the phase relation (39), which we can write as some arms of the seesaw:

$$m_0^5 = \nu_i^2 (m_e \chi_i)^3 \rightarrow m_0^2 = \nu_i^{4/5} (m_e \chi_i)^{6/5} \tag{48}$$

which is not due to the exchange of bosons or fermions referred to in this paradigm [8], but to the energy conversion process itself (fermion exchange, in this case) represented by Eq. (37), together with the conceptual and practical differentiation, which we will not discuss here, between electron and free electron.

On another level, Eqs. (44) and (46) clear any doubt about some kind of coincidence or chance in (42) and put us in the idea that truly such a particle $\mathfrak{M}$ exists, either as a particle or as a circumstantial element of energy distribution, since it is valid for all types of particles and antiparticles, charged and uncharged, or even as the foundation of all present and non-present matter (dark matter), through which the entire spectrum of particles is systematically distributed in phase units, and every particle in this spectrum defined by the phase reached, and by its mass as an evolutionary parameter. A situation that we can represent (**Figure 5**):

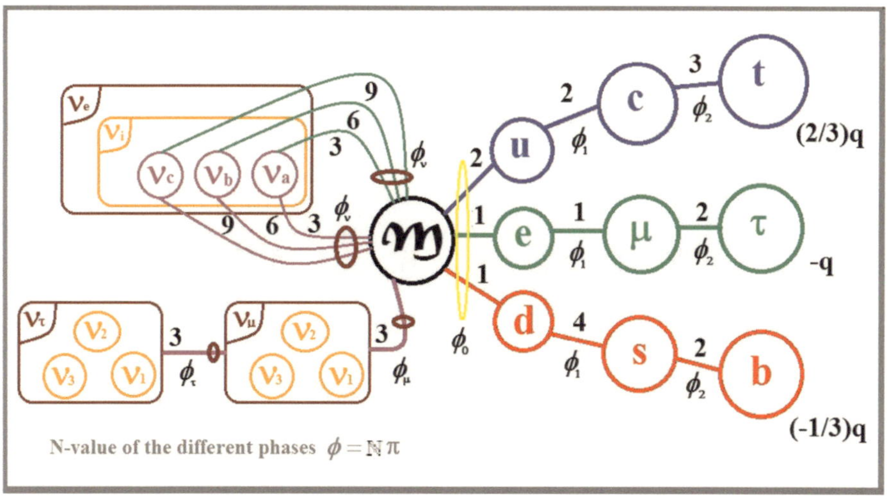

Figure 5.
*Phases diagram for 2/3 charge quarks, −1/3 charge quarks and charged leptons. Each generation of a class of particles is nothing but a phase or differentiated state of the same particle in its evolution or progress (of phase) from the fundamental state $\mathfrak{M}$ (as already shown in **Figure 4**). Likewise, charged leptons progress through $\mathfrak{M}$ by means of Eq. (39) towards the natural base of neutrinos ν_a, ν_b, ν_c. The same ones can then form the massive base of neutrinos and progress in this way through $\mathfrak{M}$ with any state of oscillation, and do so, in particular, in one of the eigenstates that we call flavours, associated to each of the charged leptons, from the initial ν_e.*

9. The massive base

We cannot address the issue of the neutrinos base without presenting the corresponding data to antineutrinos:

$$\overline{\chi}_a^\nu = \left[\left(\frac{m_0^5}{\overline{\nu}_a^2\overline{\nu}_\mu^3}\right)^{1/3}\right]^{-1} = \frac{\overline{\phi}_\mu}{\overline{\phi}_\nu} = 3/3 = \left(\chi_a^\nu\right)^{-1}, \tag{49a}$$

$$\overline{\chi}_b^\nu = \left[\left(\frac{m_0^5}{\overline{\nu}_b^2\overline{\nu}_\mu^3}\right)^{1/3}\right]^{-1} = \frac{\overline{\phi}_\mu}{\overline{\phi}_\nu} = 3/2 \neq \left(\chi_b^\nu\right)^{-1}, \tag{49b}$$

$$\overline{\chi}_c^\nu = \left[\left(\frac{m_0^5}{\overline{\nu}_c^2\overline{\nu}_\mu^3}\right)^{1/3}\right]^{-1} = \frac{\overline{\phi}_\mu}{\overline{\phi}_\nu} = 3/1 = \left(\chi_c^\nu\right)^{-1}. \tag{49c}$$

We can realize that as a consequence of the change of sign of the exponent, we switch from a multiplier phase relation to a divider phase relation, and so while for neutrinos the phase space $\phi_\mu = \pi$ implies $\phi_\nu = \pi, 2\pi, 3\pi$, for antineutrinos the sequence is inverse over the largest ϕ_ν taken as $\overline{\phi}_\mu$ ($\overline{\phi}_\mu = 3\pi$ implies $\overline{\phi}_\nu = 3\pi, 2\pi, \pi$), and that, in consequence, the phase $\overline{\chi}_b^\nu$ corresponding to the second antineutrino of the natural base, and consequently the mass, differs from that of the neutrino. Mass that can be calculated by Eq. (49b) or more precisely and generally, as we did with the neutrinos, from $\overline{\chi}_\nu^e$ corresponding to Eqs. (44). This is ($\overline{\phi}_0 = 9\pi$ implies $\overline{\phi}_\nu = 3\pi, 2\pi, \pi$):

$$\overline{\chi}_a^e = \left[\left(\frac{m_0^5}{\overline{\nu}_a^2 m_{\overline{e}}^3}\right)^{1/3}\right]^{-1} = \frac{\overline{\phi}_0}{\overline{\phi}_\nu} = 9/3 = \left(\chi_a^e\right)^{-1}, \tag{50a}$$

$$\overline{\chi}_b^e = \left[\left(\frac{m_0^5}{\overline{\nu}_b^2 m_{\overline{e}}^3}\right)^{1/3}\right]^{-1} = \frac{\overline{\phi}_0}{\overline{\phi}_\nu} = 9/2 \neq \left(\chi_b^e\right)^{-1}, \tag{50b}$$

$$\overline{\chi}_c^e = \left[\left(\frac{m_0^5}{\overline{\nu}_c^2 m_{\overline{e}}^3}\right)^{1/3}\right]^{-1} = \frac{\overline{\phi}_0}{\overline{\phi}_\nu} = 9/1 = \left(\chi_c^e\right)^{-1}, \tag{50c}$$

from what it turns out:

$$\overline{\nu}_a = 0.00551 \text{ eV}, \overline{\nu}_b = 0.01008 \text{ eV}, \text{ and } \overline{\nu}_c = 0.02863 \text{ eV},$$

which represents a natural base of antineutrinos different from that of neutrinos, which, however, allows the creation of identical massive bases:

$$\begin{aligned} \nu_1 &= \nu_a = 0.00551 \text{ eV} = \overline{\nu}_a, \\ \nu_2 &= \nu_b - \nu_a = (0.01559 - 0.00551) \text{ eV} = 0.01008 \text{ eV} = \overline{\nu}_b, \\ \nu_3 &= \nu_a + \nu_b + \nu_c = (0.00551 + 0.01559 + 0.02863) \text{ eV} \\ &= 0.04973 \text{ eV} = 2\overline{\nu}_a + \overline{\nu}_b + \overline{\nu}_c, \end{aligned} \tag{51}$$

With (as required experimentally):

$$\sum\nolimits_{\nu} m_{\nu} = \nu_1 + \nu_2 + \nu_3 = 0.06532 < 0.1 \text{ eV}. \tag{52}$$

Values that clear any uncertainty and ignorance about them, the hierarchy of particles and related issues (see Ref. [9], p. 301), being fulfilled for the oscillations of the solar and atmospheric neutrinos, respectively:

$$\begin{aligned} \Delta m_{21}^2 = 7.113 \times 10^{-5} \text{ eV}^2 \text{ and } \Delta m_{31}^2 = 2.442 \times 10^{-3} \text{ eV}^2, \\ \text{with } \Delta m_{21}^2 = m_2^2 - m_1^2 \text{ and } \Delta m_{31}^2 = m_3^2 - m_1^2 \end{aligned} \tag{53}$$

Results congruent with the experimental data [10], which oblige, on the other hand, a range of values for the masses, congruent with those reached in (51). Results that, nevertheless, can be placed in **Table 5** together with other close ones (including the experimental ones) in order to appreciate the real gap and evaluate our calculations.

First, we have obtained the table by infinitesimal variations of in (44) and adjusted as in (51) again. That said, we see that the experimental data (highlighted) are reached for masses different from the initial theoretical masses, which are additionally different for solar (R-8) and atmospheric (R-5) neutrinos, with the particularity that the latter remain with this mismatch (row) even considering their experimental error, while the former can vary $\pm 5.10^{-5}$ eV considering theirs, and may be placed with the value 7.347 in the same row R-5 (0.00560 eV) as the latter. Giving the latter as less uncertain, the former could not only occupy the same row R-5 but would have to occupy it since, from our approach, Δm_{21}^2 and Δm_{31}^2 have linked evolution through ν_1 (each row is a state defined: there is no mass that gives rise to the two experimental values or row) by $\chi_i^e\ [\nu_{b,c} = \nu(\nu_a = \nu_1)]$ in (44).

Consequently, the plausible ν_1 value for the experimental values, Δm_{21}^2 and Δm_{31}^2 is 0.00560 eV which we have likewise considered as experimental (theoretical and not exact of **Table 4**) with respect to the exact, strictly theoretical and exact value 0.00551 eV of **Table 3**. Being in any case the possible error $< 10^{-4}$eV.

R	Mass ν_a (eV)	$\Delta m_{21}^2/10^{-5}$	$\Delta m_{31}^2/10^{-3}$
1	0.00551	7.113	2.442
2	0.00554	7.189	2.468
3	0.00558	7.296	2.505
4	0.00559	7.323	2.514
5	0.00560	7.347	**2.522**
6	0.00561	7.375	2.532
7	0.00562	7.402	2.541
8	0.00563	**7.427**	2.549
9	0.00564	7.455	2.559

Table 5.
Difference of squares for solar and atmospheric mass depending on the massive value of ν_1, where we have highlighted the (approx.) best experimental fit, written in red the theoretical values, and marked (blue) the most plausible row.

From there, and taking into consideration that our theoretical calculation is practically arithmetical in Eq. (42a), it would remain to determine whether this experimental deviation is from the ideal situation (that of the model), or on the contrary is attributable to the methodology used (which contains theoretical and indirect elements) to calculate the Δm_{21}^2 and Δm_{31}^2 values

10. CP violation

A physical system with this configuration presents an asymmetry that would justify the structural nonexistence of antimatter in the known universe, or supremacy of matter over antimatter as a consequence of the different number of construction elements in both species, and the concurrence of fewer of them (the particles $\mathfrak{M}$ from which the natural base starts) in a process of mutual annihilation:

$$\begin{aligned} \sum\nolimits_{\nu} m_{\nu} &= \nu_1 + \nu_2 + \nu_3 = \nu_a + 2\nu_b + \nu_c, \\ \sum\nolimits_{\overline{\nu}} m_{\overline{\nu}} &= \overline{\nu}_1 + \overline{\nu}_2 + \overline{\nu}_3 = 3\overline{\nu}_a + 2\overline{\nu}_b + \overline{\nu}_c. \end{aligned} \tag{54}$$

In reality, this differential of elements is the extrapolation of the differential that we could already see between the element ν_b and the element $\overline{\nu}_b$, which showed that under equal conditions there is more matter than antimatter, and all of this, ultimately, due to the relation between phases resulting from the charge of the electron in Eq. (39) as opposed to that of the positron.

We could elude this asymmetry or consider an alternative configuration assuming a four-element base, with which we would obtain an asymmetric element (sterile neutrino ν_c for $\phi_1 = 3\pi$), and three symmetric ones, sufficient to create a symmetric base of matter and antimatter, that if they coexist would be annihilated and if they do not coexist would not require any type of transformation to constitute a mass base:

$$\begin{aligned} \nu_1 &= \nu_a = \overline{\nu}_a \simeq 0.0056 \text{ eV}, \\ \nu_2 &= \nu_b = \overline{\nu}_b \simeq 0.0158 \text{ eV}, \\ \nu_3 &= \nu_d = \overline{\nu}_d \simeq 0.0448 \text{ eV}, \end{aligned} \tag{55}$$

but which would nevertheless report for differences in mass squares, results that are notably distant from the experimental data:

$$\begin{aligned} \Delta m_{21}^2 &= \Delta m_{ba}^2 = 21.951 \times 10^{-5} \text{ eV}^2, \\ \Delta m_{31}^2 &= \Delta m_{da}^2 = 1.903 \times 10^{-3} \text{ eV}^2. \end{aligned} \tag{56}$$

11. Summary and discussion

In this article, we have studied the relationship (39) defined for all the material particles of the Standard Model, by which it is evident that the different generations of particles are in different phases from each other, which have their own size associated to the limit value of the velocity, that is, to the value at which the particle reaches an inertial mass equivalent to the real mass of the particle of the next generation and at which, consequently, it would reach the phase of that particle. This not only regulates the distance between one particle and another in a simple measurement pattern such as the phase, expressed in units π, but also grounds or gives meaning to the dispersed and apparently chaotic values of the masses of the whole spectrum and defines other hitherto undefined ones (neutrinos), since it is these values that culminate or resolve satisfactorily (39).

This means that even if we ignore the original physical reasons for Eq. (39), they arise spontaneously since we are applying an energetic equivalence as an ordering principle which, given the results, could be showing that there is indeed such an equivalence, that is, a higher-order identity between one and another form of energy, an idea that would be reinforced by the impossibility of energetically exceeding this limit, that is, that the same energy value can exist associated with two generations.

The idea of progression or advance through the phase to the limit leads us to another one of transmutation of some particles into others or to the identification, at least practically, of both, and the latter to that of generation, that is, to that of real and effective generation of the different generations of particles from the preceding ones, until reaching generation zero as precursor and, at the same time, candidate for dark matter, as necessary responsible for the whole spectrum and variety of particles (matter and antimatter, with charge and without it), their common nexus. That common nexus or point of convergence is what is reached here with $\mathfrak{M}$, that is, that by means of it we not only reach an organization of the whole spectrum but it completes or resolves it, revealing the unification or logical organization described through the structural, that is, of the structured, quantized and hierarchical relation of the particles from the first one through the development of a single rhythmic or jump pattern defined by Eq. (39), by which, above any other consideration, such as the generational energetic transmutation, a phasic progression of the particle $\mathfrak{M}$ is evidenced, for which the other particles would be but the expression, through the different phases, of each of its states or forms in which $\mathfrak{M}$ itself can express itself.

A. Appendix

We are going to calculate the integral over v of the integrand[2]:

$$\begin{aligned}\zeta(v) &= \frac{\hbar v \sin(\Delta k(a\gamma^{-1}))}{a^3\gamma^{-3}} - \frac{\hbar v \Delta k \cos(\Delta k(a\gamma^{-1}))}{2a^2\gamma^{-2}} - \frac{\hbar w \cos(\Delta k(a\gamma^{-1}))}{a^2\gamma^{-2}} \\ &= \frac{\hbar v \sin\left[\Delta k\left(a(1-v^2/c^2)^{1/2}\right)\right]}{a^3(1-v^2/c^2)^{3/2}} - \frac{\hbar v \Delta k \cos\left[\Delta k\left(a(1-v^2/c^2)^{1/2}\right)\right]}{2a^2(1-v^2/c^2)} \\ &\quad - \frac{\hbar w \cos\left[\Delta k\left(a(1-v^2/c^2)^{1/2}\right)\right]}{a^2(1-v^2/c^2)}\end{aligned} \tag{A.1}$$

FIRST TERM- The term (A.1a) is left as:

---

[2] We will notice the different terms of the expressions with the number of the same followed by a letter, in this case, (A.1a), (A.1b), (A.1c)

$$\begin{aligned}
&\int \frac{\hbar v \sin\left[\Delta k\left(a(1-v^2/c^2)^{1/2}\right)\right]}{a^3(1-v^2/c^2)^{3/2}} dv \\
&= \frac{\hbar}{a^3}\int \frac{v \sin\left[\Delta k\left(a(1-v^2/c^2)^{1/2}\right)\right]}{(1-v^2/c^2)^{3/2}} dv \\
&= \frac{\hbar}{a^3}\left[\sin\left[\Delta k\left(a(1-v^2/c^2)^{1/2}\right)\right]\int \frac{v}{(1-v^2/c^2)^{3/2}} dv\right]_{v_o}^{v_f} \\
&+ \frac{\hbar \Delta k}{a^2}\int \frac{v \cos\left[\Delta k\left(a(1-v^2/c^2)^{1/2}\right)\right]}{(1-v^2/c^2)} dv
\end{aligned} \tag{A.2}$$

The term (A.2a) is will be the TERM-1 of the general solution, which we can put in a more elegant way as:

$$\frac{\hbar}{a^3}\sin\left[\Delta k\left(a(1-v^2/c^2)^{1/2}\right)\right]_{v_o}^{v_f}\int_{v_o}^{v_f} \frac{v}{(1-v^2/c^2)^{3/2}} dv = \frac{\hbar}{a^3}\sin\left[\Delta k(a\gamma^{-1})\right]_v \int_{v_o}^{v_f} \frac{v}{\gamma^{-3}} dv \tag{A.3}$$

SECOND TERM- The term (A.2b) has the same form as the term (A.1b) so we add it for analysis.

$$\begin{aligned}
&\frac{\hbar \Delta k}{a^2}\int \frac{v \cos\left[\Delta k\left(a(1-v^2/c^2)^{1/2}\right)\right]}{(1-v^2/c^2)} dv - \frac{\hbar \Delta k}{2a^2}\int \frac{v \cos\left[\Delta k\left(a(1-v^2/c^2)^{1/2}\right)\right]}{(1-v^2/c^2)} dv \\
&= \frac{\hbar \Delta k}{2a^2}\int \frac{v \cos\left[\Delta k\left(a(1-v^2/c^2)^{1/2}\right)\right]}{(1-v^2/c^2)} dv
\end{aligned} \tag{A.4}$$

(A.4) we can put it as:

$$\begin{aligned}
&\frac{\hbar \Delta k}{2a^2}\int \frac{v \cos\left[\Delta k\left(a(1-v^2/c^2)^{1/2}\right)\right]}{(1-v^2/c^2)} dv \\
&= \frac{\hbar \Delta k}{2a^2}\int \frac{v \cos[\Delta k(a\gamma^{-1})]}{\gamma^{-2}} dv \\
&= -\frac{\hbar \Delta k}{2a^2} c^2 \int \frac{\cos[\Delta k(a\gamma^{-1})]}{\gamma^{-1}} d\left(\gamma^{-1}\right)
\end{aligned} \tag{A.5}$$

which will be the TERM-2 of the general solution.

THIRD TERM- The term (A.1c), is left as:

$$\begin{aligned}
&-\frac{\hbar w}{a^2}\int \frac{\cos\left[\Delta k\left(a(1-v^2/c^2)^{1/2}\right)\right]}{(1-v^2/c^2)} dv \\
&= -\frac{\hbar w}{a^2}\int \frac{\cos[\Delta k(a\gamma^{-1})]}{\gamma^{-2}} dv \\
&= \frac{\hbar w}{a^2}\int \frac{c \cos[\Delta k(a\gamma^{-1})]}{(1-\gamma^{-2})^{1/2}(\gamma^{-1})} d\left(\gamma^{-1}\right)
\end{aligned} \tag{A.6}$$

which will be the TERM-3 of the general solution. That, taking into account the three results, is:

$$\begin{aligned} I &= \int \zeta(v)dv \\ &= \left(\frac{\hbar}{a^3}\right) \sin\left[\Delta k\left(a\gamma^{-1}\right)\right]_v \int_{v_o}^{v_f} \frac{v}{\gamma^{-3}} dv \\ &- \left(\frac{\hbar \Delta k}{2a^2}\right) c^2 \int \frac{\cos\left[\Delta k\left(a\gamma^{-1}\right)\right]}{\gamma^{-1}} d\left(\gamma^{-1}\right) \\ &+ \left(\frac{\hbar w}{a^2}\right) c \int \frac{\cos\left[\Delta k\left(a\gamma^{-1}\right)\right]}{\left(1-\gamma^{-2}\right)^{1/2}\gamma^{-1}} d\left(\gamma^{-1}\right), \end{aligned} \tag{A.7}$$

B. Appendix

The adjustment to Eqs. (40), presented in **Table 1**, with the tabulated experimental data [9] is unique and incontrovertible for the case of leptons, but it is not, however, for the other charged particles. Thus, we have that the best evaluation for negative quarks is $\chi = 0.517$ with these data, i.e. with the data intervals provided by the best experimental fit of the data:

$$m_d = 4.67^{+0.48}_{-0.17}\ \text{MeV}, m_s = 93^{+11}_{-5}\ \text{MeV}, m_b = 4.18^{+0.03}_{-0.02}\ \text{GeV}, \tag{B.1}$$

while for the intervals of the positives:

$$m_u = 2.16^{+0.49}_{-0.26}\ \text{MeV}, \quad m_c = 1.27 \pm 0.02\ \text{GeV}, \quad m_t = 172.90 \pm 0.04\ \text{GeV}, \tag{B.2}$$

the intended value $\chi = 1.5$ is satisfied for some combination of values (with $m_u = 2.00$ MeV), which however we have discarded.

Inexact or indefinite adjustment as the case may be, but nevertheless sufficient to ensure that in case Eq. (39) is met, i.e. satisfied by a relation of natural numbers, it must be (plausibility criterion) with the values for χ defined in Eqs. (40).

In neither of the two cases, however, are we guaranteed certainty (neither to endorse the fit nor to discard the maladjustment) because we do not have that certainty about the experimental data which, coming from indirect measurements, are not totally reliable, and which are susceptible, therefore, to be corrected to other experimental values, less recognized, by means of some criterion methodology or tool (such as the one resulting from our proposal).

Consequently, we are not proposing, and presenting in **Table 1**, a simple and timely correction of data but an exact theoretical solution starting from (39), from which all certainty regarding the starting data will be derived.

Data that we can represent in a more visual way (**Figures 6–8**) on those already existing and presented in the Data listings (Ref. [9]), where we can verify a certain disparity with respect to some of the best fitted values for A which, however, as we have already mentioned, are in accordance, respectively, with the experimental results [11, 12], [13, 14] and [15, 16], among others, which we can also find in summary form [9].

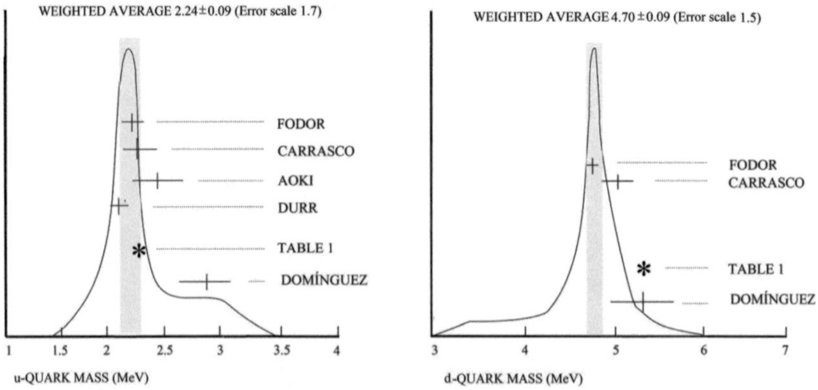

Figure 6.
*u-Quark and d-Quark. Status of our data (**Table 1**) with respect to the accredited data.*

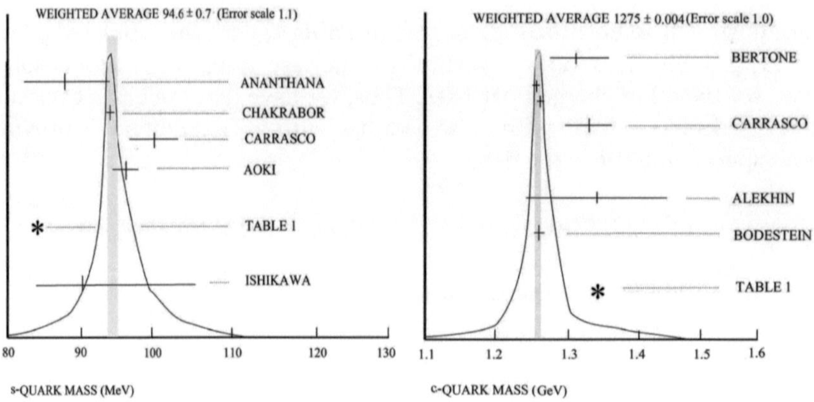

Figure 7.
*s-Quark and c-Quark. Status of our data (**Table 1**) with respect to the accredited data.*

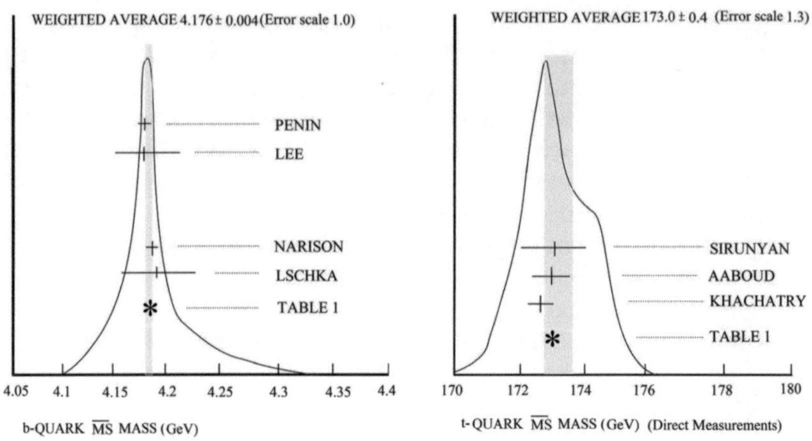

Figure 8.
*b-Quark and t-Quark. Status of our data (**Table 1**) with respect to the accredited data.*

C. Appendix

Starting from (39), we are going to express a generic Eq. (43) in an alternative form with more manageable and differentiated parameters, that is

$$\begin{aligned}\phi_i &= \left[\beta\left(\frac{1}{m_i}\right)^{1/3}\gamma_i^{-1}\right]^{-q} = \left[Z(\hbar/\pi)^{1/3}\left(\frac{1}{m_i}\right)^{1/3}\gamma_i^{-1}\right]^{-q} \\ &= \left[Z\left(\frac{\hbar}{\pi m_i}\right)^{1/3}\gamma_i^{-1}\right]^{-q} = \left[Z\delta_i\gamma_i^{-1}\right]^{-q} = \phi_0, \phi_1, \phi_2 \dots\end{aligned} \quad \text{(C.1)}$$

Parameters that remain fixed in each phase of the different transitions and that, according to (C.1), fulfil:

$$\gamma_i = \frac{m_{i+1}}{m_i} = (Z\delta_i)/\phi_i \qquad \text{for} \quad \delta_i = \left(\frac{\hbar}{\pi m_i}\right)^{1/3}, \quad \text{(C.2)}$$

which allows to determine γ_i for a phase, from the value of Z and the initial mass, and then, as we said in Section 7, the mass or starting point of the next transit (and vice versa with two masses), and allows us to progress through these transitions in a simple and reproducible way, establishing a progressive sequence of well-defined ternaries, $m_0, \delta_0, \gamma_0 - m_1, \delta_1, \gamma_1 - m_2, \delta_2, \gamma_2$, from the input values.

Input values that, consequently, serve us to modulate or see the behaviour depending on them, that is, varying Z or m_0, and check to what point they are what they should be unequivocally, as well as the initial expression that motivates all these relations and values. Verification that is done quickly and easily by means of an Excel table, such as the one represented in **Figure 9**.

On the represented table we see that, in fact, starting from the input values we reach in A9 the true initial value of the sequence, m_0, which, consequently, could be entered directly as a numerical input value (171.87). By doing it this way, we have managed to start the sequence from the value of the first-generation particle (known) and obtain γ_0 (E7) from it, that is, in reverse, although it is then recalculated, according to the equations and the progression of the transitions, in C11.

For this particular case, we have used the sequence $4\pi, 2\pi, \pi(?)$ in A15 and D15, and could have used, for another proportional value of Z, the sequence $2\pi, \pi,$ which would not allow a transition to a fourth generation. It can be seen that for the antiparticle an identical progression is reached, in which the phases change the order of the sequence $\pi, 2\pi, 4\pi(?)$ and become divisor (by the sign of the charge in the exponent) for a value $\overline{Z} = 33.0431$, which can be calculated directly by Eq. (C.2) or through its relation with Z.

$$\begin{aligned}\frac{Z\delta}{(4\pi)^3} &= \gamma = (\pi)^3\overline{Z}\delta \Rightarrow \frac{Z}{\overline{Z}} = \left[(4\pi)^3\right](\pi)^3 \\ &= \left(4\pi^2\right)^3 \Rightarrow \overline{Z} = (61528.9)^{-1}Z\end{aligned} \quad \text{(C.3)}$$

Analogously, we can establish the expression (39) in its final form $\chi = \mathfrak{R}^{-q}$ and establish its values for the two triads. Following the same case we would have an exponent (1/3) in $\mathfrak{R}$ and another (1/3) associated to $-q$, as detailed in the Excel table represented in **Figure 10**, with which we can vary any of the inputs and observe how

1	1° You must place the main table in the first column of an Excel page, matching A to the 4 row on the					
2	page. 2° Copy (insert "=") each FUNCTION into the corresponding RESULT. 3° You can place the					
3	partial tables in line (it is optional), in the way A,B,C,D,E,A,B,C...4° You can delete the first two rows.					
4	A	B	C	D	E	COLUMN
5	IMPUT VALUE	IMPUT VALUE	POWER((6,582 *9)/(PI()*B7); 1/3)	(A7*C7)/POWER(4*PI(); 3)	POWER((A7* C7/POWER(1* PI();3));3/2)	FUNCTION
6	Z	Md	δd	Ƴd	Ƴi=f(Ge)	PARAMETER
7	**2033106**	**5395000**	0,01517585	15,5483193	31390,2763	RESULT
8	A	B	C	D	E	The main table is the one delimited by the gross border
9	B7/E7	POWER((6,582 *9)/(PI()*A11) ;1/3)	(A7*B11)/POWER(PI();3)	A11*E7	POWER((6,582 *9)/(PI()*B7); 1/3)	
10	Mi	δi	Ƴi=Z*δi*fase	Md	δd	
11	171,868509	0,478723487	31390,2763	5395000	0,01517585	
12	A	B	C	D	E	The main table consists of 3 partial tables with 5 columns and 4 rows
13	(A7*E11)/POWER(4*PI();3)	D11*A15	POWER((6,582 *9)/(PI()*B15) ;1/3)	(A7*C15)/POWER(2*PI( );3)	B15*D15	
14	Ƴd	Ms	δs	Ƴs	Mb	
15	15,5483193	83883182,84	0,0060803	49,8362805	4180425830	

Figure 9.
Phases transitions for negatively charged quarks (−1/3).

	A	B	C	D	E	F	G	H
1				2,911E+13	4,153E+39	7,306E+28	1,953E-03	0,125
2	4			d^2	s^5	b^3	F1/(E1*G1)	$\Re$
3	χ		**171,86851**	**5395000**	**83883183**	**4180425830**		χ
4	$\Re$	E5/(D5*F5)	m_0^2	d^5	s^3			0,5
5	64	2,621E+05	2,954E+04	4,570E+33	5,902E+23			
In this case A,B,C…, belongs to the Excel page. Label $\mathbf{d^2}$, for example, implies POWER(F3;2) in D1								
$\Re$ is POWER(G1;1/3) in H1/ POWER(B5;1/3) in A5- χ is POWER(H4;1/3) in H2/ POWER(A8;1/3) in A2								

Figure 10.
Adjustment of Eq. (39) for negatively charged quarks (−1/3).

it affects the default values χ exactly. On which, finally, we can replace the numerical input values C3-F3 by the values A11-D11-B15-E15 of the preceding table in order to control all the masses and other variables with the input of this one.

D. The physical issue

We can think, in spite of all the above, that the expression (39) lacks validity because it is not supported by physical presuppositions or even downplay the importance of the correspondence presented and consider it simply cunning. The truth is that although these presuppositions have not been presented (for the reasons given above), they appear or ooze from the expression itself, while it could be said, as we shall see, that the correspondence is impossible without this underlying physical requirement or a mathematical one of equal importance.

Regarding the latter, we have already said, Section 5, that our analysis was as simple and unequivocal as that derived from the equation of a straight line. That is its importance.

In effect, if we think of any three straight lines defined by their respective pair of points, for example B and C, and we ask ourselves about the probability that the three converge at a single point A, or about the requirement for them to do so, we will conclude that the lines will pass through point A if the definition of the lines includes passing through that point, that is, if the lines are defined by point A and the respective point B, so that point C is as unnecessary as any other point D. Something similar happens with the point X, with respect to the different points P, although in appearance (even by the way of introducing it) it is presented as fortuitous and unexpected. That is to say, although in our case they are not straight lines but curves, and the equation therefore is not that of a straight line but the equation of the relativistic mass, defined on the Lorentz factor that defines the trajectory between B-C and C-D, if we ask ourselves the same question, the answer is identical: the points B-C-D that following the trajectory reach the points A are exclusively those that, starting from A, reach by means of the trajectory the points B-C-D for the different equations. A statement that in the present case, unlike that of the straight lines, needs to be developed, A statement that in this case, unlike that of the straight lines, needs to be developed, for which we will make use of **Figure 11**, in which we have also replaced the curves by straight lines using the logarithmic scale for better visualization, equivalence and understanding, where it is highlighted not only the confluence, or common and logically unique point of intersection of the three straight lines, but also that it occurs despite the restriction of the domain of application to those points ($\phi_0 = \pi, 2\pi, 3\pi \dots$), among all the real ones, in which it is possible, as we shall see, to find a particle, for being those that conform a phase or represent a node according to the conditions given in Eq. (39).

In a first approximation, although we have not followed this itinerary for the presentation of results, the phases are defined (the most simplified ratio of natural numbers) because χ is defined from an approximate version of χ itself, reached by means of the plausibility criterion applied to the masses of the particles (see Appendix C), which in turn defines Z for those values; or if desired by means of Eq. (C.2) and the factors γ_1 and γ_2 obtained from m_1, m_2, m_3. Having defined Z we have found ϕ_0 for each particle chain or electric charge (L, +Q, −Q) by the corresponding Eq. (C.1) or the analogous (43) as a function of β,

$$\phi_0 = \left[\beta\left(m_0^2/m_1^3\right)^{1/3}\right]^{-q} = \left[Z\delta_0\gamma_0^{-1}\right]^{-q} \qquad \text{(D.1)}$$

which gives rise to a common particle $m_0 = 171.87$ eV (that we found in (42), for $\alpha^{-q} \cong 1$, without needing to know the Z value) that holds a factor γ_0 with the first generation particle m_1 in each family.

Thus, we have expressed it, when the reality is that once defined Z, we find $\phi_0^1, \phi_0^2, \phi_0^3 \dots$ by means of (D.1) that would give rise for $\gamma_0^1, \gamma_0^2, \gamma_0^3 \dots$ to some masses $m_0^1, m_0^2, m_0^3 \dots$ in each one of those chains or electric charges, that is to say, we would reach the corresponding sequences $m_0^{L1}, m_0^{L2}, m_0^{L3} \dots, m_0^{+Q1}, m_0^{+Q2}, m_0^{+Q3} \dots$ and

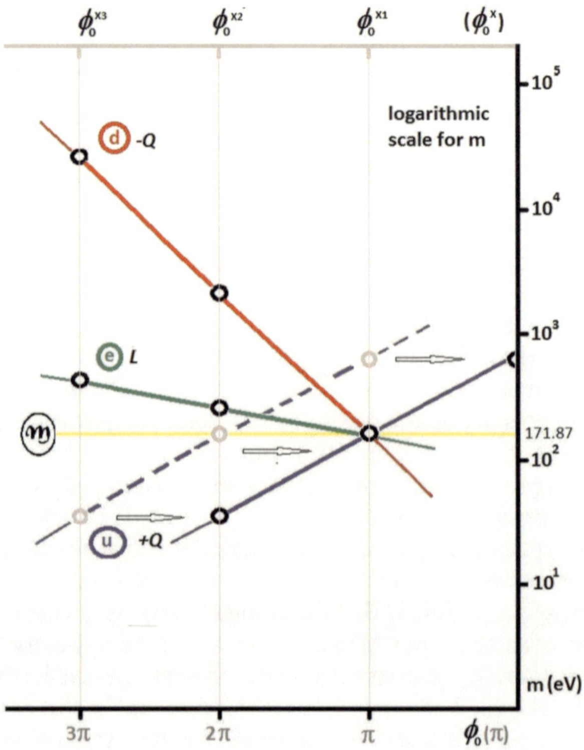

Figure 11.
Development or prolongation of the phases (ϕ_0^{xi}) prior to 𝔐 for each of the families (L, +Q, −Q).

$m_0^{-Q1}, m_0^{-Q2}, m_0^{-Q3}$ ... dispersed and discordant, that only coincide in the singular value m_0 for the phases $\phi_0^{L1}, \phi_0^{+Q2}, \phi_0^{-Q1}$ (to which we have generically called ϕ_0 obviating the sequences), from which the values follow different courses.

We can concretize this on **Figure 9** for -Q, where it is seen (Box E5-6-7) that $\gamma_0 = \gamma_0^1$ decreases in the ratio $\left[(1/n^3)^{3/2}\right] = n^{-9/2}$ as we vary n (initially to 1) in agreement with $\phi_{0-Qi} = (1, 2, 3 \ldots) \times \pi$, consequently increasing the mass (Box A9-10-11) in the form $m_0^{-Qi} = (171.87, 3888.94, 24112.48 \ldots)$ eV, the same occurring, as we will detail later, for $n^{-3/2}$ in the chain L with $m_0^{Li} = (171.87, 343.85, 515.78 \ldots)$ eV, while +Q responds to $n^{9/4}$ and would lead us $m_0^{+Qi} = (857.54, 171.87, 69.02 \ldots)$ eV for $\phi_{0+Qi} = \pi, 2\pi, 3\pi \ldots$, where we can apply an initial phase difference π, by redefining the starting "straight line", so that they all find the common value at $\phi_0 = \pi$, as a didactic adjustment (since it is not a requirement).

That is we see that there is not only one pro-precursor particle but many as a consequence of the fact that there is not only one γ_0 but γ_0^i applicable to each chain, seeing also that neither the different masses m_0^1, m_0^2, m_0^3 ... nor the generic mass m_0 chosen among them, are defined by a sort of adjustment or approximation but on the contrary they are subject to the sequence of values of a series, which we can even define in a specific way as a consequence of the fact that (and here is the key of the matter) responds to an analytical relationship between γ_1 and each γ_0^i. A relationship that we will establish initially for -Q and then generalize. From Eq. (C.1) for ϕ_1 and the d-quark we have:

$$\left[Z_d \delta_d \gamma_d{}^{-1}\right]^{1/3} = \phi_1 = 4\pi \Rightarrow \gamma_d = \frac{Z_d \delta_d}{(4\pi)^3} \tag{D.2}$$

Now we are looking for:

$$\left[Z_0 \delta_0 \gamma_0{}^{-1}\right]_d{}^{1/3} = \left[Z_d \delta_0 \gamma_0{}^{-1}\right]^{1/3} = \phi_0 = \pi \tag{D.3}$$

Given that:

$$\delta_0 = \left(\frac{\hbar}{\pi m_0}\right)^{1/3} = \left(\frac{\hbar}{\pi m_d \gamma_0{}^{-1}}\right)^{1/3} = \left(\frac{\hbar}{\pi m_d}\right)^{1/3} \gamma_0{}^{1/3} = \delta_d \gamma_0{}^{1/3}, \tag{D.4}$$

we have:

$$\left[Z_d \delta_0 \gamma_0{}^{-1}\right]^{1/3} = \left[Z_d\left(\delta_d \gamma_0{}^{1/3}\right)\gamma_0{}^{-1}\right]^{1/3} = \left[Z_d \delta_d \gamma_0{}^{-2/3}\right]^{1/3} = \phi_0 \tag{D.5}$$

From which we obtain, for $Z_d = Z_s = Z_b = Z^{-Q}$, that:

$$\gamma_0{}^{2/3} = \frac{Z^{-Q}\delta_d}{(\phi_0)^3} \Rightarrow \gamma_0 = \left(\frac{Z^{-Q}\delta_d}{\pi^3}\right)^{3/2} = 31390.28. \tag{D.6}$$

which is the result presented in **Figure 9**, with which, from m_d, we can obtain $m_0 = 171.87$ eV. From here we only have to generalize it for the three classes of charged particles (Z), for any first generation (δ_1) of reference (although it is true that it is applicable to any two contiguous generations), and for $\phi_0^1 = \phi_0, \phi_0^2, \phi_0^3 \ldots$, to finally, that was our objective, put it as a function of the Lorentz factor of that reference (γ_1), that is:

$$\gamma_0^i = \left(\frac{Z\delta_1}{\left(\phi_0^i\right)^{-1/q}}\right)^{3/2} = \left(\frac{\phi_1{}^{-1/q}}{\left(\phi_0^i\right)^{-1/q}}\right)^{3/2}\left(\frac{Z\delta_1}{\phi_1{}^{-1/q}}\right)^{3/2} = \Theta \times \gamma_1{}^{3/2} \tag{D.7}$$

Being in our starting case, for $\gamma_d = 15.548$:

$$\gamma_0^1 = \Theta \times \gamma_1{}^{3/2} = \left[\left(\frac{4\pi}{\pi}\right)^3\right]^{3/2}\left(\frac{Z^{-Q}\delta_d}{\phi_1{}^{-1/q}}\right)^{3/2}$$
$$= (64)^{3/2} \times \gamma_d{}^{3/2} = 512 \times 61.309 = 31390.28 \tag{D.8}$$

While in the other classes for $m_0 = 171.87$ eV, with $\phi_0 \equiv \phi_0^{L1} = \phi_1$ and $\phi_0 \equiv \phi_0^{+Q1} = \phi_1$, the result is $\Theta = 1$, which in all cases will be affected by taking ϕ_0^2, ϕ_0^3, affecting γ_0^2, γ_0^3 and consequently affecting m_0^2, m_0^3.

Beyond the evidence of the behaviour derived from the data, and saving the mentioned didactic adjustment, on the figure it is clearly seen that the lines -Q and L intersect at a point (as any pair of non-parallel lines) to then progress in a divergent, essentially discrete and exponentially dispersed way, in such a way that it is highly improbable that any other value m_0^{-Qi} reached by a node coincides with one m_0^{Li} reached by another node, highly improbable that it has coincided, as it happens, with a value m_0^{+Qi} (for a node or point of the application domain), and already impossible that it does so (given its decreasing sequence) with another m_0^{+Qi}, from which we conclude that the only and exceptionally possible common value of mass is $m_0 = 171.87$ eV, which is the truly relevant fact, and the one that has physical significance, being that significance the basis for such an improbable fact occurs: as we said (and now we have substantiated) when a series of straight lines intersect at a point A (and we have converted our curves into straight lines) it is because they are defined by A and their respective and differentiated slopes (governed in this case by physical criteria), everything else being a consequence: it is not possible to explain a convergence of this degree, if not as a specular process of a divergence.

In our real physical case, our straight lines are curves, and in them, in effect, the condition is only fulfilled for the curves that fulfil the above, that is to say, that in origin fulfil Eq. (C.1) for $m_i = m_0$, because these are the only ones that can count with m_0 among their points. We speak of curves and more specifically of the curves described by the Lorentz factor, that is to say, by their development. The factor γ describes the physical relation that exists between the mass of a particle and its inertial or relativistic mass, i.e., if we wanted to find another function that unites these two masses taken as points for any value of the latter (any speed) we would not find it because it is the one that describes the physical relation that is occurring. This is important because it supposes a first limitation with respect to the family of possible curves. From there, the Z value is defining a family of curves among them: the one that starting from m_1 reaches m_2 and m_3 by means of some concrete phases, according to (43), then shown by the ratio of those masses. That is the true physical condition, the one expressed in (C.1), which makes of those masses not any masses but some concrete masses, that is to say, that they are these and not others.

These are masses that have proven their real existence (although only m_1 in a natural and generalized way) while others such as m_0^1, m_0^2, m_0^3 ... only have it guaranteed within the family of curves, just as a hypothetical fourth generation m_4 would have it through Z for the corresponding ϕ_3. Apart from this, Z is defining them equally, that is, it is saying what characteristics would have some preceding and/or precursor particles subjected to the same pattern, governed by the same Z, regardless of what Z is. This said for each type of charge implies that sequence of infinite possibilities, restricting ourselves instead to the candidates that fulfil a valid pattern for the three classes of particles, it turns out that only one particle exists, that is, only one particle intersects in the projection of the three chains, only one is viable as a precursor, only one has the capacity to evolve (**Figure 4**), according to three different Z-values, in these three different ways and reach the whole material spectrum, and justify even other material forms.. That is the immediate physics of (39), which does not need to presupposes, but rather creates presuppositions or new physics environments.

Author details

Rafael Cañete Mesa
Independent Researcher, Seville, Spain

*Address all correspondence to: rafaelcanyete@protonmail.com

References

[1] Von Neumann J. Mathematical Foundations of Quantum Mechanics. 1932. DOI: 10.1515/9781400889921. Available from: https://www.degruyter.com/document/doi/10.1515/9781400889921/html

[2] Téllez AM. [Internet]. La partícula libre I. 2009. Available from: http://la-mecanica-cuantica.blogspot.com/2009/08/la-particula-libre-i.html

[3] Bombal Gordón F. Los modelos matemáticos de la Mecánica Cuántica. 1999. Available from: https://eprints.ucm.es/id/eprint/20072/

[4] Cohen-Tannoudji C. Quantum Mechanics. Vol. 1. Available from: http://www.fulviofrisone.com/attachments/article/480/Cohen-Tannoudji%20C.,%20Diu%20B.,%20Laloe%20F.%20Quantum%20mechanics,%20vol.%201(T)(887s).pdf

[5] Los Paquetes de Onda [Internet]. Física cuántica en la red. 2014. Available from: https://www.fisicacuantica.es/los-paquetes-de-onda/

[6] Casas J. ÓPTICA. ISBN: 300-2448-4. Available from: https://dlscrib.com/download/oacute-ptica-justiniano-casas_5907d627dc0d60951d959ee7_pdf

[7] Amsler C. The seesaw mechanism. In: Nuclear and Particle Physics. Bristol, UK: IOP Publishing; 2015. DOI: 10.1088/978-0-7503-1140-3

[8] Miranda OG, Valle JWF. Neutrino oscillations and the seesaw origin of neutrino mass [Internet]. Nuclear Physics B. Elsevier BV. 2016;**908**:436-455. DOI: 10.1016/j.nuclphysb.2016.03.027

[9] Zyla PA et al. Particle data group. Progress of Theoretical and Experimental Physics. 2020;**2020**:083C01. DOI: 10.1093/ptep/ptaa104

[10] Esteban I et al. The fate of hints: Updated global analysis of three-flavor neutrino oscillations. JHEP. 2020;**09**:178. DOI: 10.48550/arXiv.2007.14792

[11] Carrasco N et al. Up, down, strange and charm quark masses with $N_f = 2+1+1$ twisted mass lattice QCD. Nuclear Physics B. 2014;**887**:19-68. DOI: 10.1016/j.nuclphysb.2014.07.025. arXiv:1403.4504

[12] Abramowicz H et al. Combination and QCD analysis of charm and beauty production cross-section measurements in deep inelastic *ep* scattering at HERA. European Physical Journal C. 2018;**78**(6):473. DOI: 10.1140/epjc/s10052-018-5848-3. arXiv:1804.01019

[13] Dominguez CA et al. Up and down quark masses from finite energy QCD sum rules to five loops. Physical Review D. 2008;**79**:014009. arXiv:0806.0467

[14] Blum T et al. Determination of light quark masses from the electromagnetic splitting of pseudoscalar meson masses computed with two flavors of domain wall fermions. Physical Review D. 2007;**76**:114508. arXiv:0708.0484

[15] Mason Q et al. High-precision determination of the light-quark masses from realistic lattice QCD. Physical Review D. 2006;**73**:114501. arXiv:hep-ph/0511160

[16] Ananthanarayan B, Das D. Optimal renormalization and the extraction of the strange quark mass from moments of the τ-decay spectral function. Physical Review D. 2016;**94**:116014. arXiv:1610.08900

Chapter 3

How Black Holes Remain Neutron Stars

Martin T. Cole

Abstract

Experiments in high-energy particle physics reveal that neutrons withstand very high levels of mutual compression – the levels expected inside Neutron Stars. No limit to this elastic compressibility is indicated. A new approach to the general relativity (GR) modelling of Neutron Star cores is here suggested, utilising a robust *internal* observer. As Neutron Stars are modelled at increasing masses, upon reaching a certain critical mass suggested as ~$5M_{\odot}$, they undergo a smooth transition to Black Holes – defined as the point at which Neutron Stars become electro-magnetically invisible. At this critical mass, two concentric spherical event horizons of radius ~7 km develop simultaneously. At higher masses, these separate to form both an outer horizon (as predicted in GR), as well as a (non-Cauchy) inner horizon not predicted in GR. This inner horizon prevents any collapse of the core and the formation of a singularity. Absent any singularity, the speculated formation of wormholes is not possible.

Keywords: black hole, black star, neutron star, inner horizon, outer horizon, shell escape speed

1. Introduction

Usually, the formation of a "black hole" (BH) begins with a neutron star (NS), which is the remnant core of a supernova explosion. The NS is an extremely compact, superdense star composed almost entirely of neutrons packed together. If the size of the NS exceeds a certain "critical mass", its gravitational field is so strong (i.e. GR spacetime curvature is so great) that light itself cannot escape. The NS then behaves as a BH surrounded by an "event horizon". This event horizon marks the boundary, within which the star is completely invisible to observation at any wavelength (from gamma rays to X-rays to ultraviolet to visible to infrared and radio). Therefore, no information about the BH can be accessed from within/behind the event horizon.

It is widely accepted that general relativity (GR) field equations, based on a distant external observer, break down at this event horizon, and they cannot be used to model the likely internal structure and dynamics within a BH. It has been assumed that within the BH event horizon, this neutron star core would entirely collapse to form a singularity. A singularity has finite mass but zero size, producing infinite density and infinite GR spacetime curvature.

2. Neutrons

Simplistically, a neutron is composed of three "valence" quarks – an "up" quark with a charge of $+^2/_3$, and two "down" quarks with a charge of $-^1/_3$. The combination of these three charges produces no net charge (neutral). These quarks are held together by "gluons" which bind the quarks within the neutron. Protons are similar, except they are composed of two up quarks, with their charge of $+^2/_3$, and a down quark, with its charge of $-^1/_3$, combining to form a net charge of +1 (equal and opposite charge to an electron).

However, the composition of a neutron or proton is far more complicated than this picture presents. In addition to the three valence quarks, there are possibly fourteen quark-antiquark pairs, also held together by gluons. The external outcome of all these interactions is the short-range "strong force" (or "strong interaction") carried by the gluons, responsible for binding the neutrons and protons within an atomic nucleus (powerful enough to overcome the charge repulsion between protons).

Within a NS, the dense cluster of neutrons are prevented from collapsing under their own gravity as a result of the "Pauli Exclusion Principle", which states that no two fermions (such as neutrons) can occupy the same quantum state simultaneously. This results in a "neutron degeneracy pressure" resisting the collapse and creating a hydrostatic equilibrium. This degeneracy pressure may be understood as a force derived from the strong nuclear interaction, opposing the breakdown of a neutron into its component quarks and gluons.

It has been widely assumed that when a NS has sufficient mass to form a BH, the gravitational field becomes so strong that this degeneracy pressure is overcome, and the star collapses into a singularity.

However, no causal connection has been reported, between the moment of formation of the event horizon, and the moment of alleged collapse of the neutron core. One event does not necessarily cause the other.

A further complication is to note that, within the nucleus of an atom, the nucleons (neutrons and protons) are not necessarily "touching spheres", as might be imagined. Each nucleon may be separated by some short equilibrium distance (measured centre-to-centre), greater than the known free-nucleon diameter of ~1.68 fm. A striking example is that of the beryllium (^{9}Be) nucleus, with its 9 nucleons structured as two helium nuclei and one neutron, orbiting each-other as three distinct entities [1].

Brodsky et al. [2] have experimentally determined the form and magnitude of the strong force, which rises to a maximum at ~3 fm (~1.8 diameters) separation, in the case of isolated (free) nucleons.

Brodsky et al. [3] explain that: "So far as we know, gluons and quarks are point-like," and nucleons in isolation, behave differently from how they do in a cluster. "Overlapping nucleons would be very complex objects, possibly with no relation to the individual nucleons." To date, the magnitude of the neutron degeneracy pressure cannot be accurately quantified, theoretically or empirically.

However, it is accepted that a bound cluster of neutrons is compressible under gravity (thereby increasing its average mass density in proportion to volumetric compression). As a neutron cluster grows and gains mass, the increasing compression of the cluster under its own increasing gravity, may initially involve a reduction in the nucleon separation, until this centre separation reaches their observed free size of ~1.68 fm diameter, prior to any compression of each individual nucleon volume.

3. Neutron stars

So, what is the possible NS size range, and what could be the value of the critical mass described above? Consider NS theoretical evidence (in publishing order):

- Haensel et al. [4] say that a NS has a physical radius of 11 to 13 km containing a mass up to 2 solar masses (2 $M_\odot$ i.e. 3.98E+30 kg), with a crust density of 0.3 times the normal nuclear saturation density (ρ_{ns} = 2.7E+17 kg/m^3), rising to >2 ρ_{ns} at the core.

- Katayama and Saito [5] calculate the maximum theoretical NS mass at 2.08 $M_\odot$ (based on GR).

- Vidaña [6] says that NS's are likely the densest objects in the universe, with a central density in the range 4 to 8 ρ_{ns} and a surface radius of 10 to 12 km.

- Giddings [7] says that: "perhaps some unknown laws of physics prevent larger stars from forming BHs and instead lead them to become a kind of massive remnant – more like a NS than a BH – (but) we cannot explain what would prevent their continued collapse under gravity."

One might anticipate that such a cataclysmic and powerful event as the sudden collapse of a NS to form a BH singularity, as theorised in GR, should generate an observable energy pulse or shock wave, at least from an accretion disk (outside the event horizon). No such evidence is reported.

What may become of neutrons under extremely high compression by gravity has been a matter of conjecture. It was believed that these conditions could not be created in any Earth-bound laboratory, and that the equation of state for neutron-degenerate matter had not been determined [8].

But instead of possible "continued collapse" [7], it seems at least equally likely that neutrons (despite possibly transforming into neutron degenerate matter at some stage), can withstand any level of compression without collapsing into a singularity – being sustained by the neutron degeneracy pressure operating amongst and beyond the quarks of which neutrons are composed – a force understood to become increasingly repulsive at reducing short range and which asymptotes toward infinite repulsion at zero separation. Therefore, collapse should not be possible, and this condition is substantiated by the most recent high-energy physics laboratory evidence:

- Schmidt et al. [9] explain that strong interactions are described by the equations of quantum chromodynamics (QCD) which are not well constrained at short distances, such as in the cores of neutron stars. These interactions have been studied with the CEBAF Large Accelerator Spectrometer (CLAS Collaboration), at densities several times higher than the central density of the nucleus. "As the relative momentum between two nucleons increases and their separation thereby decreases, we observe a transition from a spin-dependent tensor force to a predominantly spin-independent scalar force ... We find even at the highest densities; the nucleons seem to keep their identities and don't turn into this bag of quarks ... The strong nuclear force creates a repulsive force between neutrons that, at a neutron star's core, helps keep the star from collapsing in on itself."

This is physical/empirical evidence that large neutron stars (beyond the critical mass), do not necessarily collapse into a singularity. It is also noted that in accordance with Haug and Gianfranco [10], the only limit to the compressibility of matter (such as the nucleons in a neutron star), is likely linked to the Planck density itself (which is not "infinite" density).

4. Analysis

Timlin [11] has produced an extensive derivation of the implied neutron degeneracy pressure in equilibrium with gravity:

$$Pn = \frac{3^{\frac{10}{3}}\,\hbar^2}{\left(15\pi^{\frac{1}{3}}\,\mathrm{m}^{\frac{8}{3}}\,R^5\right)}\left(\frac{\mathrm{M}}{4}\right)^{\frac{5}{3}} Pa \tag{1}$$

Where ħ = 1.054E-34 is the reduced Planck constant, M is the stellar mass, m = 1.674E-27 kg is the neutron mass and R is the surface radius. Substituting for the constants we obtain:

$$Pn = \frac{4.976E4\left(\frac{M}{4}\right)^{\frac{5}{3}}}{R^5} Pa \tag{2}$$

In view of Brodsky et al. [2, 3]., this equation can be regarded as a 'best guess' guide to quantifying the form and magnitude of the pressure within a neutron cluster, but this steep (5th power) inverse function illustrates how the degeneracy pressure approaches infinity at zero separation. In view of Schmidt et al. [9], this pressure is understood to *resist* degeneracy, without actual degeneracy (into quarks and gluons) necessarily ever being achieved.

Brodski et al. [3] explain that repulsion between free nucleons begins to dominate at centre separations of 0.25 fm, where quark-quark interactions remain strong.

Given that GR field equations cannot be relied upon to model the structure within a BH, a return to basic principles, with a reduction in complexity and the use of informative Figures is considered necessary to obtain a more intuitive, verifiable NS model.

Kahana [12] affirms that for a spherically symmetric object under a GR regime, in consequence of Birkhoff's theorem, the metric inside of a shell is flat Minkowski space. So, the *internal* structure of a massive object can be investigated with the Shell theorem of Newton [13] or Gauss. Bautista [14] cautions that Birkhoff's theorem may not hold for a body undergoing rotational or radial motions, which introduce velocity time dilation. Zhang [15] notes there is time dilation due to massive distant bodies. But for an isolated, non-rotating object, the Shell theorem is applicable to GR (and classical) physics, whereby the mass beyond a given shell radius has no gravitational effect.

The non-relativistic (Newtonian) gravitational potential energy U existing between two massive objects is:

$$U = -\frac{GMm}{r}\text{Joule} \tag{3}$$

where the gravitational constant $G = 6.673*10^{-11}$ m^3/kg-sec^2, M and m are masses separated by displacement r (measured from their centres of mass). By the Shell theorem, the gravitational potential energy at the centre of any isolated massive object is zero – a robust small mass placed there would hover, weightless, because all the gravitational forces in all directions cancel and nullify.

Based on Eq. (1) we first obtain the Newtonian velocity (or more-correctly, the speed) of escape from an object of mass M:

$$V_{esc} = \sqrt{2\,\Phi}\ \text{m/sec where}\ \Phi = \frac{GM}{r}\text{m}^2/\text{sec}^2 \tag{4}$$

where G is defined above, and r is the radial displacement from the object's centre of mass. Augousti [16] points out that Eq. (4) *coincides with the GR equation for the escape speed in the Schwarzchild metric*.

Obtaining a reliable *relativistic* escape speed has proven to be more controversial. Rybczyk [17] provides a derivation of the escape speed: "based on the Lorentz invariant transformation while employing the principles of GR". He equates the relativistic kinetic energy to the relativistic potential energy and in due course obtains:

$$V_{esc} = \sqrt{2\Phi - \left(\frac{\Phi}{c}\right)^2} \tag{5}$$

where c is the speed of causality, or light (2.998E+08 m/sec), and we set $\Phi \le c^2$. Eq. (3) reduces to the Newtonian escape speed when Φ/c is small. Eq. (5) is fully in agreement with an extensive derivation by Haug [18].

Alternatively, Camp [19] derives the following equation, which we then present in a form similar to Eq. (5):

$$V_{esc} = c\sqrt{\frac{2\Phi}{c^2} - \left(\frac{2\Phi}{c^2}\right)^2}$$

$$= \sqrt{2\Phi - \left(\frac{2\Phi}{C}\right)^2} \tag{6}$$

While Leinenweber [20] derives a different equation which we have re-written in a form more consistent with Eqs. (5) and (6) to assist comparison:

$$V_{esc} = c\sqrt{1 - \left(\frac{c^2}{\Phi + c^2}\right)^2} \tag{7}$$

All equations should produce results that match Newtonian escape speeds for non-relativistic objects like the Earth or Sun. However, for relativistic objects such as a 10 $M_\odot$ star, surprisingly, Eqs. (6) or (7) do not produce peak escape speeds that reach c. Whereas it will be shown that Eq. (4) can produce escape speeds that far exceed c. Overall, Eq. (5) has been found to produce the most consistent and credible data when compared with observations, and this will be relied upon for relativistic calculations here.

Commencing with a non-relativistic object such as the Earth, we apply the Shell theorem together with Eq. (5), summing small incremental steps, to compute the hypothetical "shell escape speed" at increasing shell radii (concentric with the centre of gravity). This method relies on the fact that matter beyond a shell does not exist for calculation purposes (in practice, escaping one shell would then involve the next shell, with a revised escape speed). After applying the Earth density profile data *of* Dziewonski et al. [21] to each shell, we obtain **Figure 1**.

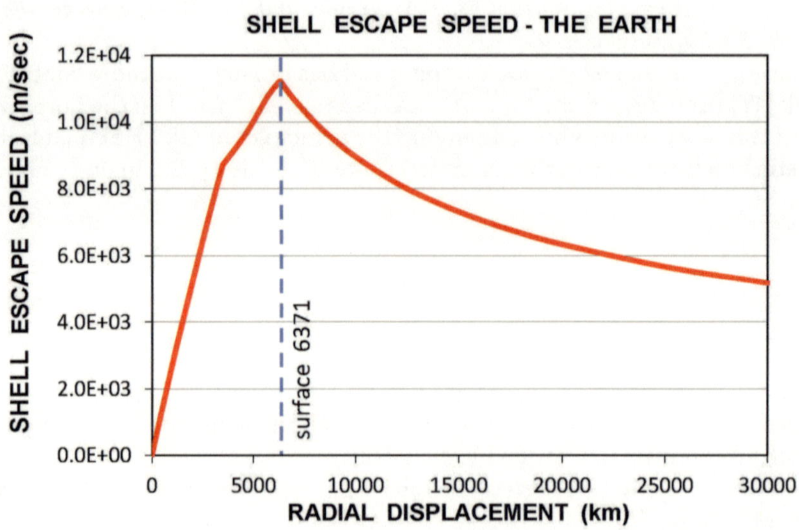

Figure 1.
Hypothetical Shell *escape speed vs. Displacement for the Earth. (Note that the* planet *escape speed applies only at or above the surface).*

Extending this analysis to a much more massive object, **Figure 2** represents the shell escape speed for the Sun, based on solar density profile data reproduced by Hathaway [22]. Here we note that the peak shell escape speed is reached far below the surface of this fluid object – this is because of the relatively high density of the core (by ~53% of the surface radius, the density has already fallen to that of drinking water [22]).

What follows is a suggested methodology, open for discussion, that could in future be adjusted according to any new findings on the observed density profile of neutron stars.

The all-important surface radius of a neutron star (to be compared with the Schwarzchild radius) depends on the uncertain density profile. In **Figure 3**, a range of plausible density profiles has been developed for the 2.08 $M_\odot$ neutron star of Katayama and Saito [5].

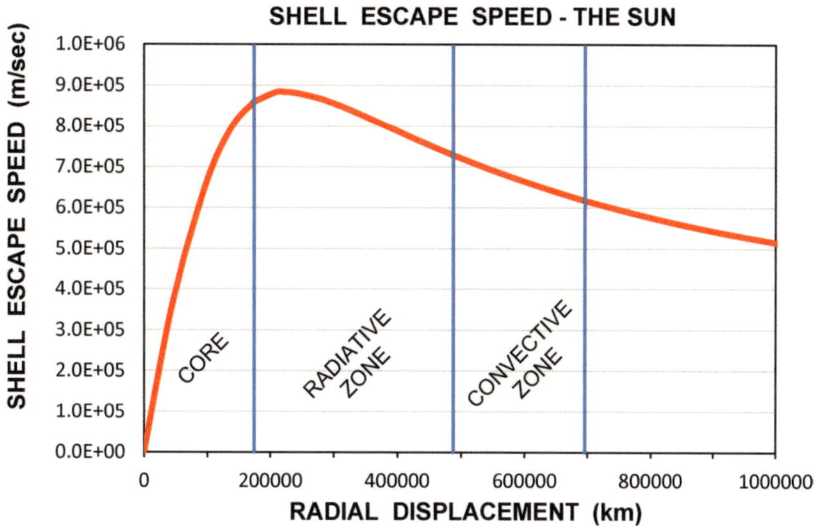

Figure 2.
Shell escape speed vs. displacement for the Sun.

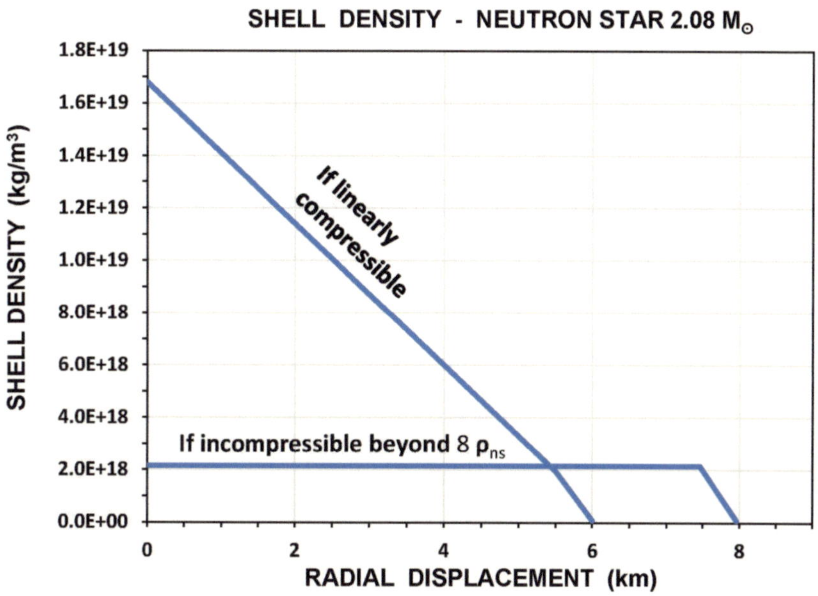

Figure 3.
Range of plausible density profiles for a 2.08 $M_\odot$ neutron star.

Assuming a crust density of 0.3 ρ_{ns} as proposed by Haensel et al. [4], **Figure 3** anticipates that all possible profiles would lie between the bounds of:

a. a maximum density of 8 ρ_{ns} [Vidaña, 6], which is achieved close to the surface – suggesting a crystalline-like structure, and

b. unlimited compressibility because the neutron degeneracy pressure asymptotes to infinity – suggesting an elastic structure.

Note that a wide range of possible NS density profiles (linear and variously curved) have been estimated and applied in this analysis (Figures not shown). Application of the various profiles was verified by summation of all the incremental shell masses, to ensure they match the specified star mass as a whole. In each case, the surface radii were determined from the density profile and the specified star mass.

It will be shown, however, that the closest match to observation data is achieved by specifying unlimited compressibility. This was also considered the most likely phenomenon, in light of Schmidt et al. [9] and Haug [10], and also because it is expected that the neutron degeneracy pressure asymptotes toward infinity at zero range.

Now, using relativistic speed Eq. (5), we calculate the *shell* escape speed (as a fraction of c), versus radial displacement for the predicted NS of 2.08 $M_{\odot}$, according to the *full span* of possible density profiles in **Figure 3**, as presented in **Figure 4**. These extremes of density profile embrace a 6.0 to 7.96 km range of possible surface radii.

In **Figure 4**, in the case of a density profile that is compressible up to 8 ρ_{ns}, we note that a peak hypothetical escape speed of 0.792 c (2.37E+08 m/sec) occurs at the surface (7.96 km radius). Alternatively, in the case of a linearly compressible density profile, the peak hypothetical escape speed of 0.886 c (2.65E+08 m/sec) occurs at 5.5 km radius which is 0.5 km below the surface, whereas at the surface (6.0 km), the escape speed is 0.876 c (2.62E+08 m/sec).

Either result for surface-escape speed is significantly below the photon speed (1.000 c), so this object should be visible.

Due to high surface temperatures, neutron stars emit strongly in X-ray (consistent with blackbody radiation), which may also contain the Fe Kα spectral line [23–25]. This ~6.4 keV line would be subject to gravitational redshift.

The line-of-sight redshift, **Z**, associated with the escape speed at the surface, is obtained from Eq. (8):

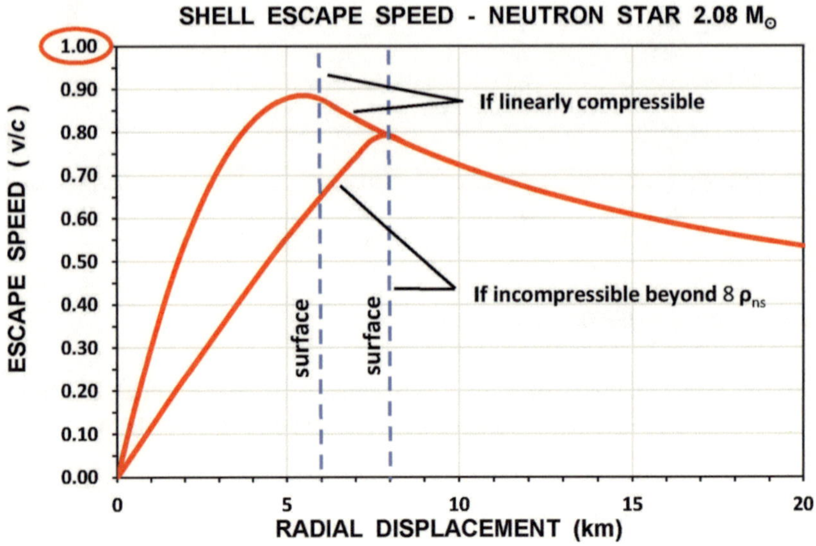

Figure 4.
Relativistic shell escape speeds for a range of plausible 2.08 $M_{\odot}$ neutron star density profiles.

$$Z = \frac{\Delta\lambda}{\lambda} = \sqrt{\frac{1+\frac{v}{c}}{1-\frac{v}{c}}} - 1 = \sqrt{\frac{1+\beta}{1-\beta}} - 1 \tag{8}$$

In **Figure 4** we found that the plausible surface escape speed ranges from 79.2% *c* at a possible 7.96 km radius, to 87.6% *c* at a possible 6.0 km radius. These determine a redshift of 1.9 to 2.9 respectively. This is a significant difference, so if the gravitational redshift of such stars can be detected and measured, this should inform the true density profile (discussed later).

Based on Katayama and Saito [4], we would expect a NS of 2.08 $M_{\odot}$ to be nearly massive enough to form a BH, but according to **Figure 4**, the formation of a BH should require substantially greater mass, or higher density (smaller surface radius). We are drawn to review the observational evidence for the size of BH's as follows (in publishing order):

- Gelino [26] initially estimated GRS 1009–45 at 5.2 $M_{\odot}$, but this has since been revised to 8.5 $M_{\odot}$ [StarDate Black Hole Encyclopedia 2012].

- Gelino and Harrison [27] report that the smallest stellar BH yet found, in GRO J0422 + 32, with a size range of 3.66 to 4.97 $M_{\odot}$. Further analysis in 2012 revised this mass to 2.1 $M_{\odot}$, which suggests this object is in fact a NS.

- Casares [28] lists 20 "confirmed cases" of stellar BH's – binaries with mass functions exceeding 5–6 $M_{\odot}$, and X-ray transients with mass estimated in the range 4–14 $M_{\odot}$.

- Remillard and McClintock [29] report that H 1705–250 has a confirmed dynamical measurement for the mass of its BH in the range of 5.6 to 8.3 $M_{\odot}$.

- Kreidberg et al. [30] report on several analyses, that indicate a significant "mass gap" between the anticipated maximum neutron star mass (3 $M_{\odot}$) and the anticipated low end of the BH mass distribution (5 $M_{\odot}$).

- Ozel et al. [31] concur with the existence of such a gap in the observed masses of these stars, occurring between 2 $M_{\odot}$ and 6 $M_{\odot}$.

- Antoniadis et al. [32] find that one of the largest neutron stars yet detected is 2.01 ± 0.04 $M_{\odot}$.

- Cromartie et al. [33] of Green Bank report discovery of a neutron star J0740 + 6620 with a mass of 2.14 $M_{\odot}$.

- Sathraprakash [34] estimates neutron star masses of up to 2.2 $M_{\odot}$ and 2.6 $M_{\odot}$ involved in mergers with BH's, based on LIGO-Virgo-KAGRA data.

- Bodensteiner et al. [35] compile a list of some 200 NS's and BH's detected using gravitational wave or electromagnetic detectors, revealing a paucity within the 2 to 5 $M_{\odot}$ range (with very wide error bars on gravitational detections).

The mass gap identified above could be due to an observational bias, or obscurity due to dimming and redshift at the higher masses, but according to the above observations, we should expect BH formation to occur at a minimum mass of ~5 $M_{\odot}$.

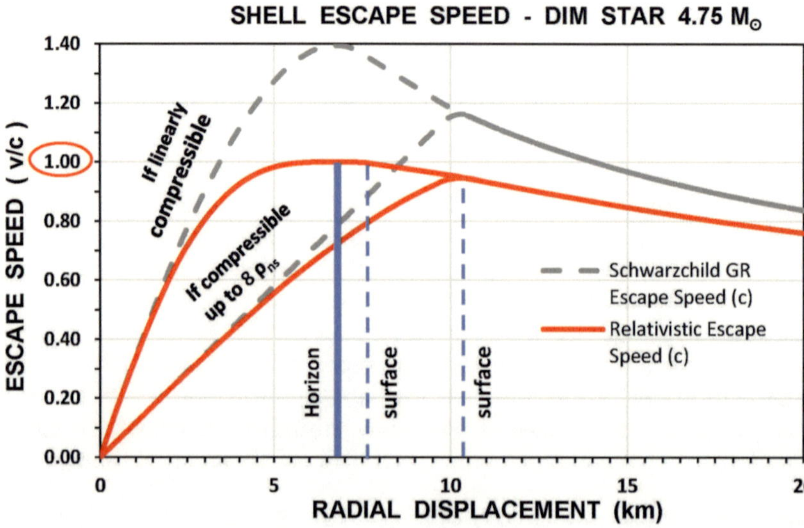

Figure 5.
Shell escape speeds for a range of plausible 4.75 M☉ neutron star density profiles, also comparing Relativistic and Schwarzchild models.

A slightly lower mass of 4.75 M☉ is initially chosen for examination. Extending the criteria of **Figure 4**, we investigate the range of plausible density profiles for a neutron star of 4.75 M☉, which obtains surface radii ranging from 7.56 km to 10.38 km.

Again using relativistic speed Eq. (5), **Figure 5** explores a range of hypothetical shell escape speeds expressed as a fraction of the photon speed, *c*, for a neutron star of mass 4.75 M☉ according to the full span of possible density profiles. In the case of a density profile that is compressible up to 8 ρ_{ns} we find the peak escape speed of 0.950 *c* occurs at the surface (10.6 km radius). Whereas in the case of the linearly compressible density profile, the peak escape speed just reaches the photon speed at a radius of 6.8 km, being 0.8 km below the surface.

We shall note that this is the first occurrence of an event horizon for a neutron star. This horizon is contained wholly within the star, a scenario that calls into question whether the GR assumed collapse of all matter within the horizon, while leaving outer layers unaffected, is even conceivable.

In this linearly compressible case, the shell escape speed falls to 0.997 *c* at the surface (7.60 km). In accordance with Eq. (8), this highest escape speed of 0.997 *c* at the *surface*, produces an extremely high redshift of 26.6. Even in the case of the lowest density profile, the surface redshift is 5.0, which is still quite substantial. This range of possible redshift would result in significant reddening and dimming of the star, but the star may remain electromagnetically visible. This 4.75 M☉ neutron star having such a high gravitational redshift, is consistent with a minimum observed BH mass of ~5 M☉.

Included for comparison in **Figure 5** are the Schwarzchild GR results according to Eq. (4) [15], for the two extremes of density profile used (dashed curves). These results clearly exceed 1.00 *c*, which cannot be valid, confirming our decision to rely upon Eq. (5).

Consider now a slightly larger neutron star of 5 M☉ for which the range of plausible density profiles, produces a range of radii from 7.65 to 10.6 km. The shell escape speeds (again using relativistic speed Eq. (5)), for the range of plausible density profiles are shown in **Figure 6**.

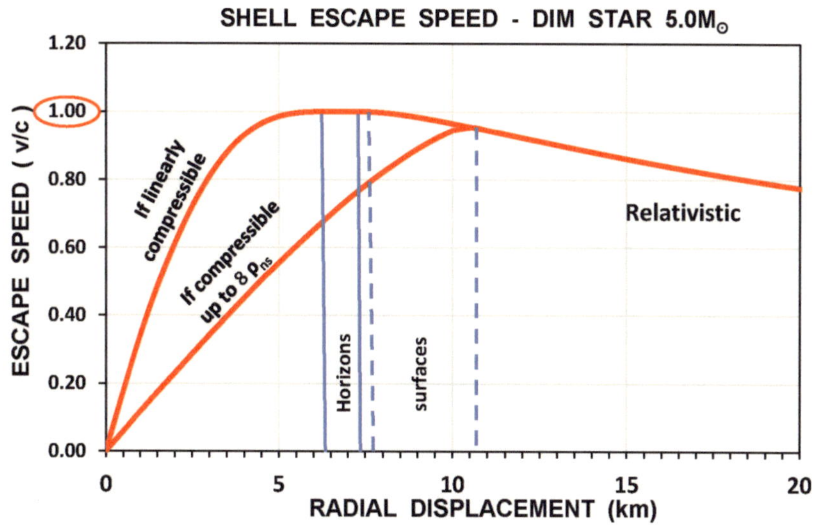

Figure 6.
Relativistic shell escape speeds for a range of plausible 5.0 $M_\odot$ neutron star density profiles.

In the case of a density profile that is compressible up to 8 ρ_{ns} we find a peak escape speed of 0.950 *c* occurring at the surface, at a radius of 10.6 km, producing a large redshift of 4.5. In the case of a linearly compressible density profile, the escape speed reaches *c* for all radii between 6.4 and 7.4 km, being just slightly below the surface of 7.65 km. The surface escape speed is 0.9994 *c* corresponding to an extreme redshift of 55.8, so this star may be nearly invisible, and may approximately represent the *critical size* at the cusp of invisibility.

Once again referring to **Figure 6**, when approaching from the direction of the core, a relativistic shell escape speed equal to the photon speed is reached at an "inner horizon", formed at a radius of 6.4 km (not to be confused with the once proposed Cauchy horizon).

When approaching from an external direction, the photon speed is reached at an "outer horizon" at a radius of 7.4 km. Apparently, a *pair* of event horizons are thereby defined. The region between the inner horizon and the outer horizon forms a "photon trap" – a broad "event zone" from which photons cannot escape.

Any internal photons released anywhere inside the event zone would (subject to scattering or absorption) circulate within the zone indefinitely, as would any external photons entering the event zone from beyond either horizon.

As noted, only the linearly compressible density profile scenario shown in **Figure 3**, has allowed invisibility in neutron stars to begin as small as 5 $M_\odot$.

If in fact there is a phase change in stellar composition occurring between 3 $M_\odot$ and 5 $M_\odot$ as suggested by Kreidberg et al. [30], then the stellar density would increase and the surface radius would correspondingly reduce, without affecting the outer event horizon radius.

Consider now the shell escape speeds for the same range of plausible density profiles in a neutron star of higher mass such as 6.5 $M_\odot$ as shown in **Figure 7**. In the case of a density profile that is compressible up to 8 ρ_{ns} we find a peak escape speed of 0.987 *c* at a surface radius of 11.5 km (redshift of 11.3).

In the case of linear compressibility, we find an inner horizon is developed at 5.5 km and an outer horizon is developed at 9.6 km. We note that at an event horizon,

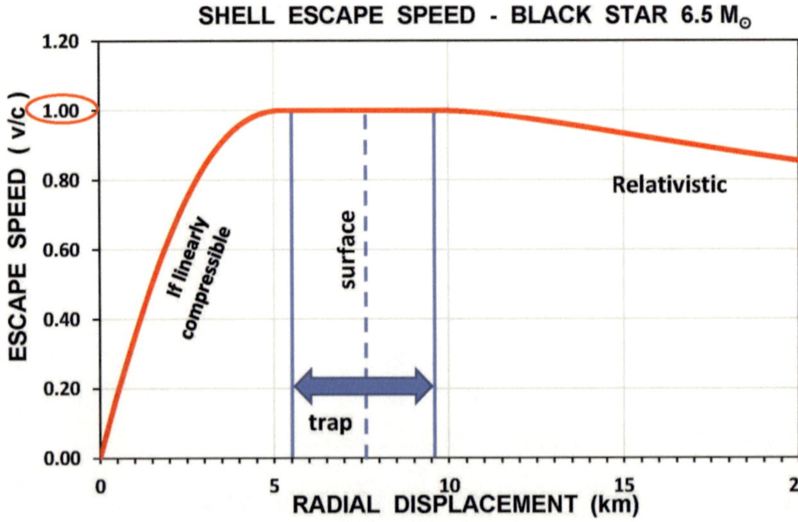

Figure 7.
Relativistic shell escape speed for a 6.5 M☉ neutron star assuming linear compressibility.

the redshift is infinite [34] (invisible). Since the outer horizon is well beyond the surface radius of 8.2 km, this star is certainly invisible. Such an invisible object is preferably termed a "black star" [34] (a star which is electro-magnetically invisible), which removes the connotation of the object necessarily containing or forming a wormhole.

If we investigate a significantly larger mass star of 13.2 M☉ for example, we find that the photon trap region is expanded. The inner horizon is at 7.0 km, and the outer horizon is at 19.5 km, being well outside the surface radius of 12.7 km (Figure not shown).

As we extend this analysis to consider possible black stars of larger mass such as 35 M☉ we obtain a nominal surface radius of 16.2 km and an expanded photon trap zone ranging from 3.5 to 52.5 km as shown in **Figure 8**.

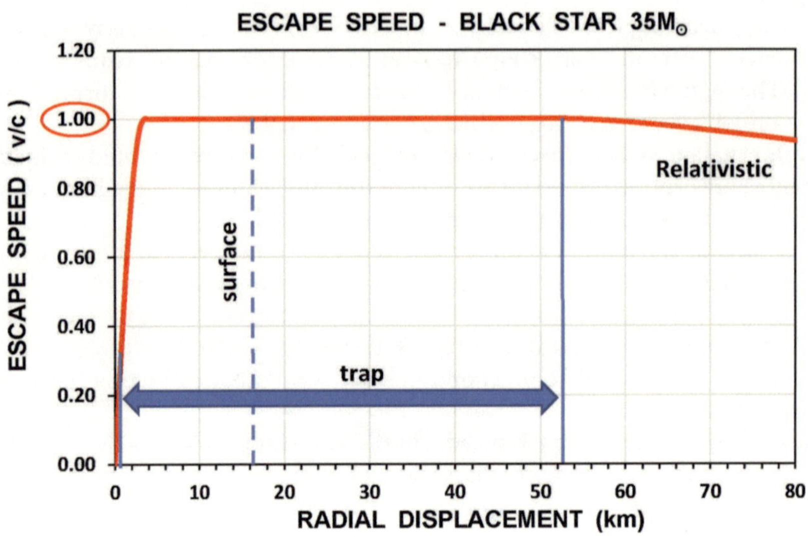

Figure 8.
Shell escape speed for a 35 M☉ neutron star of linear compressibility.

Consider the implications of this trap region (or event zone) using **Figure 7** as an example. Can matter or radiation continue to reside within this zone, including the compacted neutrons (or neutron degenerate matter) comprising the outer layers of this star?

Under GR it is widely anticipated that, in order to remain stable and avoid collapsing to the stellar centre, matter and radiation within the event horizon (the outer horizon) would need to move outward faster than the speed of light, which is accepted as impossible. It has been claimed that for this reason, collapse would be inevitable and unlimited, resulting in a singularity. This argument fails, because of the previously unknown inner horizon, which presents a barrier to any possible collapse (the shell escape speeds at lesser radii are subluminal).

If we challenge the model further, by taking a more extreme jump to 1000 $M_\odot$ we find a surface radius of 37.5 km, with the photon trap zone extending from an inner horizon of 2.5 km radius to an outer horizon of 1470 km radius (Figure not shown).

The linearly compressible density profile data proposed herein are summarised in **Figure 9** (log scale). Here we see that the surface radius and the central density, as functions of stellar mass, each conform to a one-quarter power exponential curve, fitting data with high precision ($R^2 = 1$).

The critical mass value of 5 $M_\odot$ was selected, to be consistent with the observational evidence cited earlier [26–35], but it should be regarded as an estimate which is tested using the calculations and Figures shown here. Future, more-comprehensive observational evidence may be used to refine this estimate.

If our methodology used herein holds, then based on this critical mass, if the gravitational redshift for a neutron star could be detected (after accounting for any velocity redshift), **Figure 10** could inform or confirm its likely mass and radius (e.g., a redshift of 10 indicates a mass of 4.1 $M_\odot$ and a radius of 7.3 km). **Figure 10** also demonstrates a smooth transition toward invisibility, rather than a sudden, cataclysmic event (implied by GR).

By collating all the foregoing data, we can chart the photon trap dimensions (the inner and outer horizons), together with the surface radii, for a range of stellar masses, in **Figure 11**.

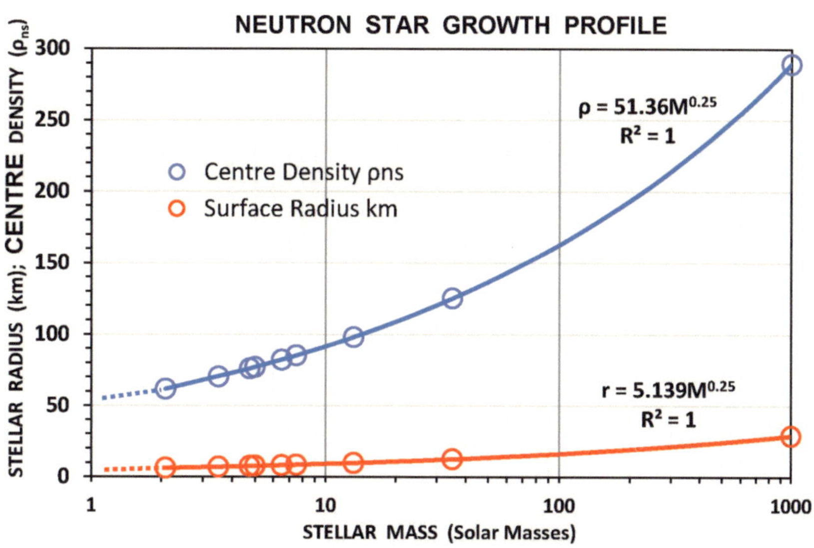

Figure 9.
Proposed surface radius and centre density of neutron stars vs. mass.

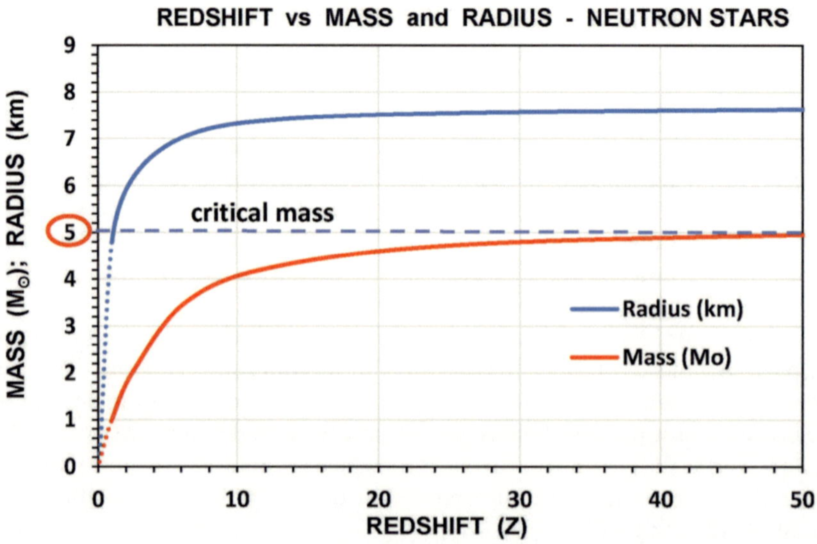

Figure 10.
Predicting NS mass and surface radius from observed gravitational redshift, based on a suggested critical mass of 5 $M_{\odot}$.

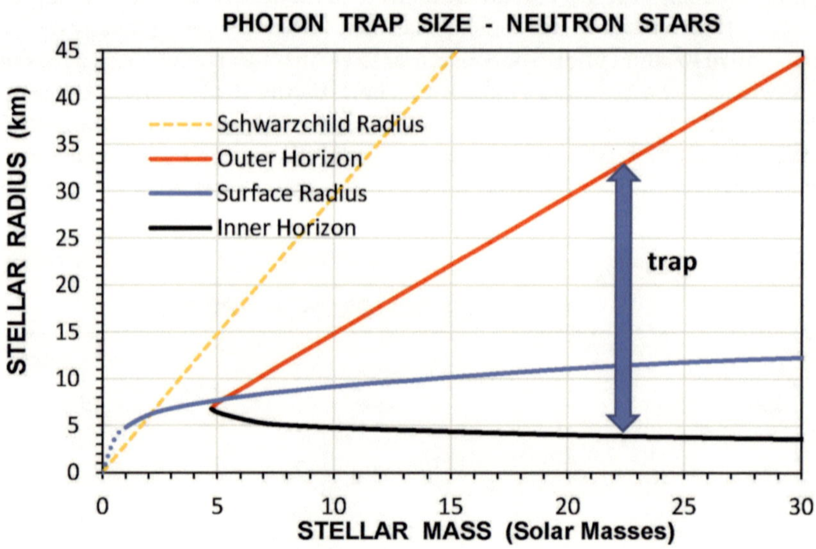

Figure 11.
Summary of results revealing the photon trap formed between the inner horizon and outer horizon, which initially form together below the surface.

The point of convergence of the inner and outer horizon radii (lower left of **Figure 11**, below the surface) represents the minimum mass for an event horizon to exist, being approximately 4.75 $M_{\odot}$. We see that the outer horizon exceeds the surface radius at masses beyond 5 $M_{\odot}$, whereby the star becomes invisible.

Importantly, when an outer horizon is developed, so too is an inner horizon. Given the existence of the inner horizon, and in light of Schmidt et al. [9], and also in light of Barcelo et al. [36] (who characterise the situation as: 'quantum matter always seems to find new ways of delaying gravitational collapse'), there appears to be *no compelling physical reason for a core collapse of an invisible (black) star that would result in the production of a singularity*.

It seems unlikely that broadening the scope of this analysis to include charged, rotating and/or revolving stars, or the influence of other distant massive objects, could result in the formation of a real singularity.

In contrast to the inner horizon radius, the outer horizon radius increases linearly with mass, and is unrelated to the density profile assumptions made here.

On the assumption of a non-rotating, spherically symmetric body, the Schwarzchild radius equation [37] has been applied to **Figure 11** (dashed line) as follows, with stellar mass M_S measured in $M_\odot$:

$$R_{sch} = 2\frac{MG}{c^2} = 2.95E{+}03\, M_S\, m \tag{9}$$

By inspection, we find that the outer event horizon radius (based on relativistic speed Eq. (5)) calculates to exactly 50% of the Schwarzchild radius, consistently across all data obtained (>5 $M_\odot$). Accordingly:

$$R_{out} = \frac{MG}{c^2} = 1.48E{+}03\, M_S\, m \tag{10}$$

This is confirmed by Haug [18] who newly identified the escape radius (where the escape speed reaches the photon speed), is: "equal to a half of the Schwarzschild radius".

5. Conclusions

The foregoing has used a methodology (based on a critical mass of ~5 $M_\odot$), that is open for discussion. This methodology could in future be adjusted according to any new findings on the real density profile of neutron stars.

1. At least within the range of stellar masses up to 1000 $M_\odot$ modelled here, isolated neutron stars fit a continuum of increasingly massive neutron stars that happen to become electromagnetically invisible once their mass exceeds ~5 $M_\odot$, described here as the "critical mass".

2. Neutron stars may appear increasingly dim and red-shifted as masses increase from 3 to 5 $M_\odot$, leading to the appearance of a mass gap in observations.

3. The closest match to observation data was achieved by specifying linear (elastic) neutron compressibility under gravity, resisted by neutron degenerate pressure. The magnitude of this pressure (based on the strong nuclear force) has not been reliably quantified.

4. A photon trap (or event zone) is developed between an inner event horizon and an outer event horizon. The inner horizon first appears below the surface at some 6.75 km radius, in a neutron star that has reached a mass of ~4.75 $M_\odot$,

reducing to 2.5 km radius at 1000 $M_\odot$. The outer horizon first appears coincident with the inner horizon at ~4.75 $M_\odot$, separating at higher masses, increasing to ~1475 km at 1000 $M_\odot$.

5. The outer horizon radius corresponds mathematically to exactly 50% of the Schwarzchild radius.

6. The existence of an inner event horizon *prevents* the physical possibility of stellar collapse with the formation of a singularity. This is supported by cited high-energy physics experimental evidence that neutrons under compression may not transition into degenerate matter, nor collapse to a singularity.

7. It is apparent that the first occurrence of an event horizon, at a neutron star mass of ~4.75 $M_\odot$, has a radius of ~6.8 km. This is below the surface radius of ~7.6 km and it seems inconceivable that the 'core' of the star (up to 6.8 km radius), could collapse to a singularity while the 'crust', from 6.8 km to 7.6 km, remains in place. Since the crust (with escape speeds less than **c**), cannot collapse, this suggests that the core does not collapse either.

8. There seems no necessary physical mechanism whereby neutron stars represent a special case within a relativistic universe, which can (at ~5 $M_\odot$) abruptly transform into a singularity with zero surface radius and infinite density, producing infinite spacetime curvature. No observed evidence of such a cataclysmic and powerful event is reported, even from an accretion disk or jet (beyond the outer event horizon).

9. Accordingly, the singularity anticipated under GR field equations, apparent to an external observer, may be a mathematical illusion. Internally, no real singularity is formed, therefore a GR black hole *per se* does not physically exist. An invisible neutron star is more appropriately termed a "black star".

10. Absent a real singularity existing within a black hole (or black star), no bridge can form between a pair of them – the wormhole has no physical reality.

11. It may be possible to inform the mass, surface radius and average density of a sufficiently luminous <5 $M_\odot$ neutron star based on gravitational redshift, observed in the Fe Kα spectral line emitted in X-ray.

Acknowledgements

This chapter represents a significant revision of the paper by Dr. M.T. Cole [38] (reproduced material is reprinted with permission). The encouragement, advice and guidance provided by Prof. M. Bailes (Swinburne University of Technology), Prof. E.G. Haug (Norwegian University of Life Sciences), Dr. K. Leschinski (University of Vienna), and Em Prof. S.J. Brodsky et al. [3]. has been very much appreciated.

Author details

Martin T. Cole
Independent Researcher, South Melbourne, Victoria, Australia

*Address all correspondence to: drcolemartin@gmail.com

References

[1] Carter K. Proton's Party Pals May Alter its Internal Structure. Newport VA: Jefferson Lab; 2009. Available from: https://phys.org/news/2011-02-bound-neutrons-pave-free.htm

[2] Brodsky SJ, Deur A, Roberts CD. Nature's Strongest Force, Scientific American. 2024. Available from: https://www.scientificamerican.com/article/physicists-finally-know-how-the-strong-force-gets-its-strength/

[3] Brodsky SJ, Deur A, Roberts CD. Private Communication, SLAC National Accelerator Laboratory, Menlo Park CA; Thomas Jefferson National Accelerator Facility, Newport VA; and Institute for Nonperturbative Physics. Nanjing University; 2025. DOI: 10.1016/j.ppnp.2023.104081

[4] Haensel P, Potekhin AY, Yakovlev DG. Neutron Stars 1: Equation of State and Structure. NY: Springer; 2006

[5] Katayama T, Saito K. Hyperons in neutron stars. Physics Letters B. 2015;**747**:43. DOI: 10.1016/j.physletb.2015.03.039

[6] Vidaña I. Hyperons in neutron stars. Journal of Physics: Conference Series. 2016;**668**:1-8. DOI: 10.1088/1742-6596/668/1/012031

[7] Giddings SB. Scientific American. 2019, p. 42. Available from: https://www.scientificamerican.com/article/escape-from-a-black-hole/

[8] Potekhin AY. arXiv 1102.5735v3 [astro.ph SR]. 2011:1-8. DOI: 10.48550/arXiv.1102.5735

[9] Schmidt A, Pybus JR, Weiss R, Segarra EP, Hrnjic A, Denniston A, et al. Probing the core of the strong nuclear interaction. Nature. 2020;**578**(7796):540. DOI: 10.48550/arXiv.2004.11221

[10] Haug EG, Gianfranco S. The European Physical Journal Applied Physics. 2022;**4**(2):40-47. DOI: 10.24018/ejphysics.2022.4.2.165

[11] Timiln J. Neutron Degeneracy Pressure. 2013. Available from: https://www.physics.drexel.edu/~bob/Term_Reports/John_Timlin.pdf

[12] Kahana D. Brookhaven National Laboratory. 2019. Available from: https://www.quora.com/Is-there-a-relativistic-shell-theorem?

[13] Newton I. Theorem XXXI, 1687. Available from:https://plato.stanford.edu/entries/newton-principia/

[14] Bautista M. Cogent Physics. Milton Park, Oxfordshire: Taylor & Francis Online; 2017. DOI: 10.1080/23311940.2017.1357325

[15] Zhang SN, Yi S. International Journal of Modern Physics. 2012;**12**:419-431. DOI: 10.1142/S2010194512006642

[16] Augousti AT, Radosz A. An observation on the congruence of the escape velocity in classical mechanics and general relativity in the schwarzchild metric. European Journal of Physics. 2006;**27**:331. DOI: 10.1088/0143-0807/27/2/015

[17] Rybczyk JA. Relativistic Escape Velocity. 2009. Available from: http://www.mrelativity.net

[18] Haug EG. Hal Archives. 2021. Available from: https://hal.archives-ouvertes.fr/hal-03240114

[19] Camp P. What Is the Formula for Escape Velocity According to General Theory of Relativity? Milledgeville GA: Georgia College; 2017. Available from: http://www.Quora.com/

[20] Leinenweber M. Physics Forums. 2012. Available from: https://www.physicsforums.com/threads/relativistic-escape-velocity.645667/

[21] Dziewonski AM, Anderson DL. Preliminary reference earth model. Physics of the Earth and Planetary Interiors (Elsevier). 1981;**25**(4):297. DOI: 10.1016/0031-9201(81)90046-7

[22] Hathaway DH. NASA/Marshall Space Flight Center. 2015. Available from: https://physics.stackexchange.com/questions/201235/density-of-the-sun

[23] Majczyna A, Madej J. ArXiv: astroph/060 1076v1; 2006, DOI: 10.48550/arXiv.astro-ph/0601076

[24] Piraino S, Santangelo A, Kaaret P, Muck B, D'Ai A, DiSalvo T, et al. Astronomy and Astrophysics. 2012;**542**:1-4. DOI: 10.1051/0004-6361/201219283

[25] NCBJ. National Centre for Nuclear Research. Swierk, Poland: Instytut Katagorli A+; 2022. Available from: https://www.ncbj.gov.pl/en/bp4/neutron-stars

[26] Gelino DM. California Institute of Technology, researchgate.net/publication/259263717 (2002), revised: StarDate Black Hole Encyclopedia - GRS 1009-45 (2012)

[27] Gelino DM, Harrison TE. The lowest mass neutron star? The Astrophysical Journal. 2003;**599**(2):1254-1259. DOI: 10.48550/arXiv.astro-ph/0308490

[28] Casares J. Instituto de Astrofısica de Canarias, 38200 – La Laguna, Tenerife, Spain. Ithaka, NY: Cornell University; 2006. DOI: 10.1017/S1743921307004590

[29] Remillard RE, McClintock JA. X-ray properties of black hole binaries. Annual Review of Astronomy and Astrophysics. 2006;**44**:49-92. DOI: 10.1146/annurev.astro.44.051905.092532

[30] Kreidberg L, Bailyn CD, Farr WM, Kalogera V. Mass measurements of black holes in x-ray tranisents: Is there a mass gap? The Astrophysical Journal. 2012;**757**:36. DOI: 10.1088/0004-637X/757/1/36

[31] Ozel F, Psaltis D, Narayan R, Villarrea AS. Mass distribution and birth masses of neutron stars. The Astrophysical Journal. 2012;**757**:55. DOI: 10.48550/arXiv.1201.1006

[32] Antoniadis J, Freire PCC, Wex N, Tauris TM, Lynch RS, Van Kerkwijk MH, et al. A massive pulsar in a compact relativistic binary. Science. 2013;**1**:340, arXiv:1304.6875. DOI: 10.48550/arXiv.1304.6875

[33] Cromartie HT, Fonseca E, Ransom SM, Demorest PB, Arzoumanian Z, Blumer H, et al. Nature Astronomy. 2019;**4**:72-76. DOI: 10.1038/s41550-019-0880-2

[34] Sathraprakash BS. Fact Sheet GW200105 and GW200115. PA: Penn State University & Wales: Cardiff University; 2021. Available from: https://www.ligo.caltech.edu/news/ligo20210629

[35] Bodensteiner J, Heida M, Abdul-Masih M, Baade D, Banyard G, Bowman DM, et al. The Messenger, No 186, European Southern Observatory. 2022. Available from: https://issuu.com/eso_observatory/docs/messenger-no186

[36] Barcelo C, Liberati S, Sonego S, Visser M. Black stars not holes. Scientific

American. 2009;**301**(4):21. Available from: https://physics.ucf.edu/~britt/AST2002/R9-Barcelo-Black%20stars,%20not%20holes.pdf

[37] Kutner M. Astronomy: A Physical Perspective. Cambridge UK: Cambridge University Press; 2003. p. 148. Available from: https://www.cambridge.org_9780521529273

[38] Cole MT. Neutron star event horizons. Physics Essays. 2022;**35**(4):364-371. DOI: 10.4006/0836-1398-35.4.364

Chapter 4

Gravitational Waves, Tokamaks, and the Pre-Planckian to Planckian era Universe

Andrew Walcott Beckwith

Abstract

Our idea for black hole physics being used for gravitational wave generation is using torsion to form a cosmological constant. Planck-sized black holes create a spin density term linked to torsion. In doing so, we review torsion-generated gravitational wave frequencies as being similar to frequency values for gravitational waves due to a tokamak.

Keywords: cosmological constant, torsion, spin density, BEC scaling of black hole physics, tokamak, wave function of the universe, wormholes

1. Introduction

The author has in prior work given the idea that a decay of millions of Planck-sized black holes as within the very early universe as in [1] could generate gravitational waves (GWs) and gravitons, due to a breakup of black holes as predicted in [1]. In our present work we consider the same problem, following [2]. The breakup of black holes commences due to what is stated in [1]. Furthermore, the breakup of black holes is also affected by [3]. Chavanis [3] addresses what happens if gravitons are acting as a Bose-Einstein condensate. We state the Bose-Einstein condensate contributes to a resulting dark energy. Either the breakup of black holes as stated in [4] or [3] would lead to GWs (and graviton frequencies) of the order of 10^{10} Hz. The upper frequency of GWs would be possibly as high as 10^{19} Hz. In doing so we look at the duration of an observed signal, its relative noisiness, and stochastic noise contributions. These problems of signal from GWs from black holes are covered in [5]. In addition, the generation of GWs in a tokamak, if commensurate with evolved laser interfermoter space antenna data after a step-down of 10^{-25} to 10^{-26} due to 60 or more e folds [6], can be reviewed for adequate polarization states for GW. The question we may analyze is if we need higher dimensions to be in fidelity to the data sets obtained as given in [7]. Having said that, what are the justifications as to using tokamaks? Justification of why we may use tokamaks for relic black hole GW radiation as similar to tokamak GWs forms the last part of our chapter.

2. The basis of torsion and relic black holes, giving GWs that are analyzed and compared to tokamaks later in this document

To do this, review how torsion may allow for understanding a quantum number n as well as primordial black holes and the cosmological constant.

Following [1, 2] we do the introduction of black hole physics in terms of a quantum number n.

$$\begin{aligned}\sqrt{\Lambda} &= \frac{k_B E}{\hbar c S_{entropy}} \\ S_{entropy} &= k_B N_{particles}\end{aligned} \tag{1}$$

And then a Bose-Einstein condensate (BEC) is given by [1, 3] as to

$$\begin{aligned}&m \approx \frac{M_P}{\sqrt{N_{gravitons}}} \\ &M_{BH} \approx \sqrt{N_{gravitons}} \cdot M_P \\ &R_{BH} \approx \sqrt{N_{gravitons}} \cdot l_P \\ &S_{BH} \approx k_B \cdot N_{gravitons} \\ &T_{BH} \approx \frac{T_P}{\sqrt{N_{gravitons}}}\end{aligned} \tag{2}$$

This is promising but needs to utilize [4], in which we make use of the following. First, a time step:

$$\tau \approx \sqrt{GM\delta r} \tag{3}$$

By use of [5] we use Eq. (3) for energy [4] for radiation of a particle pair from a black hole:

$$|E| \approx \left(\sqrt{GM\delta r}\right)^{-1} \hbar \tag{4}$$

Here we assert that the spatial variation goes as

$$\delta r \approx \ell_P \tag{5}$$

This is a Plank length, whereas we assume in Eq. (6) that the mass is a Planck-sized black hole:

$$M \approx \alpha M_P \tag{6}$$

This mass of primordial black holes is part of **Table 1**.

As to **Table 1**, we obtain, due to the quantum number n, information loading and insertion of data from the prior universe into the present universe as given per black hole. This makes use of [1, 2, 7–9].

Table 1 data will be connected to the following given consideration of spin density, as to Planck-sized black holes:

In [1, 9] we have the following, that is, we have a spin density term of [1, 9]. And this will be what we input black hole physics into with regard to forming a spin density term from primordial black holes.

End of prior universe time frame	Mass (black hole): super massive end of time BH 1.98910^{+41} to about 10^{44} grams	Number (black holes) 10^6 to 10^9, usually from center of galaxies
Planck-era black hole formation Assuming start of merging of micro black hole pairs	Mass (black hole) 10^{-5} to 10^{-4} grams (an order of magnitude of the Planck mass value)	Number (black holes) 10^{40} to about 10^{45}, assuming that there was not too much destruction of matter-energy from the pre-Planck conditions to Planck conditions
Post Planck-era black holes with the possibility of using Eq. (1) and Eq. (2) to have say 10^{10} gravitons/second released per black hole	Mass (black hole) 10 grams to say 10^6 grams per black hole	Number (black holes) Due to repeated black hole pair forming a single black hole multiple times. 10^{20} to at most 10^{25}

Table 1.
Assuming Penrose recycling of the universe as stated in [2].

$$\sigma_{Pl} = n_{Pl}\hbar \approx 10^{71} \tag{7}$$

Also, the initial energy [7] per black hole is given as

$$E_{Bh} = -\frac{n_{quantum}}{2} \tag{8}$$

We then can use for a black hole the following scaling:

$$\begin{aligned} |E| &\approx \left(\sqrt{G\cdot(\alpha M_P)\cdot \ell_P}\right)^{-1}\hbar \\ &\underset{G=M_P=\hbar=k_B=\ell_P=c=1}{\longrightarrow} (1/M_{BH})^{1/2} \approx \frac{n_{quantum}}{2} \end{aligned} \tag{9}$$

We then reference Eq. (2) to observe the following:

$$\begin{aligned} & M_{BH} \approx \sqrt{N_{gravitons}} M_P \\ & \Rightarrow (1/M_{BH})^{1/2} \approx \frac{n_{quantum}}{2} \approx \frac{1}{\left(N_{gravitons}\right)^{1/4}} \\ & \Rightarrow n_{quantum} \approx \frac{2}{\left(N_{gravitons}\right)^{1/4}} \end{aligned} \tag{10}$$

This is a stunning result. That is, Eq. (2) is BEC theory, but due to micro-sized black holes that we assume, the associated quantum number n goes increases significantly. Is this implying that corresponding increases in quantum number per black hole, n, are commensurate with increasing temperature? We start off with **Table 1** for conditions with the entropy as given in Eq. (1) and Eq. (2) for primordial black holes as brought up in **Table 1**. Whereas for the tokamaks, we eventually have

$$n|_{massive-gravitons/\,second} \propto \frac{3\cdot\hbar\cdot e_j}{\mu_0\cdot R^2\cdot \xi^{1/8}\cdot\tilde{\alpha}} \times \frac{\left(T_{\text{Tokamak}-temperature}\right)^{1/4}}{\lambda^2_{Graviton}\cdot m_{graviton}\cdot c^2\cdot(.87)^{5/4}} \sim 1/\lambda^2_{Graviton} scaling \tag{11}$$

This value of Eq. (11) as to the number of gravitons would be then related to the quantum number N (gravitons) as related to a quantum number n only in the very onset of the operation of the tokamak. That is, we would have the number of gravitons go up as we would have a shrinking graviton wavelength for a massive graviton. This will be discussed more later. However, the wavelength of the massive graviton as in Eq. (11) as related to GW frequency and tokamaks will be described when we conclude our document with respect to the wave function of the universe, a work partly drawing upon Kieffer and also on Weber. The wave function of the universe condition is heavily influenced by the similarities to Eq. (11) with the quantum number n per black hole, and the number N of black holes, as initially presented in **Table 1**. To do so we consider **Table 1** as giving a template for a wormhole connecting a prior universe to the present universe.

3. Wave function of the universe, and the assumption of connecting the prior to the present universe, via say multiple wormholes

We are looking at the Weber book to reformulate quantization imposed in general relativity (GR).

We advise readers to review [10–21] extensively before reading this section.

Using [10], a statement as to quantization for a would-be GR term comes straight from

$$\Psi_{Later} = \int \sum_{H} e^{(iI_H/\hbar)(t,t^0)} \Psi_{Earlier}(t^0) dt^0 \tag{12}$$

The approximation we are making is to pick one index, so as to have

$$\Psi_{Later} = \int \sum_{H} e^{(iI_H/\hbar)(t,t^0)} \Psi_{Earlier}(t^0) dt^0 \underset{H \to 1}{\longrightarrow} \int e^{(iI_{H_{FIXED}}/\hbar)(t,t^0)} \Psi_{Earlier}(t^0) dt^0 \tag{13}$$

This corresponds to being primarily concerned as to GW generation, which is what we will be examining in our ideas, using

$$e^{(iI_{H_{FIXED}}/\hbar)(t,t^0)} = \exp\left[\frac{i}{\hbar} \cdot \frac{c^4}{16\pi G} \cdot \int_{M} dt \cdot d^3 r \sqrt{-g} \cdot (\Re - 2\Lambda)\right] \tag{14}$$

We will use the following—that is, if Λ is a constant, do the following for the Ricci scalar [17]:

$$\Re = \frac{2}{r^2} \tag{15}$$

If so, then we can write the following: If we have an invariant cosmological constant, so we write $\Lambda \underset{all-time}{\longrightarrow} \Lambda_0$ everywhere, then Eq. (14) becomes [10].

$$e^{(iI_{H_{FIXED}}/\hbar)(t,t^0)} = \exp\left[\frac{i}{\hbar} \cdot \frac{c^4 \cdot \pi \cdot t^0}{16G} \cdot \left(r - r^3 \Lambda_0\right)\right] \tag{16}$$

Then, Eq. (12) is rewritten to be

$$\Psi_{Later} = \int \sum_{H} e^{(iI_H/\hbar)(t,t^0)} \Psi_{Earlier}(t^0) dt^0$$
$$\underset{at-wormhole}{\longrightarrow} \int \exp\left[\frac{i}{\hbar} \cdot \frac{c^4 \cdot \pi \cdot t^0}{16G} \cdot \left(r - r^3\Lambda_0\right)\right] \Psi_{Earlier}(t^0) dt^0 \tag{17}$$

4. Examining the behavior of the earlier wave function in Eq. (17)

Hartle and Hawking [13] states a Hartle-Hawking wave function, which we will adapt for the earlier wave function as stated in Eq. (6) to read as follows

$$\Psi_{Earlier}(t^0) \approx \Psi_{HH} \propto \exp\left(\frac{-\pi}{2GH^2} \cdot (1 - \sinh(Ht))^{3/2}\right) \tag{18}$$

Here, making use of Sarkar [14], if g_* is the degree of freedom allowed, we set

$$H = 1.66\sqrt{g_*} T_{temp}^2 / M_{Planck} \tag{19}$$

We assume initially a relatively uniformly given temperature, so H is constant.

Then we will be attempting to write out an expansion as to what Eq. (6) gives us while we use Eq. (18) and Eq. (19), with H approximately constant.

5. Methods used in calculating Eq. (17), with interpretation of the results

If so, then

$$\Psi_{Later} = \int \exp\left[\frac{i}{\hbar} \cdot \frac{c^4 \cdot \pi \cdot t^0}{16G} \cdot \left(r - r^3\Lambda_0\right)\right] \exp\left(\frac{-\pi}{2GH^2} \cdot (1 - \sinh(Ht))^{3/2}\right) dt^0 \tag{20}$$

Then using numerical integration [18–20],

$$\Psi_{Later} \underset{t_M \to \varepsilon^+}{\longrightarrow} \int_0^{t_M} e^{i \cdot (\tilde{\alpha}1) \cdot t - (\tilde{\alpha}2) \cdot (1 - \sinh(Ht))^{3/2}} dt$$
$$\approx \frac{t_M}{2} \cdot \left(e^{i \cdot (\tilde{\alpha}1) \cdot t_M - (\tilde{\alpha}2) \cdot (1 - \sinh(H \cdot t_M))^{3/2}} - 1\right) \tag{21}$$
$$\tilde{\alpha}1 = \left[\frac{c^4 \cdot \pi}{16G\hbar} \cdot \left(r - r^3\Lambda_0\right)\right], \tilde{\alpha}2 = \frac{\pi}{2GH^2}$$

Notice the terms for the H factor, and from here we will be making our prediction. If the energy, E, has the following breakdown,

$$\begin{aligned} & H = 1.66\sqrt{g_*}T_{temp}^2/M_{Planck} \\ & \Rightarrow E \approx k_B T_{Temp} \approx \hbar \cdot \omega_{signal} \\ & \Rightarrow \omega_{signal} \approx \frac{k_B \cdot \sqrt{M_{Planck} H}}{\hbar \cdot \sqrt{1.66\sqrt{g_*}}} \end{aligned} \tag{22}$$

the upshot is that we have, in this, a way to obtain a signal frequency by looking at the real part of Eq. (22), if we have a small t initially (small time step).

6. How to compare with a Kieffer solution and thereby isolate the cosmological constant contribution

This means looking at [21]. Eq. (11) would imply initial frequency dependence; what we are doing next is to strategize to understand the contribution of the cosmological constant in this sort of problem. The way to do it would be to analyze a Kieffer "dust solution" as a signal from the wormhole. That is, look at [21], where we assume that t would be in this case the same as in Eq. (21); namely, in this case we will write

$$\Delta\omega_{signal}\Delta t \approx 1 \tag{23}$$

If so, then we can assume that the time would be small enough so that

$$\Delta t \approx \frac{\hbar\sqrt{1.66\sqrt{g_*}}}{k_B \cdot \sqrt{M_{Planck} H}} \tag{24}$$

If Eq. (24) is of a value somewhat close to t, in terms of general initial time, we can write [21].

$$\psi_{\tilde{n},\lambda}(t,r) \equiv \frac{1}{\sqrt{2\pi}} \cdot \frac{\tilde{n}! \cdot (2\lambda)^{\tilde{n}+1/2}}{\sqrt{(2\tilde{n})!}} \cdot \left[\frac{1}{(\lambda + i \cdot t + i \cdot r)^{\tilde{n}+1}} - \frac{1}{(\lambda + i \cdot t - i \cdot r)^{\tilde{n}+1}}\right] \tag{25}$$

Here the time t would be proportional to Planck time, and r would be proportional to Planck length, whereas we set

$$\lambda \approx \sqrt{\frac{8\pi G}{V_{volume}\hbar^2 t^2}} \underset{G=\hbar=\ell_{Planck}=k_B=1}{\rightarrow} \sqrt{\frac{8\pi}{t^2}} \equiv \frac{\sqrt{8\pi}}{t} \tag{26}$$

Then a preliminary emergent space-time wave function would be

$$\begin{aligned} & \psi_{\tilde{n},\lambda}(\Delta t, r) \\ & \equiv \frac{1}{\sqrt{2\pi}} \cdot \frac{\tilde{n}! \cdot \left(2 \cdot \sqrt{8\pi} \cdot (\Delta t)^{-1}\right)^{\tilde{n}+1/2}}{\sqrt{(2\tilde{n})!}} \cdot \left[\begin{array}{c} \dfrac{1}{\left(\sqrt{8\pi} \cdot (\Delta t)^{-1} + i \cdot \Delta t + i \cdot r\right)^{\tilde{n}+1}} \\ -\dfrac{1}{\left(\sqrt{8\pi} \cdot (\Delta t)^{-1} + i \cdot \Delta t - i \cdot r\right)^{\tilde{n}+1}} \end{array}\right] \end{aligned} \tag{27}$$

just at the surface of the bubble of space-time, with $t_{Planck} \propto \Delta t$, and $r \propto \ell_{Planck}$.

This is from a section, page 239 of the third edition of Kieffer's book, regarding a quantum theory of collapsing dust shells, and so we have the following procedure to isolate out the contribution of the cosmological constant. Namely, take the real part of Eq. (27) and compare it with the real part of Eq. (21).

Another way to visualize this situation is a different way to interpret Eq. (26). To do so we examine looking at page 239 of Kieffer, namely [21], where one has an expectation value for energy we can write as

$$\langle E \rangle_{\kappa=n,\lambda} = \frac{(\kappa = n) + 1/2}{\lambda} \underset{\lambda \approx 1/\hbar\omega}{\rightarrow} \hbar\omega \cdot ((\kappa = n) + 1/2) \tag{28}$$

What we can do is to ascertain that the last step would be to make Eq. (28) in a sense partly related to the simple harmonic oscillator. But we should take into consideration the normalization using that $\hbar = \ell_P = G = t_P = k_B = 1$ is done via Plank unit normalization [14, 15]. If so, then we have that frequency is proportional to $1/t$, where t is time. Hence, if there is a value of $n = 0$, and making use of the frequency, we then would be able to write Eq. (27) as [21].

$$\Psi_{1,\kappa=n=0} \approx \sqrt{\frac{\omega}{\pi}} \cdot \left[\frac{1}{\omega + i \cdot (t+r)} - \frac{1}{\omega + i \cdot (t-r)} \right] \tag{29}$$

or

$$\Psi_{2,\kappa=n=0} \approx \frac{1}{\sqrt{\pi}} \sqrt{\frac{\sqrt{8\pi}}{t}} \cdot \left[\frac{1}{\frac{\sqrt{8\pi}}{t} + i \cdot (t+r)} - \frac{1}{+\frac{\sqrt{8\pi}}{t} i \cdot (t-r)} \right] \tag{30}$$

with, say,

$$\omega \approx \frac{\sqrt{8\pi}}{t} \tag{31}$$

And this is in a setting where we have the dimensional reset of Planck units:

$$\hbar = \ell_P = G = t_P = k_B = 1 \tag{32}$$

7. The big picture: Polarization of signals from a wormhole mouth may affect GW astronomy investigations

We will be referencing [22, 23]. That is, for [22] we have a rate of production from the wormhole mouth we can quantify as

$$\Gamma \approx \exp\left(\omega_{signal} / T_{temperature}\right) \tag{33}$$

Whereas we have from [23] a probability for "scalar" particle production from the wormhole given as

$$\Gamma \approx \exp\left(-E / T_{temperature}\right) \tag{34}$$

Whereas if we assume that there is a "negative temperature" in Eq. (34) we rewrite Eq. (34) as obeying

$$\left(\omega_{signal}/T_{temperature}\right) \approx \left(-E/T_{temperature}\right) \tag{35}$$

This is specifying a rate of particle production from the wormhole. So then, if we refer to black holes with extra dimension, n, of Planck-sized mass, we have a lifetime value of about

$$\begin{aligned} &\tau \sim \frac{1}{M_*}\left(\frac{M_{\text{BH}}}{M_*}\right)^{\frac{n+3}{n+1}} \underset{M_{\text{BH}} \approx M_{Planck}}{\longrightarrow} 10^{-26}\text{seconds} \\ &M_* \approx \textit{is the low energy scale,} \\ &\textit{which could be as low as a few TeV,} \end{aligned} \tag{36}$$

The idea would be that there would be n additional dimensions, as given in Eq. (38), which would then lay the door open to investigating [24, 25] in terms of applications, with [26] additional polarization states to be investigated as to signals from the mouth of the wormhole. We will next go into some predictions into the strength of the signals, the frequency range, and several characteristics as to the production rate of Planck-sized black holes, which conceivably could get evicted by use of Eq. (36), in terms of what could be observed via instrumentation.

8. A first-order guess as to the rate of production of Planck-sized black holes through a wormhole, using Eq. (35)

In order to do this, we will be estimating that the temperature would be of the order of Planck temperature, that is, using ideas from [25, 27],

$$\frac{\omega_p}{T_p} \equiv \frac{\sqrt{Gk_B^2}}{\hbar} \underset{\hbar=G=k_B=1}{\longrightarrow} 1 \tag{37}$$

If so, then there would be to the first-order the following rate of production of Gravitons associated with a white hole-black hole pair, with the white hole in the prior universe and the black hole in the present universe. Namely, for white hole to black hole transition per unit of Planck time, the production rate looks like

$$\Gamma_{rate-of-production} \approx e \approx 2 - 3 \tag{38}$$

9. Interpretation of Eq. (38) in light of Table 1

What we are seeing is that **Table 1** is implicitly assuming millions of white hole (prior universe) to black hole (present universe) transitions, and enormous generation of gravitons as a wormhole transition. If so, then we can then relate this to our problem via the cosmological transition by the following argument.

The reason for using **Table 1** is because of the modification of dark energy and the cosmological constant [1–4]. To begin this, look at [2].

$$\rho_\Lambda c^2 = \int_0^{E_{Plank}/c} \frac{4\pi p^2 dp}{(2\pi\hbar)^3} \cdot \left(\frac{1}{2} \cdot \sqrt{p^2c^2 + m^2c^4}\right) \approx \frac{\left(3 \times 10^{19} GeV\right)^4}{(2\pi\hbar)^3} \underset{E_{Plank}/c \to 10^-30}{\longrightarrow} \frac{\left(2.5 \times 10^{-11} GeV\right)^4}{(2\pi\hbar)^3} \tag{39}$$

In [2], the first line is the vacuum energy, which is completely canceled in their formulation of application of torsion. In our article we are arguing for the second line. In fact, by [2] we can assume we are having dark energy created by the following:

$$\frac{\Delta E}{c} = 10^{18} GeV - \frac{n_{quantum}}{2c} \approx 10^{-12} GeV \tag{40}$$

The term n_{quantum} comes from a Corda expression as to the energy level of relic black holes [7]. We argue that our application of [1, 2] will be commensurate with Eq. (39), which uses the value given in [2] as to the following; that is, relic black holes will contribute to the generation of a cut-off of the energy of the integral

$$\rho_\Lambda c^2 = \int_0^{E_{Plank}/c} \frac{4\pi p^2 dp}{(2\pi\hbar)^3} \cdot \left(\frac{1}{2} \cdot \sqrt{p^2c^2 + m^2c^4}\right) \approx \frac{\left(3 \times 10^{19} GeV\right)^4}{(2\pi\hbar)^3} \tag{41}$$

Furthermore, the claim in [2] is that there is no cosmological constant, namely that torsion is always canceling Eq. (30), which we view as incommensurate with **Table 1** with regard to [2]. We claim that the influence of torsion will aid in the decomposition of what is given in **Table 1** and will furthermore lead to the influx of primordial black holes, which we claim is responsible for the behavior of Eq. (30) discussed previously.

10. Stating what black hole physics will be useful for in our modeling of dark energy, namely, inputs into the torsion spin density term

In [2, 9] we have a spin density term of [1, 2, 9]. And this will be what we input black hole physics into regarding forming a spin density term from primordial black holes: $\sigma_{Pl} = n_{Pl}\hbar \approx 10^{71}$, as given in Eq. (7).

11. Statement of the torsion problem

Eventually, in the case of an unpolarized spinning fluid in the immediate aftermath of the big bang, we would see a Roberson Walker universe given as follows, if σ is a torsion spin term added due to [1, 2, 9]:

$$\left(\frac{\dot{\tilde{R}}}{\tilde{R}}\right)^2 = \left(\frac{8\pi G}{3}\right) \cdot \left[\rho - \frac{2\pi G\sigma^2}{3c^4}\right] + \frac{\Lambda c^2}{3} - \frac{\tilde{k}c^2}{\tilde{R}^2} \tag{42}$$

12. What does to Eq. (42) versus what we would do and why

In the case of [1] we would see σ be identified as due to torsion so that Eq. (42) [9] reduces to

$$\left(\frac{\dot{\tilde{R}}}{\tilde{R}}\right)^2 = \left(\frac{8\pi G}{3}\right)\cdot[\rho] - \frac{\tilde{k}c^2}{\tilde{R}^2} \tag{43}$$

The claim is made in [2] that this is due to spinning particles, which remain invariant so the cosmological vacuum energy or cosmological constant is always canceled. Our approach instead will yield [1, 2, 9].

$$\left(\frac{\dot{\tilde{R}}}{\tilde{R}}\right)^2 = \left(\frac{8\pi G}{3}\right)\cdot[\rho] + \frac{\Lambda_{Observed}c^2}{3} - \frac{\tilde{k}c^2}{\tilde{R}^2} \tag{44}$$

That is, the observed cosmological constant $\Lambda_{Observed}$ is 10^{-122} times smaller than the initial vacuum energy.

The main reason for the difference in Eq. (39) and Eq. (41) is mainly that the reason for the existence of σ^2 is due to the dynamics of spinning black holes in the precursor to the big bang, to the Planckian regime of space-time, whereas in the aftermath of the big bang, we would have a vanishing of the torsion spin term. That is, **Table 1** dynamics in the aftermath of the Planckian regime of space-time would largely eliminate the σ^2 term.

13. Filling in the details of the collapse of the cosmological term, versus the situation given in Eq. (33) via numerical values

First look at numbers provided by [17] as to inputs; these are very revealing:

$$\Lambda_{Pl}c^2 \approx 10^{87} \tag{45}$$

This is the number for the vacuum energy, and this enormous value is 10^{122} times larger than the observed cosmological constant. Torsion physics, as given by [17], is solely to remove this giant number. In order to remove it, the reference [1, 17] proceeds to make the following identification, namely

$$\left(\frac{8\pi G}{3}\right)\cdot\left[-\frac{2\pi G\sigma^2}{3c^4}\right] + \frac{\Lambda c^2}{3} = 0 \tag{46}$$

What we are arguing is that instead, one is seeing [2].

$$\left(\frac{8\pi G}{3}\right)\cdot\left[-\frac{2\pi G\sigma^2}{3c^4}\right] + \frac{\Lambda_{Pl}c^2}{3} \approx 10^{-122} \times \left(\frac{\Lambda_{Pl}c^2}{3}\right) \tag{47}$$

Our timing as to Eq. (47) is to unleash a Planck time interval t about 10^{-43} seconds. As to Eq. (46) versus Eq. (47), the creation of the torsion term is due to a presumed particle density of

$$n_{Pl} \approx 10^{98} cm^{-3} \tag{48}$$

Finally, we have a spin density term of $\sigma_{Pl} = n_{Pl}\hbar \approx 10^{71}$, which is due to innumerable black holes initially.

14. Brief recap of tokamak physics obtaining Eq. (11) and comparison with Grishchuk and Sachin results for obtaining GW generation count

Russian physicists Grishchuk and Sachin [29] obtained the amplitude of a GW in plasma as

$$\mathrm{A(amplitude - GW) = h} \sim \frac{\mathrm{G}}{c^4} \cdot E^2 \cdot \lambda_{GW}^2 \tag{49}$$

This is compared with [30], and we diagram the situation as follows [30]:

Note that a simple model of how to provide a current in the toroid is provided by a transformer core. This diagram is an example of how to induce the current I, used in the simple Ohm's law derivation referred to in the first part of the text. Here, E is the electric field whereas λ_{Gw} is the gravitational wavelength for GWs generated by the tokamak in our model. In the original Griskchuk model, we would have very small strain values, which we will comment upon but which require the following relationship between GW wavelength and resultant frequency. Note, if $\omega_{GW} \sim 10^6 Hz \Rightarrow \lambda_{Gw} \sim 300 meters$, we will be assuming a baseline of the order of setting $\omega_{GW} \sim 10^9 Hz \Rightarrow \lambda_{Gw} \sim .3 meters$, for GW detection above the tokamak.

15. Restating the energy density and power using the formalism of Eq.(49) directly

$$W_E \cdot V_{volume} \sim \tilde{\alpha} \cdot \lambda_{GW}^2 \cdot \frac{\xi^{1/4} \tilde{\alpha}^2 T_{Temp-plasma-fusion-burning}^2}{e_j^{\,2}} \tag{50}$$

The temperature for plasma fusion burning is then between approximately 30–100 KeV, as given by Wesson [28]. The corresponding power for the tokamak as given by Wesson [28] is then

$$P_\Omega = E \cdot J \leq \frac{E}{\mu_0} \cdot \frac{B_\phi}{R} \tag{51}$$

In a one-second interval, if we use the input power as an experimentally supplied quantity, then the effective E field is

$$E_{applied} \sim \frac{\xi^{1/8} \cdot \tilde{\alpha}}{e_j} \times T_{\text{Tokamak}-temperature} \tag{52}$$

What is found is that if Eq. (50) and Eq. (51) hold, then by Wesson [28] (**Figure 1**), pp. 242–243, if $Z_{eff} \sim 1.5, q_a q_0 \sim 1.5, (R/\tilde{a}) \approx 3$. Then the temperature of a tokamak, to good approximation, would be between 30 and 100 KeV, and then we have [28].

$$B_\phi^{4/5} \sim .87 \cdot \left(\tilde{T} = T_{Tokamak-temperature}\right) \tag{53}$$

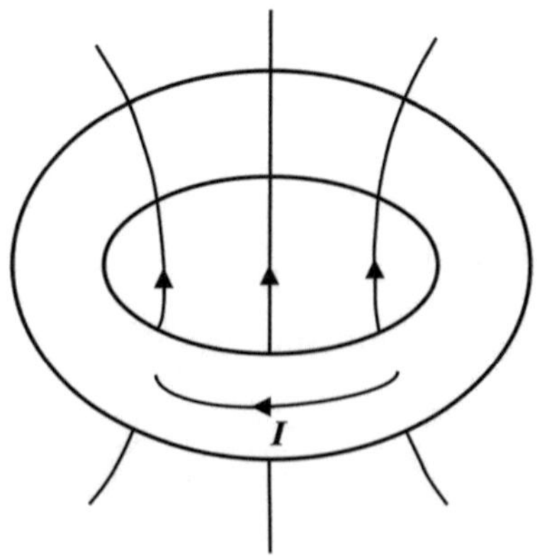

Figure 1.
We outline the direction of gravitational wave flux. If the arrow in the middle of the tokamak ring perpendicular to the direction of the current represents the z axis, we represent where to put the GW detection device as 5 meters above the tokamak ring along the z axis. This diagram was initially from Wesson [28].

Then the power for the tokamak is

$$P_{\Omega}|_{Tokamak-toroid} \leq \frac{\xi^{1/8} \cdot \tilde{\alpha}}{\mu_0 \cdot e_j \cdot R} \times \frac{\left(T_{Tokamak-temperature}\right)^{9/4}}{(.87)^{5/4}} \tag{54}$$

Then, per second, the author derived the following rate of production per second of a graviton, as brought up in Eq. (11).

16. Wrapping it all up: Some specific interconnections for future work

We can state that Eq. (11) is also tied into the quantum number n, as given in Eq. (10), which in turn is linkable to N as the black hole number. In addition, we have also stated that if we have multiple wormhole-style connections between a black hole and a white hole, with the white hole as given in the pre-Planckian section of space-time and the black hole in the present era, then we should pay attention to what Eq. (38) is saying commensurate with **Table 1**. In short, lots of interconnections, and proof of Eq. (11) by tokamak physics, may be extremely important.

This also means we can safely review the issues given in [26, 28–37] with this in mind.

17. Future project as to explicitly working in prior universe white hole linked to present universe black hole, via a special wormhole, for each wormhole linking prior to present universes

What we are doing is using the following wormhole connection. In doing this we should note that we are assuming as a future work that there would be black holes in our initial configuration, plus a white hole in the immediate preinflationary regime, likely in a recycled universe. References [7, 17] is what we will start off with, and their given metric as far as a black hole to white hole solution. That is,

$$dS^2 = -A(r,a)dt^2 + B(r,a)^{-1}dr^2 + g^2(r,a)d\Omega^2 \tag{55}$$

We can perform a major simplification by setting

$$A(r,a) = B(r,a) = f(r,a) \tag{56}$$

In doing so, [7] gives us the following stress energy tensor values:

$$\begin{aligned} T_t^t &= \frac{1}{8\pi}\cdot\left(\frac{1}{g}\cdot(f'g' + 2fg'') - \frac{1}{g^2}\cdot\left(1 - fg'^2\right)\right) \\ T_r^r &= \frac{1}{8\pi}\cdot\left(\frac{1}{g}\cdot(f'g') - \frac{1}{g^2}\cdot\left(1 - fg'^2\right)\right) \\ T_\theta^\theta = T_\phi^\phi &= \frac{1}{8\pi}\cdot\left(\frac{1}{g}\cdot(f'g' + fg'') + \frac{1}{2}\cdot(f'')\right) \end{aligned} \tag{57}$$

In doing this, we will chose the primed coordinate as representing a derivative with respect to r. Also in the case of black hole to white hole joining, we will be looking at a gluing surface connecting a black hole to a white hole, which we give as ξ. And $\tilde{\tilde{n}}$ is a quantum gravity index. Note that in [7] the authors often set it at 3; if so, then for a black hole to white hole to worm hole configuration they give

$$\begin{aligned} g(r,a) &= \left\{r^2 + a^2\left(1 - \frac{r^2}{\xi^2}\right)^{\tilde{\tilde{n}}}\right\}, when(r \le \xi) \\ g(r,a) &= \left\{r^2\right\}, when(r > \rho) \end{aligned} \tag{58}$$

We then make the following connection to energy density in a black hole to white hole system, namely

$$\begin{aligned} &\rho_{black-hole-white-hole-wormhole} \equiv -T_r^r \\ &\approx \hbar\omega_{black-hole-white-hole-wormhole}\tilde{n}_{black-hole-white-hole-wormhole} \end{aligned} \tag{59}$$

If we use Planck units where we normalize h bar to being 1, this will lead to

$$\begin{aligned} &\tilde{n}_{black-hole-white-hole-wormhole} \\ &= \frac{1}{8\pi}\cdot\left(\frac{1}{g}\cdot(f'g') - \frac{1}{g^2}\cdot\left(1 - fg'^2\right)\right)\cdot\frac{1}{\omega_{black-hole-white-hole-wormhole}} \end{aligned} \tag{60}$$

If we are restricting ourselves to quantum geometry at the start of expansion of the universe, it means that we can set these values to be compared to the inputs used to specify a quantum number n, and furthermore

$$a \approx \ell_P = Planck - length \xrightarrow[Planck-normalization]{} 1 \tag{61}$$

We get further restrictions as to the quantum number in Eq. (60) when we compare it to where we had a value of n given in the first section of our document. Furthermore, it means that we can use this to model additional work in a future project on a white hole (specified as in the prior universe). If we go back to **Table 1** of this document, there will be a join between the prior and present universes, where Eq. (61) will be subsequently modified.

18. Conclusions for this document

First, the tokamak may enable the number of gravitons generated from early universe black holes to be formally worked out. This is tricky and will require a lot of work. Second, black holes generate gravitons, and we have stated a relationship between gravitons and a quantum number n. Third, we are assuming that relic black holes have a quantum number as well. Fourth, we have tried through **Table 1** to specify regimes between prior universe to our present universe black holes, assuming a collapse and rebirth of a universe structure. Fifth, a wormhole connection between white holes, in the prior universe, to black holes in our present universe, as discussed in **Table 1**, is alluded to as a formal wormhole connection; this has to be formally worked out. Sixth, the rudiments of a wave function of the universe, as discussed by Kieffer, is also discussed and set up for further elaborations in future research. The main item to be considered is if we can get pre-Planck to Plank space-time metrics understood so as to explicitly understand the details of Section 17 fully. We also leave as a future investigation the items brought up in [38–40] as to their feasibility and application to this document.

Conflict of interest

The authors declare no conflict of interest.

Author details

Andrew Walcott Beckwith
Physics Department, Chongqing University, People's Republic of China

*Address all correspondence to: rwill9955b@gmail.com

References

[1] de Sabbata V, Sirvaram. Quantum effects and the problem of cosmological constant. In: Zichichi A, de Sabbata V, Sachez N, editors. Gravitation and Modern Cosmology, the Cosmological Constant Problem, Ettore Majorana International Science Series (Physical Sciences). Vol. 56. New York City, New York, USA: Plenum Press; 1991. pp. 19-36

[2] Beckwith AW. How torsion as presented by De Sabbata and Sivaram in Erice 1990 argument as modified may permit cosmological constant, and baseline as to dark energy. Journal of High Energy Physics, Gravitation and Cosmology. 2024;**10**:138-148

[3] Chavanis P. Self gravitating Bose-Einstein condensates. Fundamental Theories of Physics. 2012;**178**:151-194

[4] Carlip S. Black hole physics and statistical mechanics. In: Elftherious P, editor. Lecture Notes in Physics 769: Physics of Black Holes, a Guided Tour. Berlin, Federal Republic of Germany: Springer-Verlag; 2009. pp. 89-123

[5] Corda C. Black hole spectra from Vaz's quantum gravitational collapse. Fortschritte der Physik. 2023:2300028, 11 pages. arXiv:2305.02184v2 [gr-qc]. Available from: https://arxiv.org/abs/2305.02184, forthcoming

[6] Casadio R, Octavian Micu O. Quantum matter core of black holes (and quantum hair). In: Joshi P, Malafarina D, editors. New Frontiers in Gravitational Collapse and Spacetime Singularities. Singapore: Springer; 2024, arXiv: 2303.14945v1 [gr-qc]

[7] Feng Z, Ling Y, Wu X, et al. New black-to-white hole solutions with improved geometry and energy conditions. Science China Physics, Mechanics and Astrononomy. 2024;**67**: 270412. DOI: 10.1007/s11433-023-2373-0 arXiv:2308.15689v5 [gr-qc] 19 Mar 2024

[8] Ohanian H, Ruffini R. Gravitation and Spacetime. 3rd ed. United Kingdom: Cambridge University Press; 2013

[9] Will C. Was Einstein right? A centenary assessment. In: Ashtekar A, Berger B, Isenberg J, MacCallum M, editors. General Relativity and Gravitation: A Centennial Perspective. Cambridge, UK: Cambridge University Press; 2015. pp. 49-96

[10] Weber J. General Relativity and Gravitational Waves. Mineola, New York, USA: Dover Publications, Incorporated; 2004

[11] Lu HQ, Fang W, Huang ZG, Ji PY. The consistent result of cosmological constant from quantum cosmology and inflation with born–Infeld scalar field. The European Physical Journal C. 2008; **55**:329-335

[12] Sarkar U. Particle and Astroparticle Physics. New York City, New York, USA: Taylor & Francis Group; 2008

[13] Hartle J, Hawking S. Wave function of the universe. Physical Review D. 1983;**28**(12):2960. Bibcode:1983PhRvD.. 28.2960H. 10.1103/PhysRevD.28.2960. S2CID121947045

[14] DeBenedictis A, Das A. On a general class of wormhole geometries. Classical and Quantum Gravity. 2001;**18**(7): 1187-1204. arXiv:gr-qc/0009072. Bibcode:2001CQGra..18.1187D. CiteSeerX 10.1.1.339.8662. 10.1088/0264-9381/18/7/304. S2CID 119107035

[15] Einstein A, Rosen N. The particle problem in the general theory of relativity. Physical Review. 1935;**48**(1): 73. Bibcode:1935PhRv...48...73E. DOI: 10.1103/PhysRev.48.73

[16] Visser M. The Quantum Physics of Chronology Protection. Available from: https://arxiv.org/abs/gr-qc/0204022

[17] Lightman A, Press W, Price R, Teukolsky S. Problem Book in Relativity and Gravitation. Princeton, New Jersey, USA: Princeton University Press; 1975

[18] McCracken DD, Dorn WS. Numerical Methods and FORTRAN Programming with Application in Engineering and Science. Moscow: Publishing House Mir; 1977

[19] Davis PJ, Rabinowitz P. Methods of Numerical Integration. 2nd ed. New York City, NY, USA: Dover; 2007

[20] Zwillinger D. CRC Standard Mathematical Tables and Formulae. 31st ed. Boca Raton, LA, USA: Chapman and Hall/CRC; 2003

[21] Kieffer K. Quantum Gravity. 3rd ed. Oxford, United Kingdom: Oxford Science Publications, Oxford University Press; 2012

[22] Martın-Moruno P, Gonzalez-Dıaz PF. "Thermal radiation from Lorentzian traversable wormholes," Spanish Relativity Meeting (ERE 2010): Gravity as a crossroad in physics. Journal of Physics: Conference Series. 2011;**314**: 012037. Available from: https://iopscience.iop.org/article/10.1088/1742-6596/314/

[23] Gecim G, Sucu Y. Quantum gravity correction to Hawking radiation of the dimensional wormhole. Advances in High Energy Physics. 2020;**2020**: 7516789. Available from: https://www.hindawi.com/journals/ahep/2020/7516789/, https://arxiv.org/pdf/2002.03699, https://onlinelibrary.wiley.com/doi/10.1155/2020/7516789

[24] Cheung K. Black hole production and extra large dimensions. Physics Review Letters. 2002;**88**:221602. Available from: https://arxiv.org/abs/hep-ph/0110163

[25] Li F-Y, Wen H, Fang Z-Y, Li D, Zhang T-J. Electromagnetic counterparts of high-frequency gravitational waves having additional polarization states; distinguishing and probing tensor-mode, vector-mode, and scalar-mode gravitons. European Physics Journal C. 2020;**80**:879 Available from: https://arxiv.org/abs/1712.00766

[26] Li FY et al. Signal photon flux and background noise in a coupling electromagnetic detecting system for high frequency gravitational waves. Physical Review D. 2009;**80**:064013. arXiv:0909.4118v2 [gr-qc]

[27] Carr BJ. Primordial Black Holes: Recent Developments. Available from: https://arxiv.org/abs/astro-ph/0504034

[28] Wesson J. Tokamaks, International Series of Monographs on Physics. 4th ed. Vol. 149. Oxford, United Kingdom: Oxford Science Publications; 2011

[29] Grishchuk LP, Sazchin MV. Excitation and detection of standing gravitational waves. Journal of Experimental and Theoretical Physics. 1975;**68**:1569-1582

[30] Li F, Tang LM, Luo J, Li Y-C. Electrodynamical response of a high energy photon flux to a gravitational wave. Physics Review D. 2000;**62**:44018

[31] Woods RC et al. Journal of Modern Physics. 2011;**2**(6):498

[32] Li J et al. A long pulse high confinement plasma regime in the experimental advanced super conducting tokamak. Nature Physics. 2013;**9**(12):817-823. DOI: 10.1038/nphys2795

[33] Wen H, Li F, Fang Z. Electromagnetic response produced by interaction of high-frequency gravitational waves from braneworld with galactic–extragalactic magnetic fields. Physics Review D. 2014;**89**: 104025

[34] Beckwith AW. Part 2: Review of tokamak physics as a way to construct a device optimal for graviton detection and generation within a confined small spatial volume, as opposed to Dyson's 'infinite astrophysical volume' calculations. Journal of High Energy Physics, Gravitation and Cosmology. 2017;**3**:138-155. DOI: 10.4236/jhepgc.2017.31015

[35] Chongchitnan S. Inflation Model Building with an Accurate Measure of E-Folding. Available from: https://arxiv.org/pdf/1605.04871.pdf

[36] Cornish N, Robson T. Galactic binary science with the new LISA design. Journal of Physics: Conference Series. 2017;**840**(1):012024 arXiv:1703.09858. Bibcode:2017JPhCS.840a2024C. 10.1088/1742-6596/840/1/012024. S2CID 119335855

[37] Danzmann K, The eLISA Consortium. The gravitational universe. arXiv:1305.5720. 2013. [astro-ph.CO]

[38] Barrow JD. The Constants of Nature, from Alpha to Omega: The Numbers That Encode the Deepest Secrets of the Universe. New York City, New York: Pantheon Books; 2002

[39] Freeze K, Brown M, Kinney W. The phantom bounce, a new proposal for an oscillating cosmology. In: Mercini-Houghton L, Vaas R, editors. The Arrows of Time: A Debate in Cosmology, Fundamental Theories in Physics. Vol. 172. Heidelberg, Federal Republic of Germany: Springer Verlag; 2012. pp. 149-156

[40] Maggiore M. Gravitational Waves, Volume 1: Theory and Experiment. Oxford, UK: Oxford University Press; 2008

Section 2

New Boundaries for Cosmological Models

Chapter 5

On the Use of the Principle of Least Action in Cosmology

Martin Tamm

Abstract

In this article, it is argued that the solutions to some of the problems we face in cosmology may not require new physical theories as much as they require a time symmetric perspective. The Principle of Least Action is nowadays mostly used as a convenient tool for deriving the field equations. It is suggested that new insights could be obtained by restoring the original role of the Principle of Least Action as a (time symmetric) minimizing principle. A possible way to achieve this is to replace the Einstein-Hilbert Principle by another principle, which is closer to the idea of multiple histories, and use methods from statistical mechanics. As an example, the accelerating expansion is discussed. Although it is not the purpose here to discuss all possible explanations of this phenomenon, it is argued that the present framework of cosmology may force us to exclude certain types of explanations and hence limits our possibilities to find the right one. In particular, a simple model for the accelerating expansion, which cannot be treated as an initial value problem, is given. This model also has the property to explain why the rate of acceleration now seems to be decreasing with time.

Keywords: cosmology, accelerating expansion, dark energy, force, causality, multiple histories, principle of least action

1. Introduction

Perhaps the most common method in physics consists in first formulating the equations of motion and then solving these with respect to certain initial conditions. Normally, this method is used to predict the state of the system in the future. However, these equations can also be used to obtain the state of the system in the past. But in both cases, the method is in a sense time asymmetric, since the initial and final states play different roles, even if the equations of motion themselves are time symmetric. Under ordinary circumstances, this is not considered as a problem, and it is also very much in this spirit that science has been practiced since the days of Isaac Newton.

But, it may not be obvious that this is the right way to make progress in, for example, present-day cosmology. The current state of the art tends to consider the development of our universe as an initial value problem. This is certainly one possible kind of attack, and it is quite possible that in most cases it is also the best one. But to make this approach conform to what we now observe on a global scale, it has been necessary to modify the model by introducing new hypothetical components, like, for example, dark energy. And, it has turned out to be very difficult to explain where these hypothetical components come from.

One of the purposes of this article is to discuss the possibility that the development of our universe cannot always be well understood as an initial value problem, but rather should be treated in a time symmetric way. In the model to be discussed here, I will start from a modified form of the Principle of Least Action, which is based on statistical mechanics and a multiple-histories approach to physics. The model that emerges suggests a way to explain the accelerating expansion without using dark energy, and in addition, it also gives a natural explanation for why the rate of acceleration now seems to be decreasing with time. And in particular, it can also be seen as an example of why it may be wrong to consider the development of cosmos as an initial value problem.

I make no big claims here about how the possibilities sketched in this article could actually materialize as the successful explanation to all the big problems of cosmology. Rather, I hope to contribute to a broader field of non-standard perspectives as a basis for future research.

The form of this article is close to an essay, and I start from a historical perspective on the development of cosmology. It has been my explicit ambition to write this article in a non-technical way so as to make it accessible to a broader public. However, since mathematics does play an important role, I have in the appendices also included some material on the mathematical background underlying the theory. The first (Appendix A) is about the field equations of general relativity, and the second (Appendix B) is about the accelerating expansion. In both cases, the perspective differs from the mainstream perspective.

2. A historical analogy

Before the Copernican revolution, astronomers had for a long, long time developed more and more sophisticated tools for describing the motion of the planets, including cycles, epicycles, eccentrics, equants, and even epicycles on epicycles to name a few [1]. In this way, they had built a complicated machinery that could produce orbits that reasonably well mimicked the data available at the time. But in the end, the picture was drastically simplified by a change of perspective. And in addition to the simplification, this also turned out to be the way forward toward a much more precise machinery. The history of science is full of similar (although perhaps usually less dramatic) situations, where a change of perspective has led to great simplification; Mendeleev's periodic system [2] and the eightfold way [3] may serve as examples.

Can we see something that could be similar symptoms in the cosmology of today? Again, we can never know for sure beforehand if something is a symptom of a misleading perspective or just the best we can do for the moment. In this article, I will in particular be concerned with concepts like "forces" and closely related ones like "pressure" and "cause." What do these words stand for? And could it be that these are in fact the epicycles, eccentrics, and equants of our time?

To understand the forces around us and to predict what they will lead to is something that every human being has constantly been trying to do since long before the definition of a force in the scientific sense had emerged and, in fact, long before physical science had even been thought of. Since our ability to do so has obviously been crucial for our survival, it should certainly not come as a surprise that this is also the way in which our physical theories tend to be formulated. And certainly, this way of thinking has taken us very far. But the further we get from situations where our usual intuition applies, the more problematic, and perhaps even dangerous, it could be to let this way of thinking lead us.

Let us note that the scientific use of the word "force" also has a long history. Aristotle's way of perceiving the forces in nature was far more advanced than earlier animistic views. But it was still quite anthropomorphic; bodies and elements "strived" to reach their natural positions, very much like animals and humans do. And it took another 2000 years before a more mechanical view, as proposed by Descartes and realized by Newton, could replace it.

But how free from anthropomorphisms was Newton's concept of force? Clearly, the forces that were known at his time were mostly of a very human life-related type (like the force experienced when pulling a rope), with one important exception: gravity. From this point of view, Newton's conception of the force of gravity as something of the same kind as the pulling of a rope appears to have been the only natural one, even though the rope itself had disappeared. The modern view on gravity in terms of geometry, rather than as a classical force, was still far beyond reach.

Since Newton's time, the concept of force has been inherited from one generation of physicists to the next in a long and complicated process [4]. Nowadays, there is a kernel of knowledge where the concept of force (or quantum field in a more modern setting) works extremely well. This kernel includes quantum field theory for the electro-weak and probably also strong interactions. In these cases, even if quantum field theory may not be the final word, it seems very likely that this is the best we can do presently. But how about gravity? It has turned out to be very difficult to treat gravitational interaction in the same way as the other interactions. Should we still try to squeeze it into the cage of quantum field theory?

Since the nineties, the picture has been further complicated by the discovery that our universe is expanding at an accelerating rate [5, 6]. So far, the most common way to try to explain this phenomenon has been to introduce new hypothetical forms of energy, like dark energy or quintessence, that would give rise to a pressure, or a repulsive force (for an overview, see Refs. [7, 8]). However, there is so far very little independent evidence for the existence of such energy. And it does not end there. A similar problem occurs when we try to understand the very early history of our universe: The observed cosmic microwave background seems to require a short period of "inflation" immediately after the Big Bang, in order to explain the distribution of mass that we can now observe. And again, a standard way to try to understand this has been to introduce some kind of repulsive force, or "inflatory field." But also in this case, it has turned out to be very difficult to explain where this field comes from.

It is very much in the tradition of physics to interpret deviations from a current model as indicating the presence of a new force. But with several such hypothetical new fields popping up, it may be questioned if this is the right way if we are looking for a fundamental theory. Also, physicists often tend to push the problem ahead of them by assuming that these forces will somehow in the future turn out to be one and the same (at sufficiently high temperatures) and that their splitting into different manifestations is just some kind of secondary process (at lower temperatures). However, a skeptic might instead suspect that these forces are just names for various different phenomena that we can put into our formulas to make the theories conform to observations, in fact, in a way not too different from how pre-Copernican astronomers used epicycles, eccentrics, and equants to make their planetary orbits resemble the observed ones. And if so, perhaps it is time to look for a different perspective?

In the following, I will argue that a possible such perspective is given by the Principle of Least Action. This principle can be said to be the closest we have to a universal law of physics. However, in contemporary physics (at least in classical physics including general relativity), it has essentially been reduced to a technical tool for deriving the field equations. My suggestion is that new insights could be obtained from returning to its original meaning as a minimizing principle. But such a reformulation inevitably demands replacing the Einstein-Hilbert Principle by something else, since minimizing the scalar curvature in itself obviously does not make sense (see Section 7).

3. Cosmology

Cosmology is one of the oldest branches of physics, with roots that go back to antiquity and perhaps even further. It can also be said that attempts to build a consistent picture of what our universe is like were common long before anything that we can nowadays call science. The journey from the oldest mythological ideas to present-day cosmology can in fact be seen as a model for the development of science itself, where old anthropomorphic ways of thinking have successively been replaced by something that in a sense is more formal and less related to our own special human perspective.

Has cosmology reached the end of this process now? We do not know, but there are lots of examples in history where we have overestimated our own understanding, and it is quite possible that we may still be somewhere in the middle of this process. It is simply too early to tell where it will all end up.

Since the birth of general relativity a little more than a hundred years ago, cosmology has been in a phase of very rapid development, and this certainly goes for both the theoretical part and the observational part. And surely, the development will not be stopped forever by problems like the abovementioned ones. For what we know, there are lots of ways in which these problems could eventually be solved: New experimental facts could reveal to us the true nature of these new mysterious forces, and it is then also possible that solving the field equations in the traditional way will enable us to predict the future of our universe convincingly. It is also quite possible that the solutions will require new clever physical ideas that no one has thought of so far. But it is also possible that the most important thing we need could be a change of perspective.

The main example in this article will be the accelerating expansion. What I do want to argue is that it may be that our present paradigm limits our possibilities, just as the old Aristotelian concepts limited astronomy and cosmology in the sixteenth century. And that the way forward may be to take one more step away from our old anthropocentric perspective, just as Copernicus did.

4. The Copernican revolution and the standard method

Few authors have contributed more to our understanding of the development of science than Thomas Kuhn [9]. And in fact, his first major contribution was his study of the Copernican revolution ([1], this book is also a general reference for many of the historical facts touched upon in this article).

It is nowadays common to consider this revolution to be the historical process that can roughly be described by listing its main actors as follows:

- Copernicus - Tycho Brahe - Kepler - Galilei - Newton.

Copernicus was not the first one to consider the heliocentric point of view. Such ideas can in fact be traced back at least to Aristarchus from Samos in the third century B.C. But clearly, Aristarchus did not have the arguments he would have needed to convince his contemporaries, whereas Copernicus and his followers, as it turned out in the end, did. Similarly, it can be questioned if Newton is a natural endpoint, and in general, it can be argued that the revolution was not the result of individual contributions, but rather an unstoppable consequence of the scientific and technological development. However, most of us can probably agree to the following general characteristics of the process:

1. An expansion of our world view from a human-centered view, where we ourselves inhabit the center of the universe, to a view where the human civilization is just a small part of something much bigger.

2. An increasingly precise and less anthropomorphic description of the world, aiming at an explanation, not just a description.

3. The shift did not take place because of a sudden discovery, rather what happened was that gradual changes in the beliefs of the scientific community led to a promotion of ideas, previously considered irrelevant or obscure, to mainstream.

At the end of the Copernican revolution, a general way of attacking physical problems had emerged. Simplifying somewhat, and perhaps formulating it in a little bit more modern language, the method looks as follows: Determine the forces involved, and then write down the equations of motion. Having done so, the problem has essentially been transformed into a mathematical one. Solving the equations with appropriate initial conditions will then give us predictions that can be compared with reality. This recipe is still very central in modern physics. It is in fact so central that I will in the following refer to it as "the Standard Method."

Should the Standard Method be considered to be the main method for doing physics, or does it just represent our temporary understanding, which is waiting to be replaced by something better? There is absolutely no doubt that the Standard Method has served science extremely well. In fact, it may even be argued that it is the most successful scientific method that mankind has ever come up with. But of course, this does not mean that it is free from problems and that every task fits equally well into the scheme.

Sometimes, part of the procedure may so to speak disappear out of sight. For example, in statics, only the forces are important, and the time development is trivial (all bodies remain at rest). In statistical mechanics, on the other hand, the large number of particles usually makes the treatment of the equations of motion hopeless, and we are simply forced to try to use statistical methods instead. However, these examples may perhaps still be considered to be covered by the Standard Method in principle.

But more serious problems occur if we turn to quantum mechanics: There are phenomena that, from the point of view of the observer, are genuinely unpredictable, for example, the decay of a neutron into a proton and an electron. And it is not difficult to construct examples of situations where decays at different times will lead to completely different futures, even within a relatively short period of time. As an obvious example, we may let the life and death of Schrödinger's famous cat depend on the time of the decay.

Another kind of doubt about the Standard Method stems from general relativity: As mentioned above, in most cases, gravitation can be better understood in terms of the geometry of space–time, rather than as a classical force. There could also be problems of even more fundamental types: For example, from a modern point of view, it is not at all evident that time can be used as a global parameter. If the topological structure of space–time is non-trivial, it could very well be that it is impossible in principle to solve initial value problems, for example, since solving along different paths could give contradictory results.

In modern physics, there are also a growing number of other kinds of methods, based on, for example, symmetry and group theory, which can give us complementary understanding, sometimes of a more spectacular type than the Standard Method gives. However, this new understanding seldom competes with the traditional perspective of the Standard Method.

In spite of the above examples, the grip of the Standard Method seems to be firm in all parts of physical science. For example, an orthodox apologist of the Standard Method in quantum mechanics could still claim that the method applies: The Schrödinger equation is completely deterministic; it is just that we cannot know the full initial conditions, nor can we in fact know the full final state (and this is indeed so for very fundamental reasons). Thus, for an all-knowing being, the Standard Method would work just as well in quantum physics as it does in classical physics. But we are not all-knowing creatures, and hence, we have to accept things like, for example, certain kinds of unpredictability.

As for the case of general relativity, let me quote "Gravitation" (see Ref. [10]), one of the most influential textbooks of modern physics: "No plan for predicting the dynamics of geometry could at the same time be more mistaken and more right than this: Give the distribution of mass-energy; then solve Einstein's second-order equation, $G = 8\pi T$, for the geometry. Give the distribution of mass-energy and solve for the geometry? No. Give the fields that generate mass-energy and their time-rates of change, and give 3-geometry of space and its time-rate of change, all at one time, and solve for the 4-geometry of spacetime at that time? Yes. And only then let one's equations for geometrodynamics and field dynamics go on to predict for all time ... ". What the authors are essentially telling us is that general relativity should use the Standard Method and that everything else is misunderstanding. In support of this claim, it should of course be noted that the method has been very successful; most of the progress in modern astrophysics, both before and after the publication of the book, has followed this path.

Summing up, there are issues with the Standard Method. But usually, these issues have not been thought of as being serious enough to cause scientists to question the Standard Method as such. Nevertheless, sooner or later, a situation could come up where we simply have to rethink the traditional foundations. Could this be where cosmology has come right now?

5. Can the standard method solve the problems of cosmology?

Let us start by comparing the three characteristics of the Copernican revolution listed in Section 4, with what has happened afterward.

DOI: http://dx.doi.org/10.5772/intechopen.1009085

As for (1), the expansion of our world view has continued up to the present and to an extent which would have been impossible to imagine at Newton's time. Not only has the heliocentric view been replaced by a view where the sun is just a very ordinary star in a very ordinary galaxy, but also, what we traditionally perceive as our whole world may be just a very ordinary world among an enormous number of others. These ideas have led to a very extensive debate in the scientific community about parallel worlds and multiverses. In this article, however, it is not the ontological status of these parallel worlds that is central. The question is rather whether or not it is advantageous to include parallel histories in our strategies for problem-solving.

As for (2), it has already been noted that although the formulation of the Standard Method has now taken a scientific and mathematical form, its content is still based on concepts with a definite anthropomorphic history. This includes not only forces but also, for example, the closely related concept of causality, which has been a nuisance for philosophers for thousands of years. In our everyday life, we are always trying to understand the role of causes and effects, and surprisingly often, we also agree about them. But still, when it comes to trying to understand what it really means, the history of science is full of different non-compatible statements (see for instance Refs. [11, 12]). In this context, the Standard Method has come to serve as a kind of substitute in physics: If we at time t_0 know precisely the state of a system in the part of space that is contained in the backward light-cone of a certain event Ξ at time t_1, then it could be said that this information contains the "cause" of Ξ, since the Standard Method should then give a unique solution of the equations of motion leading to Ξ.

When it comes to (3), this may be the most difficult point to discuss, since we do not know where the seed to a forthcoming revolution will be found. In this article, I will be content with looking at the parallel track that is based on the idea that instead of starting from given initial values and solving the equations of motion from there on, we should consider all developments and then try to find out which ones are possible and which ones are not (or in modern language, perhaps rather try to find out which ones have non-negligible probabilities). This is very much in the spirit of what is nowadays often referred to as Constructor Theory. This theory is the most powerful when it is combined with quantum physics, but it also has historical roots in the Principle of Least Action.

To my mind, this fits well into the abovementioned scheme of the Copernican revolution in the following sense:

1. It does represent an expansion of our world view since it includes in a natural way not just one but all possible developments.

2. It does not start from (or at least tries to avoid extensive use of) concepts with a very human history, like "forces" and "causes."

3. It has in fact been around for as long as the Standard Method, if not longer, but has never been fully mainstream.

6. The principle of least action

When trying to solve such problems of cosmology as were mentioned earlier, in particular the problem with the accelerating expansion, the Standard Method does not allow for many choices: Either we have to find the sources of the forces we are using, or we have to change the equations of motion, or (perhaps) try to modify the initial conditions. But so far, these attempts have not been very successful in explaining the accelerating expansion, since there is still no agreement about why it occurs. So maybe it is time to at least consider the possibility that it is in fact the Standard Method itself which is the problem?

The Principle of Least Action is often attributed to Maupertuis who in the eighteenth century formulated the belief that the universe develops according to a principle of ultimate economy [13]. Maupertuis was not the first one to express such an idea: Already Fermat had formulated the special case for light almost a century earlier, and also Leibniz had formulated something similar and more general. When it comes to the mathematical formulation, it can rather be found in the works of Euler and Lagrange. Still, Maupertuis did contribute to the discussion by emphasizing the philosophical implications of the idea: To him, this principle showed that the universe always chooses the simplest way of developing, and according to his beliefs, this could only be so if the universe had been created with a purpose and hence by somebody. So, his conclusion was that this must be the final and indisputable proof of God's existence.

A less metaphysical and more scientific formulation of the original Principle goes as follows:

> Principle 1: Among all possible developments from a state Ξ_0 at time t_0 to a state Ξ_1 at time t_1, the one that is realized is the one that minimizes the action.

Note that this formulation makes no references to "time development," nor does it state that there should be some kind of "causal relationship" between the states at t_0 and t_1, and forces are not even mentioned. In this way, it represents a very different perspective of what physics is really about.

Strangely enough, however, under ordinary physical circumstances, the Principle of Least Action turns out to be more or less equivalent to the Standard Method: In fact, the equations of motion can in general be deduced by minimizing the action. For this reason, the Principle of Least Action in classical physics with time became a useful and common technical tool for finding these equations. On the other hand, the very existence of such an equivalence also seems to have served as a pretext to dismiss the philosophical aspects of the Principle, not only as a proof of God's existence but also as a fundamental principle of nature. Many physicists simply saw no point in questioning the Standard Method but happily accepted the Principle of Least Action as a technical tool.

> Remark 1: In a way, this can perhaps be compared to the position of Tycho Brahe. At the beginning of the Copernican revolution, he was collecting what was to become by far the most accurate collection of astronomical data of his time. He was aware of Copernicus' work, but his belief in the geocentric viewpoint was so strong that he could not accept the main claim, namely that the sun is the natural center of the solar system. On the other hand, he clearly understood the gain in simplicity achieved by considering the planets to be circulating around the sun in circular orbits, rather than describing their orbits in terms of cycles and epicycles. So, the best compromise that he could find was to consider all the ordinary planets, except Earth, to be circulating around the sun but then let the sun itself circulate around the earth. This model is sometimes referred to as the Tychonic or the geoheliocentric model. It can be worth asking ourselves if we are making

the same kind of mistake as Tycho Brahe by accepting the technical benefits of the Principle of Least Action but at the same time being unwilling to discuss its real essence?

In the early days of general relativity, one of the most urgent problems was to determine the field equations of space–time. And although the first derivation should be attributed to Einstein, the easiest and fastest derivation, as suggested by Hilbert, was to use the Principle of Least Action with the simplest possible action integral:

$$I = \int R \, dV. \tag{1}$$

In this formula, R denotes the scalar curvature of the four-dimensional space–time defined by a certain metric g, and dV denotes the corresponding volume measure. For more details about these concepts, see Ref. [10]. This action principle is nowadays often referred to as the Einstein-Hilbert principle (see Refs. [14, 15]). It was certainly a great success at the time, and very much in the tradition of classical mechanics.

But is it the right way forward? It may actually also be argued that the very fact that the Einstein-Hilbert principle fits so well into the classical tradition may perhaps at the same time be what makes it a road blocker on our way toward the unification of general relativity and quantum mechanics. The Einstein-Hilbert principle cannot be said to rest on fundamental principles. It is essentially just a technical recipe, and it was soon noted that this was not the only possible choice that could explain the field equation [16]. For this reason, it seems potentially very profitable to look for an action principle that is more natural and somehow more closely related to quantum mechanics.

In view of the close ties between the Principle of Least Action and Feynman's idea of the democracy of all histories [17], this connection also seems to be a good starting point for the search for such an action principle. However, it is the opinion of the author of this article that we may simply not yet be ready for a truly quantum mechanical implementation of Feynman's theory when it comes to general relativity but that the underlying idea is still the best tool we have for trying to find a bridge between the two theories. For this reason, I have in Appendix A included a short description of a much simpler framework, based on methods from statistical mechanics, which may still catch something essential about multiple histories.

Remark 2: The method in Appendix A may be called heuristic, but if we really want to build a bridge between two so very different theories, there may simply be no good alternative to such methods from our present starting point and even more so when using an approach based on statistical mechanics.

In fact, statistical mechanics is one of the most powerful theories in all of physics, but it is notoriously difficult to prove rigorous results with it. The most common way to do so is to pass to an infinite "thermodynamic limit" (see Ref. [18]), as in, for example, Onsager's famous treatment of the Ising model (see Ref. [19]). In this article, the universe is, although enormous, always of finite volume. In such situations, there will always be approximations and compromises between the computations on the "micro-level" and the "macro-level."

In any case, Appendix A is included in this article to motivate the following action principle, only differing from (1) by making use of the square R^2 of the curvature instead of R itself:

$$I = \int R^2 dV. \tag{2}$$

Note that I here is always non-negative, so it does make good sense to minimize this integral, whereas it does not make good sense to do so with the integral in (1) (since there is no natural lower bound).

Remark 3: In the classical theory of general relativity, the action integral in (2) is usually not accepted as a natural choice, the main reason being that deriving the Euler-Lagrange equations in the ordinary way does not reproduce the standard field equations. It is, however. important to this article that this objection no longer seems to be true when switching to a multiple history perspective; see Appendix A.

Remark 4: It should perhaps be stressed that leaving causality behind would not mean that we would have to stop using such concepts in ordinary physics or in everyday life. In fact, in Kuhn [1], an interesting comment is made that the old geocentric two-sphere model survived the Copernican revolution and actually remained the best model for navigation up to modern times.

Having come so far, it is now time to face the crucial question: Could such a change of viewpoint, from the Standard Method to a minimizing principle, lead us forward? Or would it just be another dead end for cosmology?

7. Can a change of perspective open up new possibilities in cosmology?

It is not the purpose of this article to try to present new, wonderful solutions to all the big problems in cosmology. Instead, I would like to explain why the Principle of Least Action could open up possibilities that are closed to the Standard Method and, in particular. Explain why such a change could make assumptions about additional forces superfluous. To this end, I will here in an intuitive way present a highly simplified model where it can be understood why an accelerating expansion could be a natural phenomenon without any dark energy, and in fact even in a universe of finite volume. I have tried to be as non-technical as possible. A rigorous treatment of all aspects would by necessity be very mathematical and will not be considered here; see, however, the Appendices. The reader may also consult Tamm [20–22].

A very important attempt, and perhaps also the first serious attempt, to create a model for modern cosmology based on general relativity was the Friedmann-Einstein model [23]. The idea was to assume that the only factors that we need to take into account are curvature and gravitation as seen from the perspective of general relativity. In addition, we assume that the universe, at every fixed moment of time, is isotropic and homogeneous and has the simplest possible geometry (a three-dimensional sphere). With these assumptions, to determine the global structure, we only have to find the relation between the remaining two variables t (time) and $r(t)$ (the radius of the sphere at time t). This can then be done by solving the field equations, very much in the tradition of the Standard Method, and the result is the well-known cycloidal curve in **Figure 1** (see Appendix B).

The Friedmann-Einstein model was a great breakthrough when it appeared almost a hundred years ago. Nowadays, we know that it cannot give a fully satisfactory

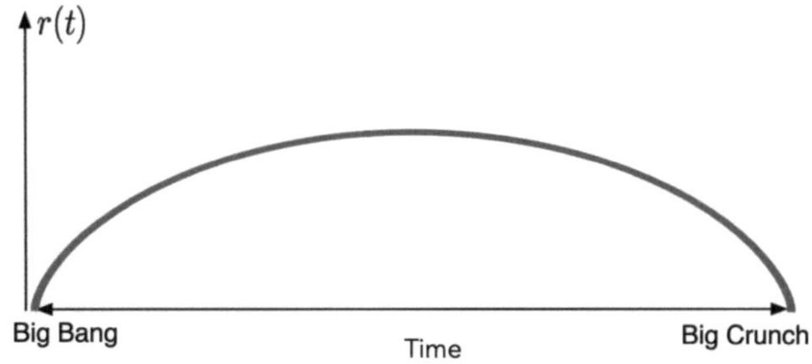

Figure 1.
The Friedmann-Einstein model. See Appendix B.

description of the history of our universe. And one of the reasons for this in particular is that it cannot explain the accelerating expansion, which again is closely connected to the reasons for introducing new hypothetical concepts like, for example, dark energy.

But how would it all look if we instead had relied on the Principle of Least Action as a time symmetric minimizing principle, for example, based on the action integral (2)? In other words, if we start with a space–time manifold with the same topology as in the Friedmann-Einstein model (essentially a four-sphere) and then try to minimize the action, what would be the form of the resulting (pseudo-Riemannian) manifold? In this case, where we have already fixed the topological shape of the universe (as a four-sphere), the end-states Ξ_0 and Ξ_1 can simply be identified with the Big Bang and the Big Crunch.

So why should this setup be able to generate something different from what the Standard Method gives? Partly, a purely technical answer is that the natural Euler–Lagrange equation for the radius $r(t)$ will be of order four, which means that the usual kind of initial conditions (using only derivatives up to order one) will simply not be enough to determine the solution if the problem is formulated as an initial value problem. Hence, other kinds of conditions are needed (see Appendix B).

When we base our theory on a minimizing principle, we may get new possibilities and a very different picture. For example, a natural additional constraint, which does not fit into the Standard Method, is given by the condition that the total space–time volume of the universe is fixed. This condition has been used by many authors, notably by Hawking [24], although from a somewhat different perspective.

Remark 5: How natural is this condition of constant space-time volume? From the traditional classical perspective, essentially shared by Newton, Maupertuis, and their contemporaries, and in which space-time is just an empty arena where particles develop according to certain laws of motion, it could appear quite unnatural to postulate that the total space-time volume should be fixed. However, in modern physics, space-time is not an empty arena but an active constituent of our universe, and this is in fact a very different situation.

Any global application of the Principle of Least Action must somehow start from specifying some parameter(s) to avoid degenerate solutions (e.g., total mass-energy, total volume, total number of particles, ...). Even Maupertuis' original principle is similar in the sense that without additional conditions, the obvious universe with the least action would be degenerate (with e.g., zero extension in time). Usually, when the

Principle of Least Action is applied locally, the need for global constraints may be hidden, because there are boundary conditions that somehow fulfill a similar function. But if we try to find a realistic (global) universe that minimizes the action, some extra condition is definitely needed to guarantee that the optimal solution would not be, for example, of zero space–time volume.

There may be many possible such conditions. But to my mind, the condition of fixed space–time volume is the most natural one. One reason is that it somehow reflects the idea that space–time is analogous to matter in the sense that it is built out of some kind of elementary building blocks and that the total space–time volume then just represents the number of such building blocks. This is also very closely related to what in statistical mechanics is referred to as a microcanonical ensemble (only now in 4 dimensions instead of 3).

There is much more that could be said about this. The minimizing principle used in this article actually originates from something that closely resembles the minimizing of the Helmholtz free energy in statistical mechanics (where constant volume is an essential condition). But I will not go further into this here; see, however, Ref. [25] and Appendix A.

> Remark 6: Some authors have tried to impose different kinds of extra conditions in the future in a more traditional setting. Typically, such conditions concern the behavior near the Big Crunch (see e.g., Ref. [26]). But there is always something strange about these attempts: In the Standard Method, there is actually no room for extra conditions, since the initial conditions and the field equations alone should determine the solution at all times.

Let us start our investigation of this alternative path by considering a completely empty universe without mass or energy but still with a fixed four-volume. Which form will it take? In this case, it turns out that the obvious solution to the problem of minimizing the action integral I in (2) is to let the geometry (not just the topology) be given by a four-sphere (except of course for the endpoints). The resulting curve can be seen in **Figure 2**, although note that, for typographic reasons, I have not used geometric units, so it looks more like an ellipsoid than a sphere. Note also that this curve is similar to, but not identical to, the cycloid in **Figure 1**.

That this curve is minimizing may be somewhat surprising, since in Euclidian geometry, a sphere gives a typical example of a manifold with positive curvature. But in Lorentz geometry, a direct computation shows that $R = 0$, which will obviously

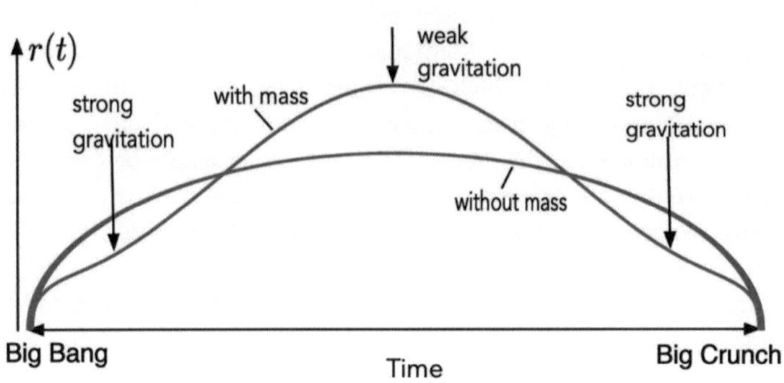

Figure 2.
The effect of mass on the radius $r(t)$, compared to the case without mass.

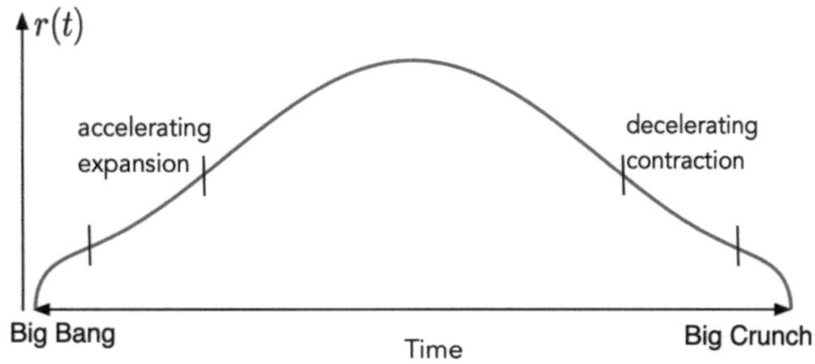

Figure 3.
A numerical computation of a solution to the Euler–Lagrange equation.

minimize I in (2). (Note, however, that $R_{ij} \neq 0$). In this case, $r(t)$ is concave, which means that the rate of expansion is decreasing with time, so there is no accelerating expansion.

But the interesting question is now: What happens if we add matter to this model? Mathematically, this is usually achieved by adding an extra term to the integral in (2). An intuitive non-mathematical analysis could look as follows:

It is clear that adding mass will give rise to something that in the traditional framework would be perceived as an (attractive) force, which is strongest close to the Big Bang and the Big Crunch where the density is high, and that this force will deform the universe by contracting the special part near the endpoints (i.e., make $r(t)$ smaller). However, since the volume is fixed, this contraction near the ends must somehow be compensated by an expansion in the middle (where the density is lower and the contracting force is weaker), thus making $r(t)$ larger. This should result in something like the behavior of the second curve in **Figure 2**.

Note that this contraction/expansion gives rise to a phase where $r(t)$ is convex, i.e., where the expansion is accelerating. It may also be worth noting that the model gives a natural explanation to the recent, much-discussed discovery that the rate of acceleration may now in fact be decreasing.

Although the analogy is not perfect, the reader, who does not wish to pursue the mathematics involved, may think about an elliptic balloon, perhaps filled with water. What happens if we squeeze the balloon at the ends? Obviously, the squeezing will make the water go away from the ends and hence make it expand in the middle in between, thus creating a form similar to the one in **Figure 2**.

The above should of course not be taken for a scientific argument; it is just an intuitive one involving the traditional way of thinking in terms of forces and so on. But the point is that the problem can also be analyzed mathematically without the concept of force. In **Figure 3** is shown a numerical computation of a solution of the corresponding Euler–Lagrange equation for the problem of minimizing the action in this case (see Appendix B). Clearly, a complete analysis of the mathematics is much more complicated than just finding an example of a suitable solution, and there are in fact many unsolved questions that remain. However, since the point here is just to show the possibilities of the method, I will not go further into this here.

Note, however, the crucial point: In the time symmetric variational problem, which determines the shape of the universe in this section, the past and the future play equally important roles, something that is completely unnatural to the Standard Method. And, as has already been remarked above, the Euler–Lagrange equation in this case is an ordinary differential equation of fourth order (see Appendix B), which means that to determine the solution uniquely by initial conditions, it would be necessary to specify the derivatives at the starting point up to order three, not just up to order one as is the case in, for example, Newtonian mechanics. No similar situation in physics, where such initial conditions have turned out to be adequate, is known to the author.

Summing up, the model in this section could give an alternative explanation for the accelerating expansion without any need to introduce extra components, like dark energy, into the field equations. But this explanation obviously does not make sense in a paradigm where the starting point is to view the development of our universe as an initial value problem and where only the behavior in the past is allowed to influence the solution.

8. Conclusions

In this article, I have tried to argue that it is possible that the situation nowadays in cosmology may have interesting similarities with the situation that led to the Copernican revolution. This should of course not be taken as a prediction about how modern cosmology will evolve in the future. Revolutions very seldom follow the paths their protagonists believe they should follow. But I do claim that it is an important part of science to try to imagine possible alternative paths into the future.

I have also tried to show that a change of perspective could open up other kinds of conditions and solutions than the Standard Method can give us access to. In particular, the Standard Method more or less forces us to build time asymmetry, in the sense of a fundamental difference between initial conditions and boundary conditions in the future, into our models from the beginning, whereas the Principle of Least Action, interpreted as a time symmetric minimizing principle, does not. However, this should only be considered as an example of how our perspective may be restricted by the Standard Method and expanded by the Principle of Least Action, although to my mind, it is a rather important one.

In this article, the starting point has been to consider universes of finite space–time volume. But it may of course also be asked if similar methods could apply to universes with infinite volume? In fact, most of the theory, including the minimizing principle (2), would make sense in such a setting, but the condition of fixed space–time volume would not. Following the traditions of statistical mechanics, a natural substitute could be to work with something similar to a chemical potential. But it is not clear how this should be interpreted physically, nor is it clear what would be gained from replacing one mysterious concept (dark energy) by a perhaps equally mysterious concept (the chemical potential). Summing up, more research is needed before any conclusions can be drawn in this case.

Finally, as already mentioned, the accelerating expansion is just one of several problems in modern cosmology. One can also ask if a shift toward a different point of view, based on the Principle of Least Action as a minimizing principle instead of the Standard Method, could offer new perspectives on, for example, the problem with cosmic inflation? My personal opinion is that the answer may very well be yes, but the arguments in this case would be quite different to the ones sketched here in connection with the accelerating expansion, so they will be discussed elsewhere.

Conflict of interest

The author declares no conflict of interest.

Appendix A. Multiple histories as a possible foundation for the field equations of general relativity

The action principle (2) does not reproduce the field equations in the usual way, using the classical calculus of variation (as does (1)). Curiously, however, this may change in a very interesting way when we switch to a more quantum mechanical point of view: Only when we apply a multiple histories perspective to general relativity, the ordinary field equations in vacuum can be obtained in the sense that practically all the probability mass will fall on metrics that satisfy $R_{ij} = 0$. A complete treatment of this approach uses methods from statistical mechanics and is far beyond the scope of the present article. But here is a very brief sketch of how the arguments would go (for the concepts from statistical mechanics, see Ref. [25]):

The starting point is to consider all possible geometries, defined by different metrics, in some region U and with certain boundary conditions. Together, these are said to constitute an Ensemble. Although there is an enormous number of such geometries on a microscopic scale (corresponding to quantum fluctuations), it can be argued that on a macroscopic scale, under suitable circumstances, there is only one geometry that has a non-negligible probability of occurring. This is perfectly analogous to what happens, for example, in the classical theory of gases.

Starting from very general assumptions and using the central limit theorem [27], one can argue that a natural Ensemble measure is given by attributing to each metric g, the (un-normalized) probability

$$\exp\left\{-\mu\int_U R_g^2 dV\right\}, \tag{3}$$

where μ in general is a very large number.

Using common procedures in statistical mechanics, we can now compute the Helmholtz free energy L, which turns out to become

$$L = \mu\int_U R_g^2 dV - \log\Omega_g, \tag{4}$$

where Ω_g is the so called "density of states" around g.

Common wisdom in statistical mechanics now tells us that the states (in this case metrics) that minimize the free energy are the ones that are realized in nature. But computing these states can be very difficult, since they depend on a complicated interplay between the two terms above. However, in the particular case when there is a metric that satisfies the field equations $R_{ij} = 0$ (and given boundary conditions), it can be argued this metric will simultaneously minimize both terms. For the first term, this is obvious. For the second, it is much more difficult but in the end boils down to the fact that in a certain (non-trivial) sense $\delta R/\delta g_{ij} = R_{ij}$, see [28, 29].

Even in this restricted case, it is not easy to make a complete and rigorous proof, but the reader is referred to Refs. [20, 21] for more information. In a sense, this is all much closer to quantum mechanics than is the Einstein-Hilbert Principle. It is in fact very much at the heart of this article that the classical Einstein-Hilbert Principle, even if it was a great step forward at the time when it was introduced, may nowadays be a road blocker on our way to a unification between general relativity and quantum physics.

Appendix B. The Euler–Lagrange Equation

The purpose of this Appendix is to briefly, and in an approximate and simplified form, illustrate the mathematics underlying the figures in this article. Similar figures have been published earlier in Tamm [20, 22], but the following is intended to give a short but self-contained summary. However, it is not intended to give a numerically realistic description of our universe, only to illustrate the mechanisms in this article.

We start from a standard description of a compact, homogeneous, and isotropic universe with the following metric in spatial, spherical coordinates r, χ, θ, φ (see Ref. [10]):

$$ds^2 = -dt^2 + r(t)^2\left(d\chi^2 + \sin^2\chi\left(d\theta^2 + \sin^2\theta d\varphi^2\right)\right). \tag{5}$$

In a closed model, the function $r(t)$ can naturally be thought of as the radius of the universe at time t, rather than as the nowadays more common scale factor. Here $-T_0 < t < T_0$, where $-T_0$ stands for the Big Bang and T_0 stands for the Big Crunch.

The field equations of general relativity are given by

$$G_{ij} = 8\pi T_{ij}, \tag{6}$$

where G_{ij} is the Einstein tensor, and T_{ij} is the stress-energy tensor (see further e.g. Ref. [10]). The most useful component of (6) is the time–time component that reads

$$\frac{6}{r(t)^2}\left(\frac{\partial r}{\partial t}(t)\right)^2 + \frac{6}{r(t)^2} = \frac{6r_0}{r(t)^3}, \tag{7}$$

where r_0 is a positive constant, which is usually considered to represent some kind of total mass of the universe. After simplifications we get

$$\left(\frac{\partial r}{\partial t}(t)\right)^2 = \frac{r_0}{r(t)} - 1. \tag{8}$$

This is the simple first-order differential equation used in drawing **Figure 1** as an illustration of the Standard Method.

If we now instead turn to the study of the consequences of the minimizing principle, we get for the geometric part:

$$I_g = \int_{\Omega} R^2 dV = \frac{\pi^2}{2}\int R(t)^2 r(t)^3 dt. \tag{9}$$

The contribution to the action from mass will be (except for an unimportant constant which can be dropped out) given by the classically potential energy, which is here coarsely estimated as one over the (average) distance between the particles, just like in a classical gas. This gives

$$I_p = -\beta\frac{\pi^2}{2}\int \frac{1}{r(t)} dt, \tag{10}$$

for some suitable constant $\beta > 0$.

Under these conditions, the problem becomes to minimize

$$I = I_g + I_p = \frac{\pi^2}{2}\int\left(R^2 r(t)^3 - \beta\frac{1}{r(t)}\right)dt, \tag{11}$$

under the condition

$$V = \int_\Omega dV = \frac{\pi^2}{2}\int r(t)^3 dt = V_0, \tag{12}$$

where V_0 is a fixed number, corresponding to the total four-volume of the universe Ω. What do the solutions to this minimizing problem look like?

The traditional method of attack is to look at the Euler–Lagrange equation for the functional $I_\lambda =$

$$I + \lambda\left(V_0 - \int_\Omega dV\right) = \frac{\pi^2}{2}\int\left(R^2 r(t)^3 - \beta\frac{1}{r(t)}\right)dt + \lambda\left(V_0 - \frac{\pi^2}{2}\int r(t)^3 dt\right). \tag{13}$$

The scalar curvature for the metric in (5) becomes, after a not too long computation:

$$R(t) = \frac{6\left(1 + r'(t)^2 + r(t)r''(t)\right)}{r(t)^2}. \tag{14}$$

From this we obtain

$$I_g = \frac{\pi^2}{2}\int R(t)^2 r(t)^3 dt = \frac{\pi^2}{2}\int \frac{36\left(1 + r'(t)^2 + r(t)r''(t)\right)^2}{r(t)}dt. \tag{15}$$

Recall that for the problem of finding the minimum of a functional of the form

$$\int F\left(r(t), r'(t), r''(t)\right)dt, \tag{16}$$

where $F(u, v, w)$ is some sufficiently regular function, the Euler–Lagrange equation is given by

$$\frac{\partial F}{\partial u}\left(r(t), r'(t), r''(t)\right) - \frac{d}{dt}\left(\frac{\partial F}{\partial v}\left(r(t), r'(t), r''(t)\right)\right) + \frac{d^2}{dt^2}\left(\frac{\partial F}{\partial w}\left(r(t), r'(t), r''(t)\right)\right) = 0. \tag{17}$$

Using Mathematica, we can now compute the Euler–Lagrange equation associated with the functional in (13) and obtain (after multiplying with $r(t)^2$):

$$\begin{gathered} 2r(t)^3 r^{(4)}(t) + 3r(t)^2 r''(t)^2 - 4r(t)r''(t) + 3r'(t)^4 + 2r'(t)^2 + \\ 4r(t)^2 r^{(3)}(t)r'(t) - 12r(t)r'(t)^2 r''(t) - 1 + \beta = 3\lambda r(t)^4. \end{gathered} \tag{18}$$

This is the equation that has been used to draw the graph in **Figure 3** (with $\beta = 3$ and $\lambda = 2.175$). No initial conditions have been used, but rather I have assumed that the universe is time symmetric. This is a very reasonable condition for the large-scale form of a closed universe, which then gives $r'(0) = r'''(0) = 0$ ($r(t)$ will be an even function, symmetric around $t = 0$). In principle, the even derivatives should be determined from something comparable to r_0 in the case off the Friedmann-Einstein model and the condition of constant volume. However, this leads to non-trivial technical complications near the endpoints $-T_0$ and T_0, where in any case the given model is far too simple to give a good description. (Note, however, that $-T_0$ and T_0 are dynamical, i.e., they are not fixed from the beginning, but rather, they are given by the solution.)

So, for the purpose of illustrating the curve in the region, which is relevant for the accelerating expansion, I have just put $r(0) = 1$ and $r''(0) = -1/2$. There are very many solutions to Eq. (18), most of which do not represent closed universes at all, so a lot of work is needed before we draw any more substantial conclusions. This example is only to indicate a possible way to explain the accelerating expansion.

Author details

Martin Tamm
Stockholm University, Stockholm, Sweden

*Address all correspondence to: matamm@math.su.se

References

[1] Kuhn TS. The Copernican Revolution. United States: Harvard University Press; 1957

[2] Scerri E. The Periodic Table: Its Story and Significance. United Kingdom: Oxford University Press; 2006

[3] Gell-Mann M. The Eightfold Way: A Theory of Strong Interaction Symmetry. United States: California Institute of Technology; 1961. DOI: 10.2172/4008239

[4] Brookes DT, Etkina E. "Force," ontology, and language. Physical Review Physics Education Research. 2009;**5**: 010110

[5] Riess AG et al. (supernova search team). Observational evidence from supernovae for an accelerating universe and a cosmological constant. The Astronomical Journal. 1998;**116**(3): 1009D38

[6] Perlmutter S et al. (the supernova cosmology project). Measurements of omega and lambda from 42 high redshift supernovae. The Astrophysical Journal. 1999;**517**(2):565D86

[7] Dark Energy Survey Collaboration: Abbott T, et al. Monthly Notices of the Royal Astronomical Society. 2016;**460** (2):1270-1299. DOI: 10.1093/mnras/stw641

[8] Li et al. Dark Energy. Communications in Theoretical Physics. 2011;**56**:525. DOI: 10.1088/0253-6102/56/3/24

[9] Kuhn TS. The Structure of Scientific Revolutions. Chicago: University of Chicago Press; 1962

[10] Misner CM, Thorne KS, Wheeler JA. Gravitation. San Fransisco: W. H. Freeman and Company; 1973

[11] Margenau H. Meaning and scientific status of causality. Philosophy of Science. 1934;**1**(2):133-148. DOI: 10.1086/286314

[12] Goldvarg E, Johnson-Laird PN. Naive causality: A mental model theory of causal meaning and reasoning. Cognitive Science. 2001;**25**(4):565-610

[13] Maupertuis P. Accord de differentes lois de la Nature qui avaient jusqu'ici paru incompatibles. Mémoires de l'Académie Royale des Sciences de Paris. 1744. pp. 417-426

[14] Hilbert D. Die Grundlagen der Physik. (Erste Mitteilung.) Nachrichten von der Gesellschaft der Wissenschaften zu Göttingen. Mathematisch-physikalische Klasse. 1915:395-407

[15] Palatini A. Deduzione invariantiva delle equazioni gravitazionali dal principio di Hamilton. Rendiconti del Circolo Matematico di Palermo. 1919;**43**:203-212

[16] Eddington AS. Relativitätstheorie in Mathematischer Behandlung. Berlin: Springer; 1925

[17] Feynman RP, Hibbs AR, Styer DF. Quantum Mechanics and Path Integrals. New York: McGraw-Hill; 1965

[18] Ruelle D. Thermodynamic Formalism. United Kingdom: Cambridge University Press; 1978. DOI: 10.1017/CBO9780511617546

[19] Onsager L. Crystal statistics. I. A two-dimensional model with an order-disorder transition. Physics Review. 1944;**65**:117. DOI: 10.1103/PhysRev.65.117

[20] Tamm M. Natural Lagrangians. Universe. 2021;7:74. DOI: 10.3390/universe7030074

[21] Tamm M. Newton's first law and the grand unification. Symmetry. 2024;**16**(12):1694. DOI: 10.3390/sym16121694

[22] Tamm M. Accelerating expansion in a closed universe. Journal of Modern Physics. 2015;**6**:239-251

[23] Einstein A. Zum kosmologischen Problem der allgemeinen Relativitätstheorie. Sitzungsberichte der Königlich Preussischen Akademie. 1931: 235-237

[24] Hawking S. General Relativity; an Einstein Centenary Survey. Cambridge: Cambridge University Press; 1980

[25] Huang K. Statistical Mechanics. second ed. New York: John Wiley & Sons, Inc.; 1987

[26] Anninos D, Ng GS, Strominger A. Future boundary conditions in de sitter space. Journal of High Energy Physics. 2012;**2012**:32. DOI: 10.1007/JHEP02(2012)032

[27] Fischer H. A History of the Central Limit Theorem: From Classical to Modern Probability Theory, Sources and Studies in the History of Mathematics and Physical Sciences. New York: Springer; 2011

[28] Schoen RM. Variational theory for the total scalar curvature functional for riemannian metrics and related topics. In: Giaquinta M, editor. Topics in Calculus of Variations. Lecture Notes in Mathematics. Vol. 1365. Berlin, Heidelberg: Springer; 1989. DOI: 10.1007/BFb0089180

[29] Besse AL. Einstein Manifolds. Berlin, Heidelberg: Springer Berlin Heidelberg; 2008

Chapter 6

The Fallacies of ΛCDM Falsifications

Alain Blanchard

Abstract

In recent years, numerous arguments have emerged suggesting that the Lambda Cold Dark Matter (ΛCDM) model may be inconsistent with observational data, requiring more or less radical revisions. Notable examples include the "Hubble tension"—the discrepancy between early- and late-universe measurements of the Hubble constant—as well as tensions in measurements of cosmic structure growth. These issues have led some to question the validity of the ΛCDM framework and consider possible modifications or alternative models. However, upon closer inspection, many of these critiques stem from methodological or interpretive disagreements rather than from clear falsifications in the strict Popperian sense. Karl Popper proposed that scientific theories must be testable and falsifiable; in other words, a theory should be rejected if it fails a specific, reproducible test. Yet, many of the alleged "inconsistencies" within ΛCDM, while statistically significant, are not definitive falsifications but rather indicators of areas needing refinement or more complex modeling within the same framework. Thus, I review the recent claims about ΛCDM's limitations and analyze why they often reflect individual biases or philosophical preferences, rather than rigorous scientific falsifications. For example, alternative cosmological models such as Modified Newtonian Dynamics (MOND) or models incorporating new physics like quintessence or modified gravity are sometimes advocated based on theoretical appeal rather than direct evidence from critical tests. In many cases, these arguments for falsifying ΛCDM reveal more about subjective interpretations of data than about concrete observational contradictions.

Keywords: cosmology, dark matter, dark energy, CDM, CMB, ΛCDM

1. Introduction

The idea of a non-baryonic matter component contributing significantly to the density of the universe dates back to the 1960s: in 1966, Gershtein and Zel'dovich [1] noticed that an upper limit to the neutrino mass comes from neutrinos' possible contribution to the density of the universe. By the 1980s, the possibility of a critical-density universe dominated by neutrinos with masses exceeding several electron volts was explored, largely driven by the work of Yakov Zel'dovich and the claim of the detection of non-zero mass [2]. It was realized that in such case the free streaming of neutrinos will smooth out primordial fluctuations on small scales. This hypothesis marked an important shift in cosmological thinking: specifying the nature of dark matter meant that the initial spectrum of density fluctuations was no longer purely determined by unknown initial conditions but instead emerged from the dynamic evolution of perturbations. This led to specific and well-defined features in the final distribution of matter, intrinsically tied to the properties of non-baryonic dark matter.

Neutrinos with masses on the order of a few tens of electron volts (eV) introduced a characteristic suppression scale that was too large to be consistent with the observed distribution of galaxies known at the time. This suppression arose because neutrinos, being relativistic and freely streaming in the early universe, erased density fluctuations below a certain scale. As a result, structure formation was severely inhibited on small scales, preventing the efficient formation of galaxies and galaxy clusters. These fundamental issues ultimately led to the exclusion of the neutrino-dominated scenario as a viable model for dark matter [3]. The inability of hot dark matter to support the observed large-scale structure prompted a shift in focus toward alternative dark matter candidates.

By the late 1970s, it was realized that a weakly interacting particle with a mass on the order of a giga-electron volt (GeV) could also contribute significantly to the total density of the universe. Unlike neutrinos, such a particle would become non-relativistic much earlier in cosmic history, allowing density perturbations to survive on small scales and facilitating the hierarchical growth of structures. This idea laid the foundation for the cold dark matter (CDM) paradigm, which would later become a cornerstone of modern cosmology.

In 1982, Jim Peebles [4] proposed an alternative scenario based on such a much more massive dark matter particle, giving rise to what is now known as the cold dark matter (CDM) paradigm. He noted that the galaxy correlation function roughly agreed with observations and that this model predicted a significantly lower amplitude for cosmic microwave background (CMB) fluctuations than in a purely baryonic universe. Around the same time, the concept of cosmic inflation was gaining traction. Inflation naturally led to a specific initial fluctuation spectrum—now known as the Harrison-Zel'dovich spectrum—providing a logical foundation for a cosmological model with very few free parameters. In this framework, the spectrum of density perturbations was determined by just a handful of cosmological parameters and the overall amplitude of fluctuations.

Furthermore, the evolution of perturbations in the early universe produced a transfer function with multiple characteristic scales. The precise values of these scales depended on the details of the cosmological model, particularly on key parameters such as the baryonic matter density, the non-baryonic dark matter density, and the Hubble constant. Initially, this cold dark matter model was developed under the assumption of a universe with critical density. However, as observational data on small-scale CMB fluctuations started to emerge, it became evident that the universe was nearly flat, yet no dynamical measurement indicated a total matter density close to the critical value. This discrepancy led to the proposal of introducing a cosmological constant (Λ) to the model, which eventually evolved into what is now known as the ΛCDM model.

One of the most significant strengths of this framework is its predictive power. Given that the matter spectrum is approximately fixed, it allows for precise predictions of both the CMB fluctuation spectrum and the large-scale matter distribution (and thus the distribution of galaxies, up to a bias parameter) with very few free parameters. This predictive capability was put to the test when full-sky measurements of CMB fluctuations became available with the launch of the Wilkinson Microwave

Anisotropy Probe (WMAP) in 2003: once normalized to the CMB, the ΛCDM model predicted the entire linear matter power spectrum (or equivalently, the correlation function). Strikingly, when data from the Sloan Digital Sky Survey (SDSS) became available in 2005, the observed large-scale structure exhibited an almost perfect match with the predictions of the ΛCDM model—an extraordinary success for what is now considered the "standard model" of cosmology.

Since then, the robustness of this agreement has been further reinforced by increasingly precise CMB measurements from WMAP and later from the Planck satellite, as well as by improved large-scale structure data from SDSS and other galaxy surveys. Today, no alternative scenario offers a comparably successful description of cosmic structures on large scales. The strong predictive power of the standard ΛCDM model continues to enable rigorous and fruitful comparisons with observational data. This makes it possible to envisage its falsification but also makes it the dominant framework for our understanding of the large-scale evolution of the universe.

2. What makes a scientific theory successful?

Scientific inquiry does not seek to unveil an absolute reality, let alone an ultimate truth. Instead, it constructs models that describe and predict observable phenomena, always remaining open to revision in light of new evidence. A defining feature of modern science is its commitment to the Popperian framework: a scientific theory must not only explain known facts but also predict new and unexpected phenomena. Crucially, these predictions should be as distinct as possible from previously established knowledge, ensuring that the theory is testable and falsifiable.

A historical example of this principle is Newton's theory of gravity, which reached its pinnacle when it successfully predicted the existence of Neptune. The discovery was based on unexplained anomalies in Uranus' trajectory, leading astronomers to infer the presence of an unseen planet. This validation of Newtonian mechanics remains a landmark in the philosophy of science, demonstrating the predictive power that distinguishes scientific theories from mere descriptions. Of course, this success was even more striking than Newton's ability to account for planetary orbits, as it showcased the theory's capacity to anticipate phenomena beyond those it was originally formulated to explain.

Similarly, it is crucial to examine how the ΛCDM model has undergone the Popperian falsification test—and passed with flying colors. However, the Popperian sequence of prediction, testing, and potential falsification should not be interpreted naively as a strict chronological process. What must prevail in the scientific process is the logical structure of the sequence of events. An unexpected discovery that is later shown to be deducible from a preexisting theory or model constitutes a successful Popperian test. The fact that the phenomenon was not initially predicted, or that it was not announced as a test of the theory, but that it can be deduced, even retrospectively, from the theoretical framework reinforces confidence in the model, because it demonstrates its predictive power. This aspect is discussed in a little more detail by Ref. [5]. This process exemplifies the self-correcting nature of science: theories are expected not only to explain known facts but also to withstand scrutiny when new, unforeseen phenomena emerge, or even show that this new observation follows naturally from the theory. An emblematic example of this situation is Hubble's discovery in 1929: the proportionality between the redshift of distant sources, as galaxies were, and their distance had been noticed and recorded in cosmological models derived from general relativity, though it had never been explicitly stated as a prediction. Nevertheless, Hubble's observations were regarded at the time—and still are today—by the scientific community as the first validation of relativistic cosmological models. [1]

In contrast, if a new fact arises and the theory is modified after the fact by introducing additional ad hoc elements solely to accommodate it, the epistemic value of the test is considerably weakened—or even nullified altogether. A theory that is excessively flexible, capable of adapting to any new data simply by incorporating arbitrary adjustments, risks becoming unfalsifiable, thereby stepping outside the bounds of empirical science. However, this does not mean that theories cannot evolve. The Popperian test remains meaningful if the revised version of the theory is not merely retroactive accommodation but instead introduces new, independently testable predictions. In such cases, the modifications are not ad hoc but part of a refined theoretical framework that remains falsifiable. The key is that any refinement should lead to novel, discriminating, and risky predictions—ones that could, in principle, prove the new formulation wrong. Only under these conditions does scientific progress remain genuinely empirical rather than purely adaptive.

That said, although Popperian falsifiability is an important methodological ideal, science operates more flexibly in practice. Many theories are accepted because they are coherent and explanatory and drive progress, even if they are not immediately testable. The robustness of a theory often relies on an accumulation of converging evidence rather than a single falsification test. For instance, string theory is mathematically elegant and consistent with known physics, but it still lacks testable predictions that can be verified by current experiments. Likewise, the theory of evolution is not easily testable through a strict Popperian sequence, yet it is strongly supported by an accumulation of independent observations.

Furthermore, the history of science teaches us that even the most successful theories have their limits. Einstein's general relativity ultimately refined Newton's framework by incorporating the curvature of spacetime, providing deeper insights into gravity. This iterative process underscores the self-correcting nature of science: theories evolve, but they are never final. The scientific endeavor remains an ongoing process of refinement, where theories must continuously prove their explanatory and predictive power to retain their status as reliable models of reality.

A final remark concerns the nature of the facts used in the Popperian criterion. An interesting historical example is the solar neutrino deficit. For the record, models of the Sun's interior predicted a flux greater than that measured by early neutrino detectors, such as the Homestake experiment. However, doubts arose regarding the validity of these stellar interior models, given their reliance on complex and sometimes untestable hypotheses. This situation is not uncommon in astrophysics: predictions often depend on intricate modeling, which can weaken their robustness—not necessarily due to inaccuracies, but because they are grounded in assumptions that may themselves lack direct empirical validation. The key takeaway is that the strength of a Popperian test involves a degree of subjectivity: it must be regarded as robust by the scientific community as a whole, which may not be the case for models of the Sun's interior. The resolution of the solar neutrino problem came not from astrophysics but from particle physics, with the groundbreaking discovery of neutrino oscillations—a phenomenon demonstrating that neutrinos can change "flavor" as they travel through

[1] There is a certain irony in this event in two ways: Einstein modified his initial theory because he was focused on a static cosmological solution; the equations of an expanding universe can be deduced from Newtonian reasoning, and they could therefore have been deduced before the advent of general relativity.

space. This insight not only solved the deficit but also provided experimental confirmation that neutrinos have a small but non-zero mass. Subsequent experiments, such as those conducted at Super-Kamiokande and the Sudbury Neutrino Observatory (SNO), validated both the phenomenon of oscillations and the accuracy of solar interior models, marking a milestone in our understanding of both particle physics and stellar processes.

In this article, I aim to examine how the ΛCDM model has been tested against observational data and how no alternative provides the same level of Popperian corroboration.

3. Successes of the ΛCDM model

There is no precise date that marks the birth of the ΛCDM model. As mentioned earlier, the ΛCDM model itself is difficult to date. The role of the cosmological constant has been studied since Einstein first introduced it, but for a long time, the lack of observational evidence made it an exotic concept in cosmology. J. Peebles later revisited this idea to explain a flat universe without requiring the matter density to be equal to the critical density [6]. He then developed what is now known as the quintessence scenario, in which the cosmological constant is replaced by a scalar field whose behavior, depending on its potential, can mimic that of a cosmological constant [7]. The introduction to this article explains that one of the motivations is to achieve a flat universe, which requires a critical density, without invoking non-baryonic dark matter.

By the late 1990s, all the key ingredients of the ΛCDM model were in place: baryonic matter, cold non-baryonic matter, and dark energy. The theoretical framework for computing the transfer function had been established and refined since the work of Peebles and Yu [8]. Once these components were specified, the full calculation of the fluctuation spectrum in the cosmic microwave background (CMB) and the matter distribution spectrum became well-defined.

A crucial element in this framework is the initial spectrum of fluctuations. From the 1970s onward, for simplicity, it was generally assumed that this spectrum was Gaussian and followed a power law. Whether the fluctuations were adiabatic or not, however, remained a subject of debate. The emergence of inflation theory provided a theoretical foundation for these assumptions. But it should be kept in mind that by the 1980s, the essential ingredients now incorporated into Boltzmann codes such as CAMB and CLASS were already in place.

It is remarkable how observations of the cosmic microwave background (CMB) and the large-scale distribution of galaxies have consistently supported the ΛCDM model over the past four decades. For the record, when Peebles wrote his article on CDM [4], no fluctuations had been measured in the CMB, no three-dimensional catalogs were available to study the distribution of galaxies (only projected distribution was well studied). Some alternative perspectives, such as the hot dark matter model and the cosmic string scenario, were quickly abandoned: hot dark matter due to galaxy distribution data ruling out the massive neutrino scenario. Models of structure formation based on topological defects, such as textures, were gradually ruled out as increasingly precise observations of cosmic background fluctuations became available.

The measurement of the large-scale galaxy angular correlation function was the first major challenge to the standard CDM model (i.e., the critical density model) [9]. Indeed, the shape of the angular correlation function obtained from the APM was inconsistent with the expectations of a CDM model with a critical density. However, it was observed that a CDM-type model with a non-zero cosmological constant provides a good fit to the data [10]. We could consider this observation as marking the operational birth of the ΛCDM model.

3.1 The fluctuations of the cosmic microwave background

The first detection of CMB fluctuations by COBE took place only a few years after the measurement of the APM correlation function. The shape and amplitude of these fluctuations, as well as their Gaussian or non-Gaussian nature, had the potential to strongly challenge—or even outright invalidate—the ΛCDM scenario. In retrospect, it is interesting to note that, although the signal detected by COBE matches precisely what is expected in the ΛCDM model, this agreement was not particularly emphasized at the time. Yet, this is one of the most astonishing results of the theory: COBE's observations reveal the amplitude of matter fluctuations on scales 100 to 1000 times larger than those accessible to robust measurements from galaxy catalogs at the time, and their amplitude agrees to within a factor of order unity with theoretical predictions.

The fundamental role of CMB data in validating the ΛCDM model cannot be overstated here: the calculation of CMB fluctuations is highly robust, as it is performed within the linear perturbation regime. This ensures that no ad hoc or overly simplifying assumptions have been introduced. Moreover, the various existing codes—some of them independent and competing—agree to an extremely high degree of accuracy. The measurement of these fluctuations is also highly reliable: Galactic foregrounds, while present, are not dominant and can be efficiently subtracted due to their frequency-dependent characteristics[2]. Finally, after more than 30 years of improvements, Planck's results are fully consistent with those of WMAP, as well as with the most recent ground-based experiments, such as ACT and the South Pole Telescope.

After the first version of this paper was submitted, the ACT team published the DR6 data and constraints. This is an excellent Popper test: once the model is fitted to previous CMB data, it predicts very accurately the properties of the power spectrum of the fluctuations C_l. ACT provides new measurements at small scales, increasing by nearly a factor of two the scales at which accurate measurements are available. These new ACT data are fully consistent with ΛCDM [11], constraining severely current alternatives that modified the small-scale behavior of the C_l [12].

3.2 The large-scale structure of the universe

The measurement of the large-scale structure of the universe in a quantitative way is primarily performed through the correlation function of galaxies. As we have already mentioned, Peebles [4] pointed out that the shape of the power spectrum was in *qualitative* agreement with what was known at that time. Forty years on, observations of the distribution of galaxies on linear scales, particularly at the BAO scale, remain in perfect agreement with theoretical predictions, even though they could have invalidated the model, as was the case for massive neutrinos. The DESI data provides the latest tight constraints [13].

---

[2] It remains an interesting question whether oversimplistic assumptions have been made about the properties of astrophysical foregrounds, such as dust, which would lead to a (slight) bias in the cosmological parameters.

3.3 The global picture

The fact that the ΛCDM model is consistent with both the CMB and the large-scale structure (LSS) distribution of galaxies, while measurements reach an accuracy on the order of 1%, is a remarkable achievement of modern science that should not be underestimated. The ability to determine cosmological parameters independently from these two datasets, yielding consistent values with an accuracy of a few percent, is an impressive test of consistency.

4. Criticism of the ΛCDM model

Many articles deal critically with the Standard Model. But the criticisms are of different kinds. There's also a low-level skepticism among some physicists, stemming from the fact that in cosmology, there are virtually no laboratory experiments to back up the model, to the constraints of, say, atomic or particle physics. The impression is therefore that the assertions are based on a small number of observations that provide only tenuous support for the representations that are elaborated.

4.1 Early universe and initial conditions

In order to make fruitful comparison of ΛCDM model with observations, it is necessary to assume some initial power spectrum, in practice a power law:

$$P(k) \propto k^{n} \quad (1)$$

with a free index n close to 1. Furthermore, initial fluctuations are assumed to be adiabatic and Gaussian. These ingredients are easily produced by inflation models. By way of an intellectual shortcut, it is often claimed that the ΛCDM model relies on inflation—considered by some to be a speculative theory with limited observational validation, and some even argue that it is contradicted by the data [14]. ΛCDM would therefore suffer from the same weakness [15]. However, key ingredients of the model were already in use before the formulation of inflation, such as adiabatic fluctuations with a power-law spectrum and a Gaussian distribution. Whether or not inflation is correct is therefore irrelevant to the ΛCDM framework, as the defining characteristic of Cold Dark Matter (CDM) lies in the transfer function—that is, in how fluctuations evolve over time from their initial state. The introduction of a cosmological constant became necessary when observations of the cosmic microwave background indicated a flat universe, even before the supernova Hubble diagram provided direct evidence of cosmic acceleration.

4.2 The large-scale anomalies

4.2.1 (in)homogeneity on large scale

The standard cosmological model is based on Einstein's cosmological principle, which states that the universe is homogeneous on large scales, allowing the use of the Robertson-Walker metric. However, this remains a qualitative statement of a question that must ultimately be addressed in quantitative terms. To achieve this, it is relevant to examine the amplitude of metric perturbations. As long as these perturbations remain very small, they can be treated to linear order, with higher-order corrections included if observational precision warrants it. In the solar system, for example, the Sun and planets introduce only weak perturbations, and relativistic effects can be accounted for by considering higher-order terms. The Sun represents a perturbation of the metric of the order of 10^{-5}, even though the density contrast is enormous. On the scale of a galaxy, the perturbation is of the same order. On larger scales, galaxy catalogs can be used to assess density contrast. It has long been known that:

$$\frac{\delta\rho}{\rho}\left(10h^{-1}Mpc\right)\sim 1 \tag{2}$$

At larger scales, the density field is expected to be linear (i.e., fluctuations are expected to be less than one). However, visual inspection reveals structures that appear much larger, giving the impression that very large inhomogeneities might exist. These types of features are sometimes perceived as challenges to the standard picture [16], but they do not hold up to further quantitative analysis. This argument has been reiterated recently to invalidate the ΛCDM [17], but comparisons with numerical simulations remain ultimately inconclusive [18].

4.2.2 Dipole

Several surveys of distant sources have identified a dipole anisotropy in their distribution [19]. If this is interpreted as a kinematic dipole, it exceeds that of the CMB. However, to my knowledge, this argument never considers the possibility that the dipoles in sources are just due to the randomness of the clustered process. Given the above fluctuation, assuming that the universe is filled with random spheres of $10h^{-1}Mpc$, one would expect fluctuations of the order of few 10^{-3}, comparable or larger than the CMB dipole. A second problem that arises is that, in order to be confident that the dipole is real, one must control the selection function of objects across the sky with an accuracy of at least 10^{-3}, which is highly challenging.

4.3 The S_8 tension

The amplitude of matter fluctuations in the present-day universe is traditionally measured by σ_8. The abundance of clusters and the weak lensing measurements are two traditional ways to evaluate this amplitude σ_8 and it is cosmology rephrased in term of S_8:

$$S_8=\sigma_8\left(\frac{\Omega_m}{0.3}\right)^{1/2}, \tag{3}$$

which is less sensitive to Ω_m. Lensing measurements have long yielded lower values than those inferred from the CMB in the ΛCDM framework [20]. The most emblematic measure is the one obtained from KiDS 3×2 pt. [21]:

$$S_8=0.766\pm 0.02 \tag{4}$$

Indeed, the amplitude of matter fluctuations at redshift ~ 1100 can be inferred from CMB fluctuations. Planck provides the most accurate value, consistent with other CMB measurements. In order to compare this to present-day values, one must extrapolate it to $z = 0$. This is straightforward in the ΛCDM model. However, it is important to note that the inferred S_8 value is model dependent. The tension with KiDS was 3σ.

The situation has changed in recent years. Planck's latest value obtained from the PR4 release [22] is slightly lower than the 2018 value:

$$S_8 = 0.819 \pm 0.014 \tag{5}$$

lowering the tension down to 2.2σ. Moreover, the DES and KiDS teams have reached a joint conclusion, finding a value regarded as consistent with Planck [23]:

$$S_8 = 0.790^{+0.018}_{-0.014} \tag{6}$$

the two S_8 now differ by only 1.3σ. In fact, several recent measurements based on low-redshift data consistently yield a value compatible with that inferred from the CMB. In a study by Ref. [24] using a combination of different local measurements, we obtained:

$$S_8 = 0.801 \pm 0.012 \tag{7}$$

I think it's fair to say that the tension in S_8 appears to no longer exist. Ironically, the eROSITA collaboration's analysis of X-ray clusters' abundance suggests an S_8 higher value than that derived from the CMB, potentially indicating a new tension with an S_8 value higher than what is inferred from the CMB.

4.4 "Crises" on non-linear scales

On small scales (i.e., non-linear scales), numerous claims suggest that the predictions of the ΛCDM model are inconsistent with observations. These include the core-cusp problem, the too big to fail problem, the missing satellites problem, the satellite planes problem, and the baryonic Tully-Fisher relation. An extensive body of literature has explored these issues (see, for instance, the comprehensive review [25]).

Assessing the significance of these problems is challenging for several reasons. The first issue is the reliance on a posteriori statistics: a striking pattern is identified and then judged to be highly improbable. For example, in the case of the satellite planes problem, several satellites of the Milky Way and Andromeda were observed to be located within a thin plane [26]. However, such configurations were later found to be not so uncommon in numerical simulations and consistent with expectations in ΛCDM [27, 28]. The second problem is that numerical simulations are required to address the issue, as in the previous example. Indeed, the core-cusp problem, the too big to fail problem, and the missing satellites problem are based on non-linear features that can only be addressed through numerical simulations. Because we are on non-linear scales, the interpretation of data may be misleading [29]. A final difficulty is that, most of the time, the "predictions" of the ΛCDM model are derived from pure dark matter simulations. It is well known that hierarchical models of structure formation are subjected to the "overcooling" problem: that is, a simple estimation of the amount of primordial gas that should end in a "cold form" (i.e., condensed in structures below 10^4 K) is far in excess to what is seen in the universe [30]. This is a strong indication that feedback mechanisms were strong in the past universe, and the role of baryonic physics could be critical for the interpretation of some data [31]. This essentially means that small-scale problems should be discussed in a ΛCDM + baryonic physics, and conclusions drawn from these scales without considering the diversity of possibilities induced by baryonic physics are open to question and cannot, on their own, seriously disqualify the ΛCDM model.

4.5 The Hubble constant tension

The Hubble tension has now become widely recognized and extensively discussed within the scientific community. This discrepancy, which arises from the different values of the Hubble constant inferred from early- and late-universe measurements, remains one of the most significant challenges in contemporary cosmology. While probes based on early-universe physics, the cosmic microwave background (CMB), and the baryonic acoustic oscillations (BAO) scale predict a lower value of around 67 km/s/Mpc, local measurements based on the distance ladder method, including Cepheids and Type Ia supernovae, yield a significantly higher value, around 73 km/s/Mpc. Readers seeking a more comprehensive overview of the Hubble tension, its implications, and the various proposed resolutions can refer to the numerous dedicated reviews and research articles on the subject [32, 33]. As stated in Ref. [32], the resolution of the tension is either in new physics or in as-yet unrecognized uncertainties.

There have been numerous efforts to identify possible extensions to the standard ΛCDM model in an attempt to resolve the Hubble tension. In contrast, comparatively less work has focused on investigating potential systematic biases in measurements.

Most theoretical approaches to addressing the Hubble tension involve extensions of the ΛCDM framework. This is largely because this framework is sufficiently well-defined to enable precise and testable calculations that can be directly compared with observations. Moreover, the accuracy of these observational constraints, notably on the Hubble constant, has now reached the percent level, making it possible to rigorously assess the viability of any proposed modifications to the model. There is a long list of such different models that are proposed to reduce the Hubble tension [34]. A statistical comparison of their merits has led to rank them in order of success [35].

The Hubble tension is undoubtedly the issue attracting the most attention from the community. If confirmed, it would necessitate a revision of the ΛCDM model. Conversely, if the locally measured Hubble constant ultimately aligns with the value inferred from CMB observations, the ΛCDM model would receive a strong Popperian corroboration.

The key question here is what conclusions can legitimately be drawn. While it seems widely accepted that the Hubble tension is real, I would argue that it is still too early for definitive conclusions. The measurement of the Hubble constant has undergone numerous revisions since the initial work in 1929, often due to previously unidentified systematic effects. Despite the remarkable efforts of the SH0ES team [36], the possibility of such systematics influencing current measurements cannot be ruled out. The competing team from the Chicago-Carnegie Hubble Program (CCHP) concluded that their measured value of the Hubble constant was consistent with the ΛCDM model [37].

Indeed, unidentified systematics remain the key issue. For instance, a recent study on the determination of Hubble constant using time delays in gravitational lenses suggests that previous analyses were biased. The new central value obtained is consistent with the ΛCDM model, but the uncertainties remain too large to resolve the debate [38].

One reason for skepticism toward alternative theories is that they generally introduce additional parameters, which broaden the range of admissible values for the Hubble constant. However, when the SH0ES measurement is not included in the analysis, the central value typically remains close to that obtained from Planck alone. In a recent work, we have followed a statistical approach: we compared the merits of the Cepheid calibration bias hypothesis, which is applied to several probes, with those of alternative models. The statistical gain is comparable to that of the best models available [24]. The interpretation is that a bias in the SH0ES distance scale due to a miscalibration of Cepheids is as likely as the best alternative models. Recently, Ref. [39] pointed out that distance ladder measurements systematically differ from methods that do not rely on them, a conclusion in the same vein.

4.6 The matter density (ω_m) tension

The tension in a cosmological model with data can obviously be assessed using different observables or a combination of them. A common way to express the Hubble tension, for instance, is to compute the following quantity (the Gaussian tension):

$$\frac{\bar{H}_0 - H_0)_{SHOES}}{\sigma^2 + \sigma^2_{SHOES}} \tag{8}$$

This quantity expresses, in terms of standard deviation, the disagreement between the best estimate $\bar{H}_0$ obtained from a given dataset and the value measured by the SH0ES team $H_0)_{SHOES}$, with σ and σ_{SHOES} representing the standard deviations for the model and the SH0ES data, respectively. The latest SH0ES measurement [36] and the combined Planck + DESI data [13] result in a 4.56σ tension. (This can be converted into a probability, assuming the statistics follow a Gaussian distribution.)

However, under the assumptions of standard physics in the early universe, certain quantities can be derived from the CMB independently of the late-time evolution of the universe. Among these quantities is the density parameter ω_m [40]:

$$\omega_m = 0.14143 \pm 0.0013 \tag{9}$$

As we have the measurement of Ω_m in the present-day universe, we can compare values of ω_m for a given value of the Hubble constant [24]. Combining with DESI Ω_m, we have:

$$\omega_m = 0.17 \pm 0.005 \tag{10}$$

in a 4.56 σ tension with Eq. (9).

4.7 Nature of dark matter

The ΛCDM model has two major implications for fundamental physics: the existence of a dark matter component and the existence of a source of the acceleration of the expansion of the universe. While the nature of dark energy remains an open question for which only astronomical observations seem likely to provide a better understanding, the status of dark matter is somewhat different. Many experts in the field favor the idea that dark matter is a hitherto unknown massive particle, a WIMP, as originally proposed by J. Peebles [4]. This prospect has been and remains a strong driving force for particle physics research. However, it must be remembered that this is a possibility for which there is no evidence. The non-detection of dark matter particles over the last 50 years [41, 42] cannot be considered as a falsification of the model: the nature of dark matter is not an explicit ingredient of the model. It could therefore be a question of particles that escape all possibility of detection [43] or of primordial black holes [44] or generalized dark matter [45] or some other possibility yet to be discovered.

4.8 Dynamic dark energy?

There is now some evidence for the existence of dynamic dark energy [46], a result that has been confirmed by the analysis of DESI's BAO signal [47]. If confirmed, this would imply a falsification of the ΛCDM. Dynamic dark energy would be a significant discovery, but rather than fundamentally challenging the structure of the model—as the H_0 tension does—it would refine one of its least understood components, namely the source of the accelerated expansion.

5. Conclusions

The ΛCDM model has demonstrated remarkable success in explaining a wide range of cosmological observations, from the cosmic microwave background to large-scale structure formation. Its predictive power, minimal number of free parameters, and internal consistency make it the dominant framework in modern cosmology. However, challenges remain, particularly at small scales, where certain observational features appear difficult to reconcile with theoretical expectations, and in the context of global tensions such as the discrepancy in Hubble constant measurements.

While these issues have sparked debates on potential modifications to ΛCDM or alternative models, they do not yet constitute definitive falsifications in the Popperian sense. Many of the reported tensions could stem from methodological choices, observational systematics, or an incomplete understanding of baryonic physics, which remains a complex and evolving field. Before interpreting them as fundamental inconsistencies of the model, these aspects must be thoroughly examined to assess whether they reflect genuine discrepancies or limitations in current analyses. A deeper understanding of astrophysical processes and improvements in observational techniques will be crucial in clarifying the significance of these tensions.

Among the various challenges, the Hubble constant tension stands out as the most serious. The persistence of a discrepancy between early- and late-universe measurements has led to speculation about new physics beyond ΛCDM. However, nearly a century of revisions to Hubble constant measurements has shown that systematic uncertainties have often played a key role in shaping the inferred values. Given this history, resolving the current discrepancy definitively requires an independent, alternative measurement that is not directly tied to traditional distance ladder. Promising avenues, such as gravitational wave standard sirens, offer the potential for such a test [48], assuming that unforeseen systematics do not arise. A robust, unbiased determination of H_0 from these new methods would provide a decisive assessment of whether the tension signals a genuine failure of ΛCDM or remains within the realm of observational uncertainties.

Author details

Alain Blanchard
Université de Toulouse, UPS-OMP, IRAP, CNRS, Belin, Toulouse, France

*Address all correspondence to: alain.blanchard@irap.omp.eu

References

[1] Gershtein SS, Zel'dovich YB. Rest mass of muonic neutrino and cosmology. Journal of Experimental and Theoretical Physics Letters. 1966;**4**:174

[2] Lubimov VA, Novikov EG, Nozik VZ, Tretyakov EF, Kosik VS. An estimate of the ν_e mass from the β-spectrum of tritium in the valine molecule. Physics Letters B. 1980;**94**(2):266-268

[3] White SDM, Frenk CS, Davis M. Clustering in a neutrino-dominated universe. The Astrophysical Journal. 1983;**274**:L1-L5

[4] Peebles PJE. Large-scale background temperature and mass fluctuations due to scale-invariant primeval perturbations. The Astrophysical Journal. 1982;**263**:L1-L5

[5] Ferreira PG, Wolf WJ, Read J. The spectre of under determination in modern cosmology. arXiv e-prints, page arXiv:2501.06095. 2025

[6] Peebles PJE. Tests of cosmological models constrained by inflation. The Astrophysical Journal. 1984;**284**:439-444

[7] Ratra B, Peebles PJE. Cosmological consequences of a rolling homogeneous scalar field. Physical Review D. 1988;**37**(12):3406-3427

[8] Peebles PJE, Yu JT. Primeval adiabatic perturbation in an expanding universe. The Astrophysical Journal. 1970;**162**:815

[9] Maddox SJ, Efstathiou G, Sutherland WJ, Loveday J. Galaxy correlations on large scales. Monthly Notices of the Royal Astronomical Society. 1990;**242**:43

[10] Efstathiou G, Sutherland WJ, Maddox SJ. The cosmological constant and cold dark matter. Nature. 1990;**348**(6303):705-707

[11] Louis T, La Posta A, Atkins Z, Jense HT, Abril-Cabezas I, Addison GE, et al. The Atacama cosmology telescope: DR6 power spectra, likelihoods and ΛCDM parameters. arXiv e-prints, page arXiv:2503.14452. 2025

[12] Calabrese E, Hill JC, Jense HT, La Posta A, Abril-Cabezas I, Addison GE, et al. The Atacama cosmology telescope: DR6 constraints on extended cosmological models. arXiv e-prints, page arXiv:2503.14454. 2025

[13] DESI Collaboration, Adame AG, Aguilar J, Ahlen S, Alam S, Alexander DM, et al. DESI 2024 VII: Cosmological constraints from the full-shape modeling of clustering measurements. arXiv e-prints, page arXiv:2411.12022. 2024

[14] Ijjas A, Steinhardt PJ, Loeb A. Inflationary schism. Physics Letters B. 2014;**736**:142-146

[15] Kroupa P, Gjergo E, Asencio E, Haslbauer M, Pflamm-Altenburg J, et al. The many tensions with dark-matter based models and implications on the nature of the universe. In: School and Workshops on Elementary Particle Physics and Gravity. 2023. p. 231. Available from: https://data.desi.lbl.gov/doc/papers/

[16] de Lapparent V, Geller MJ, Huchra JP. A slice of the universe. The Astrophysical Journal. 1986;**302**:L1

[17] Lopez AM, Clowes RG, Williger GM. Investigating ultra-large large-scale structures: Potential implications for cosmology. arXiv e-prints, page arXiv:2409.14894. 2024

[18] Sawala T, Teeriaho M, Frenk CS, Helly J, Jenkins A, Racz G, et al. The Emperor's New Arc: Gigaparsec patterns abound in a ΛCDM universe. arXiv e-prints, page arXiv:2502.03515. 2025

[19] Aluri PK, Cea P, Chingangbam P, Chu M-C, Clowes RG, Hutsemékers D, et al. Is the observable universe consistent with the cosmological principle? Classical and Quantum Gravity. 2023;**40**(9):094001

[20] Loureiro A, Whittaker L, Spurio Mancini A, Joachimi B, Cuceu A, Asgari M, et al. KiDS and Euclid: Cosmological implications of a pseudo angular power spectrum analysis of KiDS-1000 cosmic shear tomography. Astronomy & Astrophysics (A&A). 2022;**665**:A56

[21] Heymans C, Tröster T, Asgari M, Blake C, Hildebrandt H, Joachimi B, et al. KiDS-1000 cosmology: Multi-probe weak gravitational lensing and spectroscopic galaxy clustering constraints. Astronomy & Astrophysics (A&A). 2021;**646**:A140

[22] Tristram M, Banday AJ, Douspis M, Garrido X, Górski KM, Henrot-Versillé S, et al. Cosmological parameters derived from the final Planck data release (PR4). Astronomy & Astrophysics (A&A). 2024;**682**:A37

[23] Dark Energy Survey and Kilo-Degree Survey Collaboration, Abbott TMC, Aguena M, Alarcon A, Alves O, Amon A, et al. DES Y3 + KiDS-1000: Consistent cosmology combining cosmic shear surveys. The Open Journal of Astrophysics. 2023;**6**:36

[24] Blanchard A, Héloret J-Y, Ilić S, Lamine B, Tutusaus I. ΛCDM is alive and well. The Open Journal of Astrophysics. 2024;7:32

[25] Perivolaropoulos L, Skara F. Challenges for ΛCDM: An update. 2022;**95**:101659

[26] Ibata RA, Lewis GF, Conn AR, Irwin MJ, McConnachie AW, Chapman SC, et al. A vast, thin plane of corotating dwarf galaxies orbiting the Andromeda galaxy. Nature. 2013;**493**(7430):62-65

[27] Sawala T, Cautun M, Frenk C, Helly J, Jasche J, Jenkins A, et al. The Milky Way's plane of satellites is consistent with ΛCDM. Nature Astronomy. 2023;**7**:481-491

[28] Uzeirbegovic E, Martin G, Kaviraj S, Jackson RA, Kraljic K, Dubois Y, et al. New tools for studying planarity in galaxy satellite systems: Milky way satellite planes are consistent with ΛCDM. 2024;**535**(4):3775-3784

[29] Rhee G, Valenzuela O, Klypin A, Holtzman J, Moorthy B. The rotation curves of Dwarf galaxies: A problem for cold dark matter? The Astrophysical Journal. 2004;**617**(2):1059-1076

[30] Blanchard A, Valls-Gabaud D, Mamon GA. The origin of the galaxy luminosity function and the thermal evolution of the intergalactic medium. Astronomy & Astrophysics (A&A). 1992;**264**:365-378

[31] Aricò G, Angulo RE, Zennaro M, Contreras S, Chen A, Hernández-Monteagudo C. DES Y3 cosmic shear down to small scales: Constraints on cosmology and baryons. Astronomy & Astrophysics (A&A). 2023;**678**:A109

[32] Freedman WL. Correction: Cosmology at a crossroads. Nature Astronomy. 2017;**1**:0169

[33] Verde L, Treu T, Riess AG. Tensions between the early and late universe. Nature Astronomy. 2019;**3**:891-895

[34] Di Valentino E, Mena O, Pan S, Visinelli L, Yang W, Melchiorri A, et al. In the realm of the Hubble tension-A review of solutions. Classical and Quantum Gravity. 2021;**38**(15):153001

[35] Schöneberg N, Abellán GF, Sánchez AP, Witte SJ, Poulin V, Lesgourgues J. The H_0 Olympics: A fair ranking of proposed models. Physics Reports. 2022;**984**:1-55

[36] Riess AG, Yuan W, Macri LM, Scolnic D, Brout D, Casertano S, et al. A comprehensive measurement of the local value of the Hubble constant with 1 km s^{-1} Mpc^{-1} uncertainty from the Hubble space telescope and the SH0ES Team. The Astrophysical Journal. 2022;**934**(1):L7

[37] Freedman WL, Madore BF, Jang IS, Hoyt TJ, Lee AJ, Owens KA. Status report on the Chicago-Carnegie Hubble program (CCHP): Three independent astrophysical determinations of the Hubble constant using the James Webb space telescope. arXiv e-prints, page arXiv:2408.06153. 2024

[38] Li T, Collett TE, Marshall PJ, Erickson S, Enzi W, Oldham L et al. Correcting for selection biases in the determination of the Hubble constant from time-delay cosmography. arXiv e-prints, page arXiv:2410.16171. 2024

[39] Perivolaropoulos L. Hubble tension or distance ladder crisis? Physical Review D. 2024;**110**(12):123518

[40] Lemos P, Lewis A. CMB constraints on the early universe independent of late-time cosmology. Physical Review D. 2023;**107**(10):103505

[41] Misiaszek M, Rossi N. Direct detection of dark matter: A critical review. Symmetry. 2024;**16**(2):201

[42] Aalbers J, Akerib DS, Al Musalhi AK, Alder F, Amarasinghe CS, Ames A, et al. New constraints on ultraheavy dark matter from the LZ experiment. Physical Review D. 2024;**109**(11):112010

[43] Fairbairn M, Kainulainen K, Markkanen T, Nurmi S. Despicable dark relics: Generated by gravity with unconstrained masses. Journal of Cosmology and Astroparticle Physics. 2019;**2019**(4):005

[44] Carr B, Kuhnel F. Primordial black holes. arXiv e-prints, page arXiv:2502.15279. 2025

[45] Wayne H. Structure formation with generalized dark matter. The Astrophysical Journal. 1998;**506**(2):485-494

[46] Rubin D, Aldering G, Betoule M, Fruchter A, Huang X, Kim AG, et al. Union through UNITY: Cosmology with 2,000 SNe using a unified Bayesian framework. arXiv e-prints, page arXiv:2311.12098. 2023

[47] Adame AG, Aguilar J, Ahlen S, Alam S, Alexander DM, Alvarez M, et al. DESI 2024 VI: Cosmological constraints from the measurements of baryon acoustic oscillations. Journal of Cosmology and Astroparticle Physics. 2025;(2):021

[48] Muttoni N, Laghi D, Tamanini N, Marsat S, Izquierdo-Villalba D. Dark siren cosmology with binary black holes in the era of third-generation gravitational wave detectors. Physical Review D. 2023;**108**(4):043543

Chapter 7

Predicting the Future Using Pantheon+TRGB Data: Better Analytical Techniques with Friedmann-Lemaître-Robertson-Walker Models

Michael L. Smith, Abdullah Alsakka and Ahmet M. Öztas

Abstract

The Standard Model (ΛCDM) of cosmology neither estimates the current Milky Way location nor predicts any future position, failing the scientific test of accurate prediction. We analysed data from the 1701 SNe Ia Pantheon+ collection combined with the Tip of the Red Giant Branch (TRGB) emissions to estimate the Hubble constant. We wrote Python codes and saved our results, the SNe and TRGB data as files in GitHub, github.com/Abdullah-Alsakka/About-Transforming-Data. We present the shortcomings of correlations employing the distance modulus (μ) often used for analyses supporting ΛCDM models. The associated standard deviations are also fatally flawed; incorrectly estimating Ωm and ΩΛ using the ΛCDM correlation; we better estimated parameters using DL and σ values. We investigated the Rh=ct model, the Einstein-deSitter model and our ArcTanh model comparing results against the ΛCDM model. There was little difference in the goodness-of-fit between the Rh=ct and our ArcTanh model; both correlated the data better than the ΛCDM model. The ArcTanh model presented a H0≥ 70 km s−1Mpc−1 agreeing with the Riess group. The ArcTanh model presented the smaller values for Ωm, than the ΛCDM model agreeing with BBN calculations. The ArcTanh model also supports the quasi-Euclidean not strictly Euclidean spacetime geometry of the ΛCDM model. Our results and the ability to predict the future location of our Milky Way are evidence that our analytical technique is more valuable than many current methods.

Keywords: cosmological parameters, Hubble constant, spacetime, FLRW, TRGB, SNe Ia, Standard Model, dark energy, dark matter, Big Bang

1. Introduction

Determination of an accurate Hubble constant, H_0, is a primary goal of astronomical investigations. This is the current value for the expansion velocity of the Universe. An important question which H_0 measures—is the expansion velocity increasing, constant or decreasing? The value of H_0, now wandering about 70–74 km s^{-1}Mpc^{-1}, is also important because it allows estimates of the Universe age. The value is calculated ≈72–74 when the supernovae type Ia (SNe Ia) data [1] and the James Webb Space Telescope (JWST) date are used [2], but ≈70 using the tip of the giant red branch (TGRB) data [3]. Planck measurements of the cosmic microwave background (CMB) anisotropies [4] and an independent analysis of recent data from the JWST [5] yield a $H_0 < 70$. The difference between the two schools is ≥2.5 standard deviations, which is taken to mean the chances of these values being the same are < 85%. This situation is often termed the *Hubble tension* and is considered a crisis by many cosmologists [6, 7]. We shall outline a better procedure for estimating H_0 than currently used by most cosmologists using the supernova type Ia (SNe Ia) and tip of the red giant branch (TRGB) data.

The SNe Ia distances are typically calibrated with nearby Cepheid variable stars establishing a distance ladder [8]. The accuracy of the Cepheid observations has recently been questioned [9] but data from the JWST support Cepheid use [10]. Observations are typically made using several different light filters and many different telescopes, which are the probable sources of systematic errors. The light curves (luminosity decline over time) cannot be translated directly into linear distances but must be calibrated through laborious processes. After adjustment for "nuisance" parameters, which are the initial emission amplitude and decay constant, the emission distances are estimated but significant systematic and observational errors remain [11, 12]. These two highly variable parameter values, along with their standard deviations, are ignored during modelling process and hence are considered nuisance parameters.

The quantity and quality of the SNe Ia data have increased over time, from 50 events in 1998 with an uncertainty of H_0 of 7% [13] to over 1700 in 2023, with an estimated error in H_0 of 3.9% [1]. These data are considered the best evidence for an accelerating Hubble flow and measurable Λ, the latter being a major component of the current Standard Model sometimes labelled ΛCDM (Λ Cold Dark Matter). It has been suggested that signals from the Tip of the Red Giant Branch (TRGB) can also be used as independent standard candles to calibrate the SNe Ia data [14].

Some calculations used to estimate H_0 and Ω_m demand too many free parameters to deliver accurate results. For instance, analysis using CMB data requires six or more free parameters [4]. The difficult process of finding a solution with so many parameters is typically aided by fixing several as constants (priors) before regression. To this end, CMB analysis often presumes Euclidean geometry, acceptable for first-order approximation but counter-intuitive to gravitational physics and unjustifiable [15]. Investigations using the CMB data have been justly criticised due to the low signal-to-noise ratio and only considering the Standard Model (ΛCDM) but not other models [16]. The result from all this is parameters with very large estimated errors, and one might ask what worthwhile result has been gained from all the labour.

There are several Standard Model variants currently under investigation, one including an equation of state. This was first presented as a term modifying the Standard Model as Eq. (1) in Garnavich et al. [17]. This relationship considers all varieties of matter and Λ, accounting for the decreasing matter densities with increasing Universe volume. It has been used with SNe Ia data as support for the Standard Model [18]. The second version types are the relationships used to model the CMB data. These employ from 6 to 10 free parameters to consider our Universe consisting of baryonic, dark matter and the cosmological constant [4]. A third variation models the dependence of the cosmological constant as a function of time in an expanding Universe [19]. These all require evaluation of at least four parameters, but we restrict our investigations to models requiring three parameters or fewer, thus avoiding the use of fixed parameters (priors).

A large contribution from cold dark matter (CDM) is often presumed, but failures over past decades to detect DM on earth are troubling; sensitive observations at indirect detection have **all** failed. The search for this supposed large fraction of our Universe continues [20].

The geometry of space is also currently considered a *crisis*; is it flat and unending, or is the Universe quasi-Euclidean, as suggested by Einstein [15]? The interpretation of the relationship between Ω_k and space geometry has changed significantly since the reintroduction of the cosmological constant three decades ago [21]. Some think the evidence from analysis of SNe Ia events points to Euclidean geometry when Ω_k is ≈ 0 [22], some when $\Omega_k = 1$ [23]. Results from some SNe Ia modelling point to an $\Omega_k \approx 1$, the current spacetime geometry being very slightly elliptical; Einstein employed the term quasi-Euclidean, so do we. We previously reported our investigations following several models of evolving dark energy which present troubling and non-physical results that cannot be easily explained [19]. A significant non-zero value for Ω_k and the additional failure to detect several possible forms of dark energy in a laboratory experiment [24] question the Standard Model being the best; descriptions without Λ and CDM should be considered. Then, there is the cosmological constant problem referring to the difference of many orders of magnitude between the observed vacuum energy density and the value calculated from quantum field theory [23]. The field of cosmology and particle physics has been recently reviewed [25].

Here, we analysed the Pantheon+ SNe Ia data combined with the two TRGB collections using both the *distance mag* (μ) *vs. z* and the D_L *vs. z* correlations [26]. We do not use priors to constrain free parameters, so do not consider more than three parameters. Priors meaning a fixed value; this invariably introduces evaluation bias, but that fact is ignored by those evaluating the CMB data [4]. We find the calculated parameters and σs from the *distance mag vs. z* correlations do not often match those calculated employing the D_L *vs. z* correlation. Using the values of reduced χ^2 and the F statistic to judge model worthiness, our results do not support the Standard Model any better than some other models not requiring Λ and CDM, such as the $R_h = ct$ model and our ArcTanh model. Our analyses also support the notion that often an $\Omega_k \approx 1$, rather than ≈ 0, is a good fit, describing a universe with quasi-Euclidean spacetime geometry [15, 27]. Some results with the ArcTanh model suggest the baryonic matter density is a small fraction of the critical matter density, consistent with Big Bang calculations [28]. We have not investigated more complicated, detailed models, such as those requiring the dark energy equation of state, slowly evolving dark energy or modified gravity. We have previously reported results from some more exotic models with smaller data ensembles [29]. Those models often require more than three free parameters thus reducing the goodness of fit, increasing parameter σs, and making model preference a difficult choice.

We estimate the relative current and future positions of the Milky Way, for the first time, using the D_L *vs. z* relationship; this being impossible using the *distance mag vs. z* correlation. We also introduce the F statistic in which we found a good measure of model worthiness and find the Bayes Information Criteria (BIC) of limited usefulness. We list several cosmological models, we explain several statistical tests, and present tabulated results and figures in our GitHub repository, https://github.

com/Abdullah-Alsakka/About-Transforming-Data. Our results conform with a Hubble constant of 72–73 km $s^{-1}Mpc^{-1}$ but do not support the existence of dark matter nor dark energy.

2. Data, models and analyses

We use the SNe Ia 1701, Pantheon+ data from Brout et al. [26], the 16 signals of the Tip of the Red Giant Branch (TRGB) data from Anand et al. [30] and the 18 from Freedman et al. [31]. We use all listed data even though some TRGB are listed twice. The distance values from Anand are matched to galactic redshifts (z) listed in the NASA/IPAC Extragactic Database located at `https://science.nasa.gov/astrophysics/data/nasa-ipac-extra-galactic-database-ned/`. We use the Python 3.9 family of programming tools with both the Jupyter Notebook 6.5 and Spyder 4 and 5 environments, where all codes are publicly available. To make certain our results can be duplicated, we performed analyses using older model Apple® Macintosh and PC (Windows® 11) portable computers. Data and Python 3 programs are available at our GitHub repository, `MLSResearchGroup`, *via* `github.com/Abdullah-Alsakka/About-Transforming-Data`. The D_L and σ values calculated from the *distance mag* data are stored as comma separated files (`.csv`) in the folders `Data`.

2.1 Data handling

The first plot of cosmology, presented by Hubble as evidence for our expanding universe, was of galactic recession velocity as a function of distance, inversely related to our D_L *vs.* z correlations [32]. We invert the axes to plot D_L *vs.* z because the z measurements are more accurate than distances. This is typical of physics and engineering, where velocity is usually determined with better accuracy and precision than distance [33]. Nowadays, correlations between *distance mag* and z and are preferred by most astronomers with *distance mag* being the dependent variable even though *distance mag* does not have linear correspondence to D_L. We use the Python 3 regression routine `curve_fit` for modelling, incorporating the σ values of the dependent variable for the regression. The data quality is incredibly dependent on the observatory and filter selected; signals and baselines from distant SNe Ia are almost always noisy.

2.2 Data preparation

Estimating luminosity distances from SNe events is difficult. Intervening dust scatters the signal; contaminating light from other sources and gravitational lensing from intervening masses both add noise. Even matching the SN with the proper galaxy is often difficult.

The *distance mag*, μ, is not the luminosity distance but is related to D_L by

$$\mu = 5\log_{10}(D_L) + 25 \text{ or } D_L = 10^{[(\mu - 25)/5]} \quad (1)$$

and because μ is the result of the $\log_{10}$ function, it has no metric, admitted as Eq. 47 in the Carroll monogram [21]. Because of this transformation, plots using *distance mag* values lack a true origin and are not simply proportional estimates of the distance. The use of *distance mag* was institutionalised a century ago to conform with eyeball estimates of stellar luminosity being approximately logarithmic rather than linear and was standardised in 2015. Unfortunately, the use of this convenience persists despite being totally unnecessary in our computerised age. There is really no valid reason to not use the real distance values.

To calculate the σ in terms of distance, we add and subtract the σ from the *distance mag* values and then calculate the resulting D_L values

$$D_{L-hi} = 10^{((\mu+\sigma)-25)/5}, D_{L-low} = 10^{((\mu-\sigma)-25)/5}. \tag{2}$$

The σ of D_L is the average of the difference of the above as

$$\sigma_{D_L} = (D_{L-hi} - D_{L-low})/2. \tag{3}$$

For the Hubble correlations, we use the measured galactic redshifts, z, to calculate the recession velocities as cz. Since we know the position and z of the earth exactly (0,0), we use 1702 SNe Ia data pairs and either 16 or 18 TRGB data for evaluations with D_L, σ *vs.* z. All data available at our GitHub location.

2.3 Regression method

We correlated the D_L *vs.* z and *distance mag vs.* z using the Python 3 regression, `scipy` routine `curve_fit`. The `curve_fit` routine allows two analytical methods employing the observed σ during minimisation. If the user selects `absolute_sigma = False` from the argument list, the σ values are normalised for weighing the dependent data, that is, data with larger σ do not contribute as much to the regression as those with smaller σ. If `absolute_sigma = True` is selected, the σ values are used as the regression weights. The results are similar irrespective of the choice and we present only those where the normalised values of σ are used (`absolute_sigma = False`). We allowed the parameter, H_0, to float freely between 60 and 80, which are the lower and upper values reported the last 50 years. We allowed Ω_m, Ω_Λ or Ω_k to float between 0 and 1; values outside this range cannot describe our Universe (an $\Omega_m < 0$ describes anti-gravity matter). The results are reported in **Tables 1** and 2.

The regression method is much too lengthy to detail here. The reader is referred to a book [34] or a video overview of how to use this method with Python [35]. Evaluation of each function demands individual routines, all differing slightly but are available on our GitHub site `github.com/Abdullah-Alsakka/About-Transforming-Data`.

2.4 Statistics

We used common statistical tests: first, the r^2 correlation; second, the weighted F statistic; and third, the reduced χ^2 test—a speciality used by astronomers to compare goodness of fit. This is different from other χ^2 tests and does not present the typical p value. Each test is described, including the math and variable definitions, in the GitHub site. The larger the F statistic value, the better the model describes the data. The lower the value for the reduced χ^2 test, the better the model. We also use the Bayesian information criteria (BIC), only sometimes being useful across different model types. Details are presented in various routines stored on `github.com/Abdullah-Alsakka/About-Transforming-Data`.

Model	H_0	Ω_m	Ω_Λ	Ω_k	r^2	F(10^5)	χ^2_{red}	BIC
D_LStandard‡	75.8±0.3	0.23±0.02	0.77±0.02	—	0.9895	26.5	1.60	14.77
Standard‡	73.0±0.2	0.35±0.01	0.65±0.01	—	0.9974	3.51	0.44	14.37
Hubble§	66.5±1.4	—	—	—	0.891	0.39	14.02	2.52
Hubble†	85.1±6.6	—	—	—	0.4274	0.62	91.69	2.48
D_LE-DS	68.2±0.3	—	—	≈1	0.9038	8.34	2.89	7.35
D_LENDS	75.8±0.3	0.23±0.02	0.77±0.02	0	0.9895	26.8	1.58	14.78
$D_L\Lambda$E-DS	75.6±0.3	≈0	0.49±0.04	0.51±0.04	0.9709	17.9	1.61	14.79
*mag*E-DS	65.8±0.2	—	—	≈1	0.993	2.26	2.67	7.04
*mag*ENDS	73.2±0.3	0.34±0.02	0.66±0.02	0	0.997	2.40	1.67	14.41
magΛE-DS	72.7±0.3	≈0	0.24±0.05	0.76±0.05	0.9969	2.35	1.68	14.41

‡*model with only Pantheon+ data from [3].*
§*with all 18 TRGB signals from [27].*
†*with all 16 TRGB signals of [26].*

Table 1.
Regressions results. Einstein-de Sitter models tested with Pantheon+Anand data. H_o as km $s^{-1}Mpc^{-1}$.

Model	H_0	Ω_m	Ω_Λ	Ω_k	r^2	F(10^5)	χ^2_{red}	BIC
magR_h	70.9±0.2	—	—	—	0.9968	4.44	1.75	6.97
Standard§	73.2±0.3	0.34±0.02	0.66±0.02	—	0.9970	2.40	1.67	14.41
*mag*3PStnd§	73.2±0.4	0.28±0.23	0.61±0.19	0.11±0.30	0.9970	1.60	1.67	21.86
*mag*2PAtanh*	71.6±0.2	1e^{-5}±0.01	—	> 0.99±0.01	0.9969	2.31	1.70	14.41
D_LR_h	72.6±0.2	—	—	—	0.9840	35.6	1.79	7.34
D_LStandard§	75.8±0.3	0.23±0.019	0.77±0.02	—	0.9895	26.8	1.58	14.78
D_L3PStnd§	76.1±0.4	0.52±0.22	1.00±0.18	−0.52±0.28	0.9908	18.6	1.57	22.23
D_L2PAtanh*	73.1±0.2	1e^{-5}±0.02	—	> 0.99±0.02	0.9846	21.4	1.72	14.79

**sinn evaluated as sinh(x).*
§*sinn evaluated simply as x.*

Table 2.
Results from regressions with some currently important models using 1701 (or 1702) SNe Ia data from 25 and 16 TRGB from [30]. H_o as km $s^{-1}Mpc^{-1}$.

2.5 Models

2.5.1 Hubble

Hubble first reported the recession velocities of most galaxies are correlated with their distances as

$$cz = H_0 D_L. \tag{4}$$

This represents a fundamental relationship in cosmology and provides a fairly accurate model of nearby emissions [31]. Note, the Hubble equation is uninformative about matter density, Λ, the extent of space or spacetime geometry (**Table 1**).

2.5.2 Einstein-de Sitter models

The E-DS model describes a universe modelled at the lower extreme of matter density, DE and spacetime. It can be expressed as three relationships, depending on the assumed values: Ω_m, normalised matter density; Ω_Λ, normalised cosmological constant; Ω_k, influence of empty space. Cosmological models are most easily solved when these unit-less parameters are normalised to 1 as

$$1 = \Omega_m + \Omega_r + \Omega_\Lambda + \Omega_k. \tag{5}$$

The Ω_r term, for the influence of radiation, is rightly judged too small to be estimated with data and dropped from further consideration.

With $\Omega_m, \Omega_\Lambda \approx 0$. For evaluation with the D_L *vs.* z correlation, we use

$$D_L = \frac{c(1+z)}{H_0} \, sin\,\mathrm{n}\left(\frac{z}{1+z}\right). \tag{6}$$

where c is the lightspeed and H_0 is the local Hubble constant. We allow $sin\mathrm{n}(\mathrm{x})$ to become $sinh(\mathrm{x})$ when $\Omega_k \approx 1$, presuming quasi-Euclidean geometry, and $sin(\mathrm{x})$ when $\Omega_k < 0$ and x when $\Omega_k = 0$, and this is our D_LE-DS model. For modelling the *distance mag vs.* z data with the E-DS model, Eq. (6) becomes

$$distance\ mag, (\mu) = 5\,log_{10}\left[\frac{c(1+z)}{H_0} \, sinh\left(\frac{z}{1+z}\right)\right] + 25 \tag{7}$$

which is our *mag* E-DS model.

With $\Omega_\Lambda > 0$, $\Omega_m \approx 0$. Presuming a real Λ, we derive the expression

$$D_L = \frac{c}{H_0}\frac{(1+z)}{\sqrt{\Omega_k}} \sinh\left\{ \operatorname{arctanh}\left(\frac{\sqrt{|\Omega_k|}}{\sqrt{\frac{\Omega_\Lambda}{(1+z)^2} + \Omega_k}}\right) - \operatorname{arctanh}\left(\sqrt{|\Omega_k|}\right)\right\}. \tag{8}$$

A derivation is presented in our GitHub folder. We designate the $D_L\Lambda$E-DS model and the related *mag*ΛE-DS model for use with *distance mag* and z data.

With $\Omega_k = 0$, $\Omega_m > 0$. One variant recently proposed by Yershov [36] as a good model for SNe Ia data is

$$D_L = \frac{c(1+z)}{H_0}\int_0^{z'} \frac{dz'}{\sqrt{1+\Omega_m\left[(1+z')^3 - 1\right]}} \tag{9}$$

with a brief derivation in our GitHub folder. We term this the D_L-ENDS model using D_L data and the *mag*ENDS model with *distance mag* and z data.

2.5.3 The $R_h = ct$ model

The $R_h = ct$ model, or simply the R_h, was first introduced by Melia [37] in 2007 and presented later in more detail [38]. This is a version of the FLRW model complying with Birkhoff's theorem and the Weyl postulate requiring Euclidean geometry. These constraints mean the R_h universe is incompatible with Λ, so we cannot estimate that contribution, neither Ω_m nor Ω_k.

For use with the D_L *vs.* z data, we employ the relationship

$$D_L = \frac{c(1+z)}{H_0} ln\,(1+z) \tag{10}$$

where *ln* is the natural logarithm and we designate this as the $D_L R_h$ model. For the *distance mag vs. z* data, Eq. (10) is transformed into

$$distance\ mag, (\mu) = 5 log_{10}\left[\frac{c(1+z)}{H_0} ln\,(1+z)\right] + 25, \tag{11}$$

and we designate this the $magR_h$ (model).

2.5.4 A general FLRW model

The FLRW model includes parameters for matter density, Ω_m, the cosmological constant, Ω_Λ and empty space, Ω_k. We present several derivations with explanations in our GitHub folder. A FLRW solution for use with D_L and z, where $\xi = 1/(1+z)$ is

$$D_L = \frac{c}{H_0 \xi \sqrt{|\Omega_k|}} \left[sin\,\mathrm{n}\left(\sqrt{|\Omega_k|} \int_{\xi_1}^{1} \frac{d\xi_1}{\xi_1 \sqrt{\frac{\Omega_m}{\xi_1} + \Omega_\Lambda \xi_1^2 + \Omega_k}} \right) \right]. \tag{12}$$

here $sin\mathrm{n}(\mathrm{x})$ is evaluated as in 2.5.2. We note the selection of the $sin\mathrm{n}(\mathrm{x})$ function, as presented in [13, 21], is confusing around $\Omega_k \approx 0$, so we follow the convention suggested in the work by Garnavich et al. [22]. Variations of this model can be solved using two free parameters (2P), and versions employing three free parameters are labelled with 3P.

2.5.5 Standard models of cosmology

One version of the Standard Model includes Ω_Λ with $\Omega_k = 0$, where normalisation simplifies with

$$1 = \Omega_m + \Omega_\Lambda. \tag{13}$$

Another condition presumes $sin\mathrm{n}(\mathrm{x}) = \mathrm{x}$ and Eq. (12) is reduced to

$$D_L = \frac{c}{H_0 \xi} \int_{\xi_1}^{1} \frac{d\xi_1}{\xi_1 \sqrt{\frac{\Omega_m}{\xi_1} + \Omega_\Lambda \xi_1^2}}. \tag{14}$$

Allowing $\xi = 1/(1+z)$, we derive this from Riess et al. 1998 [13].

$$D_L = \frac{c(1+z)}{H_0}\int_0^{z'} \frac{dz'}{\sqrt{(1+z')^2(1+\Omega_m z') - z'(2+z')\Omega_\Lambda}}. \tag{15}$$

Models referencing the Ω_Λ term cannot be integrated analytically, so we coded for numerical integration. By substituting $1-\Omega_m$ for Ω_Λ, the above relationships are reduced to evaluation with only two free parameters, H_0 and Ω_m, and we designate Eq. (15) as the D_L Standard Model.

For use with the *distance mag vs. z* data, we must transform the right-hand sides of Eq.(15) by applying

$$distance\,mag, (\mu) = 5\log_{10}(Eq.15) + 25. \tag{16}$$

Eq. (16), Standard Model, is the version of Eq. (12), which is used to first detect accelerating universe expansion [13]. Eq. (16) is a transformed relationship and not Eq. (15) directly. Note that $z = 0$, the position of the observer, cannot be employed since Eq. (16) cannot be evaluated at 0. The origin values are discarded without explanation by many cosmologists.

Evaluating the three parameter versions of the Standard Model is complicated by the choice of the *sin*n function. We tested all three variations, that is when *sin*n = *sin*h (x) or *sin*(x) or simply (x). For completeness, we use regressions with three parameters to include Ω_k. The version we use is

$$D_L = \frac{c(1+z)}{H_0\sqrt{|\Omega_k|}}\, sin\,\mathrm{n}(A) \tag{17}$$

where A is

$$A = \sqrt{|\Omega_k|}\left[\int_0^{z'} \frac{dz'}{\left((1+z')^3\Omega_m + (1+z')^2\Omega_k + \Omega_\Lambda\right)^{1/2}}\right]. \tag{18}$$

We designate each model specifying the number of free parameters, the version of *sin*n selected and the data format, for instance, sinhD_L3PStnd designates the Standard Model with three free parameters, *sinh(A)* and correlates D_L *versus* z.

2.5.6 ArcTanh model

Eq. (12) reduces to Eq. (19) in a straightforward manner by allowing $\Omega_\Lambda = 0$ for a universe with elliptical or quasi-Euclidean geometry with $\Omega_k > 0$ as

$$D_L = \frac{c}{H_0\xi\sqrt{|\Omega_k|}}\,\sin\mathrm{n}\left(\sqrt{|\Omega_k|}\int_{\xi_1}^{1} \frac{d\xi_1}{\xi_1\sqrt{\frac{\Omega_m}{\xi_1} + \Omega_k}}\right). \tag{19}$$

This is similar to Eq. (14) except for the replacement of the $\Omega_\Lambda\xi^2$ term by Ω_k. Eq. (19) can be integrated to a solution successfully modelling the SNe Ia data with a significant Ω_m and Ω_k terms as

$$D_L = \frac{c(1+z)}{H_0\sqrt{|\Omega_k|}} \sin\mathbf{n}\left[2\left(\text{arctanh}\left(\sqrt{|\Omega_k|}\right) - \text{arctanh}\left(\frac{\sqrt{|\Omega_k|}}{\sqrt{\Omega_m(1+z)+\Omega_k}}\right)\right)\right]. \quad (20)$$

This is the ArcTanh model and the version used with the *distance mag vs. z* correlation is the *mag*Arctanh model as

$$\textit{distance mag}, (\mu) = 5\log_{10}(Eq.20) + 25. \quad (21)$$

where two free parameters are normalised as

$$1 = \Omega_m + \Omega_k. \quad (22)$$

This allows $1 - \Omega_m$ to substitute for Ω_k, reducing the parameter count and simplifying solution. We test four versions of this model and our labelling; for example, *mag*3PArctanh designates three free parameters, H_0, Ω_m and Ω_k, employing the *distance mag vs. z* data. Since space is curved in the presence of matter, sin**n** in Eq. (20) is set to sinh.

3. Results

3.1 Hubble correlation

The Hubble correlation of TRGB emissions is shown as **Figure 1a** and **b** with the H_0 values and statistics presented in **Table 1**. The data of the more distance TRGB emissions in both plots are of unusual quality since the standard derivations are considerably smaller than their variance from the regression lines, which is possibly due to underestimating noise. The data of Anand and coworkers display much larger dispersions than those of the Freedman group, reflected by the smaller r^2 and F statistic. The H_0 calculated with the Anand data is a high 85.1±6.6 compared with the H_0 of 66.5 ±1.4 $km^{-1}s^{-1}Mpc$ with the Freedman data, which is closer to results with the Planck data and our Universe being older than 15 Gyrs.

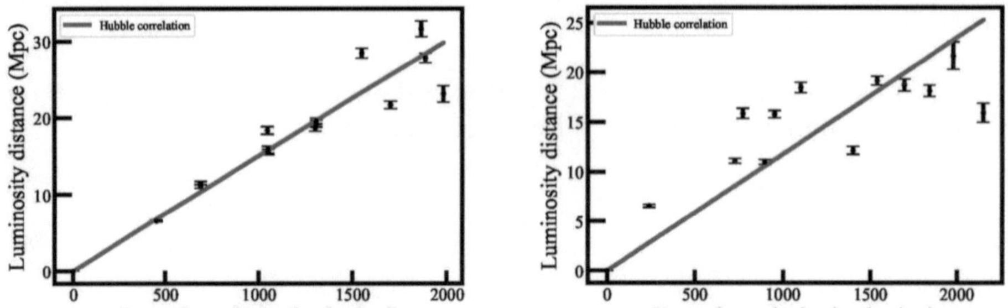

(a) Hubble correlation with TRGB data

(b) Hubble correlation with TRGB data

Figure 1.
Hubble (inverse) correlations with TRGB data from Freedman et al. [31] (left) and Anand et al. [30] (right).

3.2 Einstein-de Sitter models

The results from three variations of the E-DS model, including the ENDS model, Eq. (9), recently described by Yershov [35], are presented in **Table 1** and **Figure 2a** and **b**. Results from modelling the Pantheon+Freedman data are presented in the GitHub folder. According to the r^2, F statistic and χ^2_{red}, the D_LENDS and *mag*ENDS models are the best among the Einstein-de Sitter variants and **Figure 2b** displays a curve of the E-DS model which does not conform well with the data at distances. The σ values of distant objects displayed by the pseudo-Hubble plots presented in our GitHub folder are small and questionable because σ values of *distance mag* data are compressed by the log transformation. This also means that proportionally greater weight is given to distant signals when the *distance mag* rather than D_L values are modelled, inflating r^2 and F statistics and deflating χ^2.

The H_0 results from regressions of the *distance mag vs. z* and D_L *vs. z* correlations of the ENDS models are instructive. While the *mag*ENDS model presents a value for Ω_m of 0.34, close to some recent results for the Standard Model (**Table 2**), it is very different from the value of 0.23 calculated for the related D_LENDS model. This should not be the case if data transformation is innocuous, that is, if the resulting parameters and associated standard deviations are identical, which is definitely not the case here.

The H_0 is <70 km s^{-1}Mpc^{-1} for both E-DS variants though not unbelievable, being in the range estimated from Planck data [4] and earlier published values by the Riess group [13]. The small BIC values for both the *distance mag vs. z* and D_L *vs. z* models suggest the E-DS variant is a good fit (**Table 1**) but this is not born out in the plots and the smaller values of r^2 and larger χ^2_{red}. This is simply an example of the BIC strongly reflecting the free parameter number rather than the overall goodness of fit. Results using the Pantheon+18 TRGB emissions of the Freedman group are presented in our GitHub folder. The calculated parameters are dissimilar to those of **Table 1** though the total data number is only slightly larger. It seems the data from the Freedman group provide a better "anchor" for regression analyses as evidenced by the two inverse-Hubble plots.

3.3 R_h model

The results from the regressions of the R_h model are presented in **Table 2** and **Figure 3a** and **b**. The regression curves follow the data closely with H_0 between 71 and

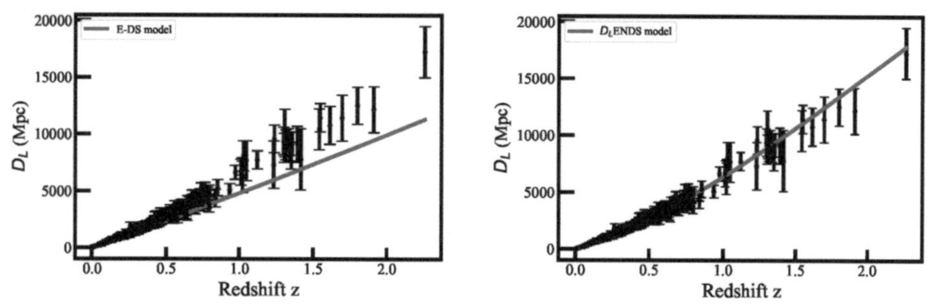

(a) The D_LE-DS model (b) The D_LENDS model

Figure 2.
Plots with Pantheon+ TRGB data from Anand et al. [30]. The distant σ values are large compared to nearby.

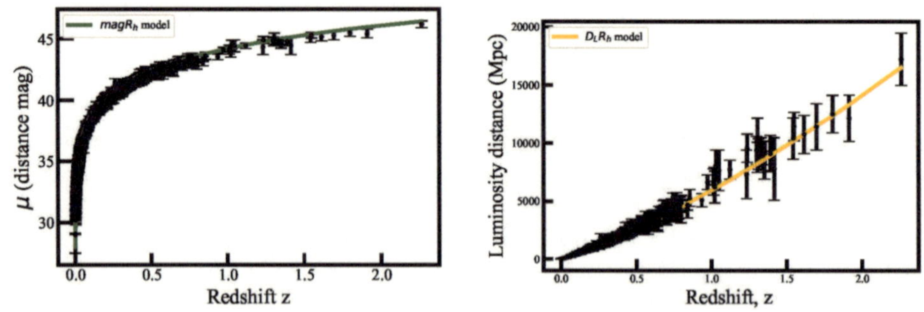

(a) The magR_h model

(b) The $D_L R_h$ model

Figure 3.
Both plots using the Pantheon+ data from [3] and TRGB data from [26]. The distant σ values of 3b are large compared with nearby events.

73 km s^{-1}Mpc^{-1}. In general, this model tends to perform well in most statistical tests with the r^2 and χ^2_{red} values similar to those of more detailed models and best overall according to the F statistic. Because estimates of parameters, Ω_m, Ω_Λ and Ω_k, are not possible along with the requirement for Euclidean geometry, the R_h model is not very informative. The results from the Pantheon+Freedman data are presented in our GitHub folder with the $H_0 \approx 70$. We found the statistics similar when either TRGB datasets are combined with the Pantheon+ data since the total number of data pairs is only slightly larger.

3.4 Standard Models

3.4.1 The Standard Model

This is modelled by Eq.(15) with two free parameters, no Ω_k and $sin(x) = x$. The presumption is that when $\Omega_k = 0$, the Universe exhibits Euclidean geometry but another appreciation is that these parameter values model a universe where empty space does not effect the Hubble flow. This universe consists of only common matter, CDM and DE, and this model is uninformative about spacetime geometry. The results are presented in **Tables 1–3** and **Figure 4a** and **b**. Our values for H_0 of 73.0 and 73.2 km s^{-1}Mpc^{-1} are close to the value recently published (73.5), likewise the values for Ω_m and Ω_Λ [3]. The value for H_0 calculated with the combined Pantheon+Freedman *distance mag* data is smaller at 70.7 ± 0.2 with a significantly larger Ω_m of 0.49 as presented in our GitHub folder.

The other form of the Standard Model correlating D_L *vs.* z, the D_LStandard, presents a larger value for H_0 of 75.8 km s^{-1}Mpc^{-1}. This value is problematic because simple change from *distance mag* to actual distance, D_L, should not present a significantly different H_0 value and, 75.8 is much too high, demanding a young Universe. The r^2, F statistic and χ^2_{red} values for the D_L Standard Model are more favourable than the Standard Model, and the Ω_m is a significantly smaller 0.25. Comparing **Figure 4a** with **Figure 4b,** we note the σ values for the *distance mag* data are much smaller and barely increasing with increasing lookback compared with the σ values displayed by the D_L data. This is even more striking as presented in **Figure 4** of of Brout et al. [25] as a supposed "Hubble diagram", but being a log-log plot is certainly not a Hubble-

Model	H_0	Ω_m	Ω_Λ	Ω_k	r^2	$F(10^5)$	χ^2_{red}	BIC
D_L2PAtanh	73.1±0.2	$1x10^{-5}$±0.02	—	>0.99±0.02	0.9846	21.4	1.72	14.79
D_L3PAtanh	73.1±0.3	0.001±0.04	—	0.99±0.04	0.9847	14.2	1.72	22.24
mag2PAtanh	71.6±0.2	$1x10^{-5}$±0.01	—	>0.99±0.01	0.9969	2.31	1.79	14.41
mag3PAtanh	72.7±0.3	0.001±0.03	—	0.99±0.03	0.9968	1.56	1.48	21.87
D_LStandard*	75.8±0.3	0.23±0.02	0.77±0.02	—	0.9895	26.8	1.58	14.78
Standard*	73.2±0.3	0.34±0.022	0.66±0.02	—	0.9970	2.40	1.67	14.41
D_L3PStnd§	76.1±0.4	0.52±0.22	1.00±0.18	−0.52±0.28	0.9908	18.6	1.57	22.23
mag3PStnd§	73.2±0.4	0.28±0.23	0.61±0.19	0.11±0.30	0.9970	1.60	1.67	21.86
D_L3PStnd†	76.2±0.33	0.42±0.07	1.00±0.09	−0.42±0.12	0.9898	18.0	1.57	22.23
mag3PStnd†	73.0±0.3	0.34±0.09	0.66±0.11	0.00±0.14	0.9970	1.60	1.67	21.86
D_L3PStnd‡	75.9±0.3	0.69±4.31	1.00±2.19	−0.69±4.84	0.9890	17.6	1.58	22.23
mag3PStnd‡	73.2±0.4	0.32±0.09	0.63±0.11	0.05±0.14	0.9970	1.60	1.67	21.86

**Standard Model,* sinn(x) = x.
§*3PStandard Model,* sinn(x) = x.
†*3PStandard Model,* sinn(x) = *sin(x).*
‡*3PStandard Model,* sinn(x) = *sinh(x).*

Table 3.
Results from analyses of all 1701 (or 1702) SNe Ia Pantheon+16 TRGB [30] emission data. H_0 *as* $km\ s^{-1}Mpc^{-1}$.

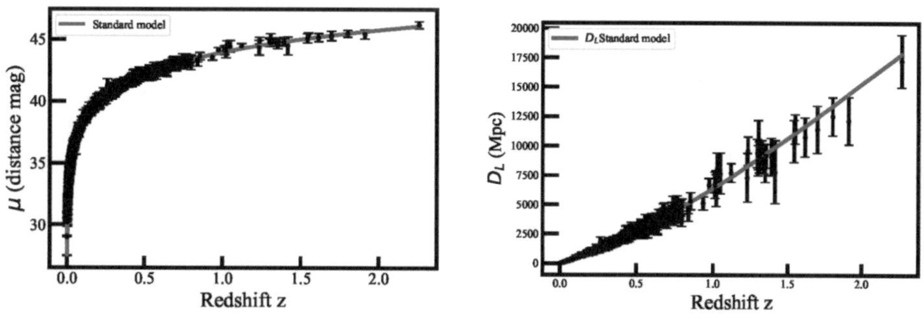

(a) The Standard Model

(b) The D_LStandard Model

Figure 4.
Both plots using the Pantheon+ data from [3] and TRGB data from [26]. The distant σ values of 4b are large compared with nearby events.

type diagram. Using transformed data, as *distance mag*, for analysis of astronomical data creates problems concerning the usefulness of calculated parameters.

3.4.2 sinn(x) = x

The results allowing freedom for H_0, Ω_m and Ω_Λ, as the three-parameter (3PStnd) versions, are presented in **Tables 2** and **3**. These models do not conform to the data significantly better than the two Standard Model as judged by r^2, F statistic and χ^2_{red}, with both versions presenting $H_0 > 73$ km s^{-1} Mpc^{-1}. The value of Ω_k for the *mag* 3PStandard Model is significant (0.11), suggesting an influence of space on H_0.

The results from the D_L *vs. z* correlation are too dissimilar to those of the *distance mag vs. z* correlation to ignore. The value of Ω_k for the D_L3PStnd model is negative – 0.52, requiring space to retard the Hubble flow or hyperbolic space geometry, both of which are ridiculous. The value for Ω_Λ of 1.00, the upper limit of the allowed regression range, describes a universe consisting primarily of dark energy. Thus, the three-parameter Standard Model allowing $sin\mathbf{n}(x) = x$ tested with the untransformed data of Pantheon+Anand fails reality. The results using the Pantheon+Freedman data, as listed in our GitHub folder, do not have this problem. Regression analysis presents our Universe primarily consisting of empty space ($\Omega_k = 0.83$) with very little matter and a bit of DE. These results seem more realistic than those from the Pantheon+Anand data.

3.4.3 sin$\mathbf{n}(x) = sin(x)$

We address two versions of the FLRW model by presuming hyperbolic space geometry employing $sin\text{n}(\text{x}) = \sin(\text{x})$. The evaluations are presented in **Table 3** as the sin*mag*3PStandard and sinD_L3PStandard Models with the Pantheon+Anand data. The major player for the D_L *vs. z* correlation is obviously $\Omega_\Lambda = 1.00$, the upper regression limit, with the negative Ω_k requiring hyperbolic space geometry and $-\Omega_m$; negative gravity. These results differ from the *distance mag vs. z* version for which all parameters are positive, with $\Omega_k > 0$ indicating a positive influence on H_0. Negative Ω_m and Ω_k are not the results from evaluations of the Pantheon+Freedman data listed in our GitHub folder, suggesting the Anand collection is flawed. This situation would remain undetected if only the *distance mag vs. z* correlation was evaluated. With that data, Ω_k is calculated as 0 with the Universe consisting of nearly equal parts matter and DE.

3.4.4 sin$\mathbf{n}(\text{x}) = sinh$

We address the FLRW model presuming quasi-Euclidean space geometry by evaluating $sin\text{n}(\text{x}) = \sinh(\text{x})$. The results from evaluating both the *distance mag* and D_L versions are presented in **Table 3** as the sinh*mag*3PStandard and sinhD_L3PStandard Models. Employing the Pantheon+Anand *distance mag* data, the values for Ω_k are positive, similar to the result for the *mag*3PStandard Model with $sin\mathbf{n}(x) = x$.

On the other hand, when the D_L data are modelled, the Ω_k is again negative. Such results are not found using the Pantheon+Freedman combination so something amiss with the Anand TRGB data. The F statistic for all D_L3P type Standard Models are similar to those of the 3PArcTanh models but none are as good as the $D_L R_h$ model. In general, the χ^2_{red} values for evaluations of the Standard Models with the Pantheon +Freedman data are all much smaller and the F statistic larger, than those using the Pantheon+Anand data.

3.5 ArcTanh models

The results from the ArcTanh models, employing the Pantheon+ *distance mag* data with the Anand TRGB data ensembles, are presented in **Tables 2** and **3**. The statistics from these regressions are similar to those of the Standard Models. The interesting results are the small Ω_m and large Ω_k values compared with any version of the Standard Model. The ArcTanh model has a very different meaning from any Standard Model since it employs the Ω_k term for space and does not consider Ω_Λ. These results suggests our Universe is primarily empty space with very little matter, consistent with astronomical observations, Big Bang calculations [24] and our previous proposals [26, 28].

The results for the ArcTanh model employing the D_L *vs.* z data are presented in **Tables 2** and **3**. These two models fit the data with similar statistics to the Standard Models. The values of H_0, both ≥ 73.1 km s^{-1} Mpc^{-1}, are similar to those calculated for the Standard Model. The Ω_k values again approach 1 for a Universe without significant DE or DM, containing only normal matter, providing evidence for quasi-Euclidean space geometry [15]. Overall, these results are fairly similar for all four ArcTanh variants, except the BIC scores which are significantly higher for the three-parameter models. The BIC statistic leans heavily on the number of free parameters and is not very informative with respect to model fitness.

4. Discussion

4.1 Predictions

The value of the D_L metric becomes clear when considering our present and future locations. The *distance mag vs. z* correlations cannot locate the present nor predict the future because log(≤ 0) has no unique future value, but only $-\infty$. Some avoid this problem by slyly publishing a discontinuity at the supposed origin [39]. We locate the present and predict our future using the D_L *vs.* z correlation, **Figure** 5. We have selected the 2PD_LArcTanh model to illustrate this, though other models display similar results. We predict that, several Gyrs from now, the relative position of the Milky Way will move by many hundreds of Mpc compared to today.

There remains no reason to transform a curved plot into a linear graphical display by employing the log function as *distance mag*. Modelling the D_L data using high-speed computers is now convenient, and programming languages with useful routines are freely available.

4.2 Problems with transformed data

For calculations of cosmological distances, astronomers often use *distance mag*, μ, a logarithmic transformation of the luminosity distance, D_L (Eq. 1). Transforming the

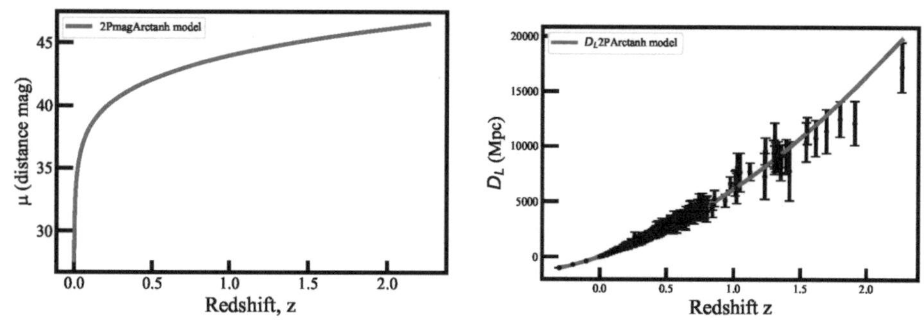

(a) The 2PmagArctanh model

(b) The D_L2PArctanh model

Figure 5.
Both plots using the Pantheon+ data from [3] and TRGB data from [26]. Plot ***5a*** *cannot be extended towards zero, only* ***5b*** *can estimate the future location of our Milky Way.*

metric to linearise data for presentation is a common practice in theoretical physics, astrophysics and even economics. Unfortunately, the log transformation is well known to introduce artefacts, especially with noisy data and is incapable of handing results at zero or less. Using the *log* or *ln* transformation is strongly discouraged in biology and biophysics [40]. Statisticians even caution medical investigators to avoid this mistake [41]. Modelling the correlation between the luminosity distance and redshift is the correct analytical method with advantages for astronomers, but only occasionally used.

The results for a model differ depending on the correlation chosen which should not be the case if the metrics are immaterial. The figures of the μ *vs.* z correlations show that the σs are not increasing much with lookback compared to those associated with D_L. This is contrary to reality, where increasing error with increasing distance is the rule in engineering and physics [32]. Valid σ estimates are required for proper regression analysis, best parameter estimates and statistics.

Our results are inconclusive with respect to model preference. The values for r^2, F statistic and χ^2_{red} of several models are similar except for some E-DS variants. The R_h view models the data well, but is uninformative with respect to Ω_m, Ω_Λ and Ω_k; hence, not very useful. We find important differences, though, between the other models. The Standard Models suggest plentiful matter and the Λ with little empty space in the Universe, while the ArcTanh models suggest the Universe is mostly empty space, which is realistic.

4.3 Spacetime geometry

Some have decided that a model of our Universe with Euclidean geometry that Ω_k must be zero; we have previously suggested otherwise. Here, we present a brief argument referencing the Friedmann equations. The **k** of Friedmann, related to Ω_k, is present in one of his two equations, without Λ, as

$$H^2 = \left(\frac{\dot{a}}{a}\right)^2 = \frac{8\pi G}{3}\rho - \frac{\mathbf{k}c^2}{a^2} \tag{23}$$

where ρ is the matter density, G is the Newton's constant, c is the lightspeed and a is the expansion factor. We do not reference Λ to keep the argument simple; the Λ is an independent term in the Friedmann relationships not directly affecting **k**.

The k of the Robertson-Walker (R-W) line element is commonly presented as

$$ds^2 = -dt^2 + a^2(t)\left[\frac{dr^2}{(1-kr^2)} + r^2 d\theta^2 + r^2 \sin^2\theta d\phi^2\right] \tag{24}$$

where s is the distance, t is the time, a is the expansion factor, θ, ϕ are the elements of spherical geometry and r is the radial distance. The k of the R-W relationship is usually assigned a range of -1 to 1 and has the metric (m^{-2}), but this range is assigned without a reality check. This same range has assigned to Ω_k and **k** by several investigators. However, neither has a metric, and the selection of this range is *obviously a definition*, introduced without explanation by Carroll [21]. Examining Eq. 23 reveals that as ρ approaches 0 in a dilute universe, **k** must become large. Likewise as Ω_m becomes small, Ω_k must approach 1 for H to remain real, in a universe without DE.

The **k** of the Friedmann equation informs us of the influence of empty space on H but nothing directly about geometry. The reason why **k** probably represents something about empty space and indirectly spacetime geometry is because on the cosmological scale, these are all that remain for assignment. Similarly, unit-less Ω_k really only represents that portion of the universe which is not matter nor dark energy. The k of the R-W equation informs us something about spacetime geometry but not the extent of empty space nor the influence of empty space on H. While k and **k** (or Ω_k) can be mathematically related after some derivation, the meanings are not exactly the same as evidenced by the different metrics. As a rule in science when one changes the metric, one unavoidably changes the meaning of that variable.

Examining Eq. (23) as **k** approached 0, the influence of space on the Hubble expansion becomes negligible. This does not mean that space disappears or the geometry becomes Euclidean or elliptical, only that it has a negligible effect on H. Conversely, a large **k** means that spacetime strongly influences H. Considering Eq. (24), as k approaches 0, spacetime geometry becomes Euclidean but k does not directly effect H. Decades of observing Einstein rings tells us that spacetime must be curved in the presence of matter so k cannot be exactly 0. More matter demands more curvature and this curvature exists even in the presence of the cosmological constant as per Einstein.

Here, we find many models presenting a positive Ω_k, (and some with -Ω_k), except for the initial condition requirement of $\Omega_k = 0$. The values of Ω_k for several ArcTanh models approach 1, all with very small Ω_m values. In a universe with even a small ρ, spacetime geometry must reflect some matter; however, lightly sprinkled, so the Friedmann model and our experience in our dilute Universe describe a gently elliptical (quasi-Euclidean) spacetime geometry and the large Ω_k tells us empty space is abundant. Evaluating Eqs. (5) and (12) when $\Omega_\Lambda = 0$ and $\Omega_m = 1$ or *vise-versa*, Ω_k is necessarily 0 as required by the Standard Model supposedly indicating Euclidean spacetime geometry [18, 19]. A universe with matter, however, must display elliptical geometry or at least quasi-Euclidean geometry. So even when $\Omega_k = 0$ and $\Omega_m = 1$, or $\Omega_m + \Omega_\Lambda = 1$, space must be highly curved but exerting little influence on H. Continuing with this reasoning in a universe without matter or DE, $\Omega_k = 1$ and spacetime geometry must be Euclidean. We expect our Universe with only a small Ω_m to exhibit a large value for Ω_k; an average m^3 portion of our Universe is empty space with a lone H atom or a couple of neutrinos just passing through.

4.4 Model fitness

The current Standard Model is not any better than the conservative ArcTanh model and the $R_h = ct$ model. We do not favour the R_h model, even though it fits the data well, because it cannot present values for Ω_m, Ω_Λ or Ω_k. The model of Yershov does well with the statistical tests but is also uninformative with respect to Ω_k and Ω_Λ. There are clear differences between the Standard and ArcTanh models around the Ω_m and Ω_k values. Given the failure to directly observe dark energy [24], we cannot favour the current Standard Model, and arguments for DE are simply special pleading. Repeated failures to detect dark matter on the subatomic and atomic scales are a persistent problem for the ΛCDM Universe [42].

All models evaluated here, except E-DS, predict a H_0 of 70–74 km s^{-1} Mpc^{-1} similar to a recent report [43]. The lower H_0 values calculated using the E-DS model are similar to those reported earlier, 62–63 [44], and by SNe Ia investigators [13]. Using the distance, D_L *vs.* z correlation, we can predict, for the first time, the future location of our Milky Way and our current location in the Universe.

5. Conclusions

An important difference between science and astrology is the ability of science to make concrete, accurate predictions.

- The analytical method employed by most cosmologists to model SNe Ia emissions cannot predict the future position of the Milky Way; therefore, it should be termed unscientific.
- Using the Hubble correlation with a detailed model, we can predict the present position and future location of our Milky Way at any time.
- The Hubble constant of 70–74 km s^{-1} Mpc^{-1} calculated here is within the currently popular range (≥ 70 to ≤ 74 km s^{-1} Mpc^{-1}).
- Our results indicate the Universe is primarily spacetime, containing only a little baryonic matter.
- Proper analyses of a recently published, large ensemble of SNe Ia data do not support speculations that dark matter or dark energy is important.

The reader might confirm our contention that spacetime is vast and overwhelming by simply viewing the heavens on a dark night.

Conflict of interest

The authors declare no conflict of interest.

Author details

Michael L. Smith[1*†], Abdullah Alsakka[2†] and Ahmet M. Öztas[3]

1 Independent Researcher, Umeå, Sweden

2 Department of Physics and Astronomy, Uppsala University, Uppsala, Sweden

3 Physics Engineering, Hacettepe University, Ankara, Turkey

*Address all correspondence to: mlsmith55@gmail.com

† These authors contributed equally.

References

[1] Galbany L et al. An updated measurement of the Hubble constant from near-infrared observations of type Ia supernovae. Astronomy & Astrophysics. 2023;**679**:A95. DOI: 10.1051/0004-6361/202244893

[2] Pascale M et al. SN hope: The first measurement of Ho from a multiply-imaged type Ia supernova, discovered by JWST. arXiv 2024. DOI: 10.48550/ARXIV.2403.18902

[3] Freedman WL et al. Calibration of the tip of the Red Giant branch. The Astrophysical Journal. 2020;**891**:57. DOI: 10.3847/1538-4357/ab7339

[4] Aghanim N et al. Planck 2018 results. VI cosmological parameters. Astronomy & Astrophysics. 2020;**A6**:641. DOI: 10.1051/0004-6361/201833910

[5] Boylan-Kolchin M. Stress testing LCDM with high-redshift galaxy candidates. Nature Astronomy. 2023;7: 731-735. DOI: 10.1038/s41550-023-01937-7

[6] DESI Collaboration. Cosmological constraints from the measurements of Baryon acoustic oscillations. ArXiv 2024. DOI: 10.48550/arXiv.2404.03002

[7] Ananthaswamy A. The Hubble tension is becoming a Hubble crisis. Scientific American. 2025. Available from: https://www.scientificamerican.com/article/the-hubble-tension-is-becoming-a-hubble-crisis/

[8] Riess AG et al. The accuracy of the Hubble constant measurement verified through cepheid amplitudes. The Astrophysical Journal. 2020;**896**:L43. DOI: 10.3847/2041-8213/ab9900

[9] Majaess D. The concerning SH_0ES Hubble constant. arXiv 2024. DOI: 10.48550/arXiv.2401.12964

[10] Riess AG et al. JWST observations reject unrecognized crowding of cepheid photometry as an explanation for the Hubble tension at 8σ confidence. The Astrophysical Journal Letters. 2024;**962**: L17. DOI: 10.3847/2041-8213/ad1ddd

[11] Toy M et al. Rates and properties of type Ia supernovae in galaxy clusters within the dark energy survey. Monthly Notices of the Royal Astronomical Society. 2023;**526**:5292-5305. DOI: 10.1093/mnras/stad2982

[12] Di Valentino E et al. Cosmology intertwined IV: The age of the universe and its curvature. Journal of Astroparticle Physics. 2021;**131**:102607. DOI: 10.1016/j.astropartphys.2021.102607

[13] Riess AG et al. Observational evidence from supernovae for an accelerating universe and a cosmological constant. The Astrophysical Journal. 1998;**116**:1009-1038. DOI: 10.1086/300499

[14] Freedman WL et al. Progress in direct measurements of the Hubble constant. Journal of Cosmology and Astroparticle Physics. 2023;**11**:50. DOI: 10.1088/1475-7516/2023/11/050

[15] Einstein A. Relativity, the Special and General Theory. Boca Raton, FL, USA: Routledge Classics; 1954. 158 pp. ISBN 0-415-25384-5

[16] Kovacs A et al. A common explanation of the Hubble tension and anomalous cold spots in the CMB. Monthly Notices of the Royal Astronomical Society. 2020;**499**:320. DOI: 10.1093/mnras/staa2631

[17] Garnavich PM et al. Supernova limits on the cosmic equation of state. The

Astrophysical Journal. 1998;**509**:74. DOI: 10.1086/306495

[18] Wood-Vasey WM et al. Observational constraints on the nature of dark energy. The Astrophysical Journal. 2017;**666**:694. DOI: 10.1086/518642

[19] Oztas AM et al. The varying cosmological constant. Monthly Notices of the Royal Astronomical Society. 2018; **476**:451-458. DOI: 10.1093/mnras/sty221

[20] Misiaszek M, Rossi N. Direct detection of dark matter: A critical review. Symmetry. 2024;**16**:201. DOI: 10.3390/sym16020201

[21] Carroll SM et al. The cosmological constant. Annual Review of Astronomy and Astrophysics. 1992;**30**:499-542. DOI: 10.1146/annurev.aa.30.090192.002435

[22] Garnavich PM et al. Supernova limits on the cosmic equation of state. The Astrophysical Journal. 1998;**509**:74-79. DOI: 10.1086/306495

[23] Carroll SM. The cosmological constant. Living Reviews in Relativity. 2001:4. DOI: 10.12942/lrr-2001-1

[24] Sabulsky DO et al. Experiment to detect dark energy forces using atom interferometry. Physical Review Letters. 2019;**123**:061102. DOI: 10.1103/physrevlett.123.061102

[25] Turner MS. The road to precision cosmology. Annual Review of Nuclear and Particle Science. 2022;**72**:1-35. DOI: 10.1146/annurev-nucl-111119-041046

[26] Brout D et al. The pantheon+ analysis: Cosmological constraints. The Astrophysical Journal. 2022;**938**:110. DOI: 10.3847/1538-4357/ac8e04

[27] Öztas AM, Smith ML. Re-evaluation of Omega_k of the normalised Friedmann-Lematîre-Robertson-Walker model. New Astronomy. 2021;**88**: 101609. DOI: 10.1016/j.newast.2021.101609

[28] Burles S et al. What is the big-bang-nucleosynthesis prediction for the baryon density and how reliable is it? Physical Review D. 2001;**63**:063512. DOI: 10.1103/physrevd.63.063512

[29] Öztas AM, Smith ML. Space-time curvature and the cosmic horizon: Derivations using the Newtonian world and the Friedmann-Robertson-Walker metric. Monthly Notices of the Royal Astronomical Society. 2015;**449**: 1270-1274. DOI: 10.1093/mnras/stv346

[30] Anand GS et al. Comparing tip of the Red Giant branch distance scales. The Astrophysical Journal. 2022;**932**:15. DOI: 10.3847/1538-4357/ac68df

[31] Freedman WL et al. The Carnegie-Chicago Hubble program. VIII The Astrophysical Journal. 2019;**882**:34. DOI: 10.3847/1538-4357/ab2f73

[32] Hubble E. A relation between distance and radial velocity among extra-galactic nebulae. PNAS. 1929;**15**:168-173. DOI: 10.1073/pnas.15.3.168

[33] Khoshelham K, Elberink SO. Accuracy and resolution of kinetic depth data for indoor mapping applications. Sensors. 2012;**12**:1437-1454. DOI: 10.3390/s120201437

[34] Massaron L, Boschetti A. Regression Analysis with Python. UK: Packt Publishing Ltd; 2016. 312 pp. ISBN 9781785286315

[35] Carlsson B. Curve Fitting in Python with curv_fit. 2024. Available from: https://www.youtube.com/watch?v=Jl-Ye38qkRc

[36] Yershov VN. Fitting type Ia supernova data to a cosmological model based on Einstein-Newcomb-de Sitter space. Universe. 2023;**9**:204. DOI: 10.3390/universe9050204

[37] Melia F. The cosmic horizon. Monthly Notices of the Royal Astronomical Society. 2007;**382**: 1917-1921. DOI: 10.1111/ j.1365-2966.2007.12499.x

[38] Melia F, Shevchuk ASH. The $R_h = ct$ universe. Monthly Notices of the Royal Astronomical Society. 2012;**419**: 2579-2586. DOI: 10.1111/ j.1365-2966.2011.19906.x

[39] Kirschner RP. Hubble's diagram and cosmic expansion. PNAS. 2003;**101**:8-13. DOI: 10.1073/pnas.2536799100

[40] Packard GC. A perspective on polyphasic loglinear allometry. Journal of Experimental Zoology Part B: Molecular and Developmental Evolution. 2016;**326**:168-175. DOI: 10.1002/jez. b.22673

[41] Feng C. Log-transformation and its implications for data analysis. Shang Arch Psychiatry. 2014;**26**:105-109. DOI: 10.3969/j.issn.1002-0829. 2014.02.009

[42] Aprile E et al. Search for coherent elastic scattering of solar 8B neutrinos in the XENON1T dark matter experiment. Physical Review Letters. 2021;**126**: 091301. DOI: 10.1103/ physrevlett.126.091301

[43] Riess AG et al. A comprehensive measurement of the local value of the Hubble constant with 1 km uncertainty from the Hubble space telescope and the SH0ES team. The Astrophysical Journal Letters. 2022;**934**:L7. DOI: 10.3847/ 2041-8213/ac5c5b

[44] Sandage A et al. The Hubble constant. The Astrophysical Journal. 2006;**653**:843-860. DOI: 10.1086/508853

Chapter 8

Dark Matter and Dark Energy in Galaxies Via Gravitational Field Self-Interaction

Firmin Oliveira

Abstract

The gravitational redshift theory (GRST) is defined in terms of the gravitational and cosmological redshifts of the gravitational field by means of the self-interaction of the field. The concept is applied to spiral galaxy rotation and to elliptical galaxies acting in the form of gravitational lensing. These field effects which are due solely to the baryonic mass in the galaxy are sufficient to account for the missing mass needed to match the observed rotation velocities without the need for other forms of matter. It is claimed that these effects which are described by the GRST, being purely relativistic in nature, are non-electromagnetic and act by gravity via the relativistic increase of the galaxy mass, and are responsible for the manifestations that have been attributed to dark matter and dark energy.

Keywords: gravitational self-interaction, gravitational redshift, cosmological redshift, spiral galaxy, elliptical galaxy, gravitational lens, lens mass, gravitons

1. Introduction

The gravitational redshift theory (GRST) is based on the premise that the gravitational field potential energy is increased (becomes more negative) by gravitational and cosmological effects through self-interaction. This is relevant in the study of galactic phenomena. In previous manifestations of this concept [1], gravitons were used as intermediary agents in producing these effects. The reliance on gravitons as agents of force transmission is conducive to picture the localized action and movement of the gravitational field. However, as the gravitational field propagates at fixed speed c in vacuum, where c is also the speed of light, then the field itself, through self-interaction, can fulfill the role of the agent to produce gravitational and cosmological effects. Thus, hereafter, the action of gravitons is absorbed into the concept of the self-interaction of the propagating gravitational field.

This process is investigated further in relation to spiral galaxy rotation velocities and to gravitational lenses. It is notable that, with this theory, only the baryonic mass of the galaxy is a free parameter, with other parameters obtained from the observation. In the case of a spiral galaxy, essentially just the radial velocity distribution $\nu(r)$, with the final radial distance r_f and velocity $\nu(r_f)$ used to determine the gravitational field self-interaction gravitational redshift coupling constant k_f and the cosmological redshift coupling constant k_e. The SPARC database [2, 3] is used in the spiral galaxy NGC 3198 analysis. It provides the baryonic mass and rotational velocities for the galaxy.

Gravitational lensing is studied by observations of Einstein rings [4]. It is found that nothing new needs be added to the galaxy version of the GRST, except that, since galactic rotation velocities are not available in many cases, the gravitational redshift coupling constant k_f and the cosmological redshift coupling constant k_e are derived from the lens mass. Analysis will be made of the galaxy lens systems JWST-ER1, COSMOS 5921 + 0638, and COSMOS 0018 + 3845.

Numerous, extensive searches for massive dark matter particles have produced constraints and limits on detectability but failed to produce any definite dark matter particle thus far [5, 6]. And, the search for dark energy has also produced null results [7].

This theory provides an explanation for the dark matter and dark energy in galaxies and the universe as due to the relativistic mass associated with gravitational and cosmological redshift energy losses of the gravitational field itself. No massive particles are required.

For the layout of this chapter, in the next section, the rotation velocity relation for a spiral galaxy is described, and comparisons are made to the predictions of the standard theory. In the following sections, the gravitational lens relations are described, and the connection to the galaxy rotation energy with the gravitational field velocity boost is established. These results are discussed in the final section. This is a development of the recent publication [8].

2. Action of gravitational self-interaction in galaxies

The gravitational field acts at constant speed c in vacuum, where c equals the speed of light. The gravitational field traveling in its own field between a source baryonic mass M and a test mass m should experience an average energy loss of $\delta\xi_{dm}$ due to motion in that field, over a short time period $\delta t = \delta r/c$, where the acceleration a at a point r in the field is given by $a = -GM/r^2$. The energy loss is expressed as follows:

$$\begin{aligned}\delta\xi_{dm} &= -mc^2\left(\frac{\delta\nu}{c}\right) = -mc\left(-a\frac{\delta r}{c}\right) = \\ &= -\left(\frac{GMm}{r^2}\right)\delta r,\end{aligned} \tag{1}$$

where m is the relativistic mass of the portion of the field involved in the interaction, where m equals the test particle mass, $\delta\nu = -a\,\delta t$ is the change in the free fall velocity of the system observed from an inertial reference frame, G is Newton's gravitational constant, M is the baryonic mass of the field source, r is the distance between the center of the source and the location of the field interaction, and $\delta r/c = \delta t$ is the short travel time of the field at speed c over distance δr. The energy change is a loss (negative) because the velocity change $\delta\nu$ is in the same direction as the motion of the field, so that for an inertial observer moving in the same direction as the field, the energy of the gravitational field is redshifted. This effect of energy loss is a gravitational redshift which, as it is described, resembles a Doppler effect, a relativistic effect, and is a form of self-interaction of the gravitational field [9].

Integrating Eq. (1) from the origin of the system of masses to the position r of the interacting field and multiplying by a coupling coefficient $K_g(r)$ of the gravitational field self-interaction to the mass at position r yields the total energy loss $\Delta\Xi_g(r)$ given by

$$\begin{aligned} \Delta\Xi_g(r) &= -K_g(r)\int_0^{\nu} mc^2\left(\frac{du}{c}\right) = \\ &= -K_g(r)\int_0^{r} m\left(\frac{GM(s)}{s^2}\right) ds, \end{aligned} \tag{2}$$

where the gravitational coupling coefficient,

$$K_g(r) = k_f\frac{M(r)}{M_b}, \tag{3}$$

where k_f is the gravitational redshift coupling constant (gravitation coupling constant), $M(r)$ is the baryonic mass within radial position r, and M_b is the total baryonic mass of the galaxy. Eq. (2) describes the gravitational redshift of the energy of the field as it travels from a lower, more negative potential to a position r of higher, less negative potential and is consistent with energy conservation.

In addition to the gravitational field experiencing an increase of its energy, that is, becoming more negative due to its own gravity, it is hypothesized that even in galaxies, the field is also susceptible to the expansion of the universe and undergoes an energy loss $\delta\xi_e$ due to cosmological redshift of its energy, which is assumed to take the form:

$$\delta\xi_e = -mc^2k_e\frac{\delta r}{R_0}, \tag{4}$$

where m is the relativistic mass of the portion of the field involved in the interaction, which equals the mass of the test particle, c is the speed of the field (speed of light) in vacuum, k_e is the gravitation cosmological redshift coupling constant (expansion coupling constant), δr is the change in distance from the galaxy center, and $R_0 = c/H_0$ is the radius of the universe where H_0 is the Hubble constant at the epoch of interaction. Integrating eq. (4) from the origin of the system of masses to the position r of the field yields the total energy loss $\Delta\Xi_e(r)$ given by

$$\begin{aligned} \Delta\Xi_e(r) &= \int_0^{\Delta\Xi_e(r)} \delta\xi_e = \\ &= -\int_0^{r} mc^2k_e\frac{\delta s}{(c/H_0)} = -mk_ecH_0r. \end{aligned} \tag{5}$$

The gravitational potential $\Phi(r)$ at radial position r in the field of a galaxy of total baryonic mass M_b, including field energy losses due to gravitational and cosmological redshifts, Eqs. (2) and (5), is defined by

$$\begin{aligned} \Phi(r) &= -\frac{GM(r)}{r} + \frac{\Delta\Xi_g(r)}{m} + \frac{\Delta\Xi_e(r)}{m} = \\ &= -\frac{GM(r)}{r} - k_f\left(\frac{M(r)}{M_b}\right)\int_0^{r}\frac{GM(s)}{s^2}\,ds - k_ecH_0r, \end{aligned} \tag{6}$$

where $M(r)$ is the baryonic mass within r and k_f and k_e are the gravitational and cosmological redshift coupling constants, respectively.

In terms of a total galaxy mass $\mathscr{M}(r)$ at radial position r, the gravitational potential Eq. (6) can be put into the form:

$$\Phi(r) = -\frac{G\mathscr{M}(r)}{r} = = -\frac{G}{r}\left(M(r) + \frac{k_f M(r)\, r}{M_b}\int_0^r \frac{M(s)}{s^2}\, ds + \frac{k_e c H_0 r^2}{G}\right), \tag{7}$$

where the total galaxy mass $\mathscr{M}(r)$ is defined by

$$\mathscr{M}(r) = M(r) + \frac{k_f M(r) r}{M_b}\int_0^r \frac{M(s)}{s^2}\, ds + \frac{k_e c H_0 r^2}{G}. \tag{8}$$

The equation for the total energy of the test mass m in the galaxy is given by the sum of the kinetic and potential energies

$$\frac{1}{2} m\nu^2(r) + m\Phi(r) = \frac{1}{2} m\Phi(r), \tag{9}$$

where $\nu(r)$ is the velocity at r and where the total energy on the right hand side is due to the kinetic energy being minus one half the potential energy (a result of the virial theorem.)

The equation for circular motion in the field, from Eq. (9), is given by

$$\nu^2(r) = \frac{GM(r)}{r} + k_f\left(\frac{M(r)}{M_b}\right)\int_0^r \left(\frac{GM(s)}{s^2}\right) ds + k_e c H_0 r. \tag{10}$$

3. Obtaining the formulas for k_f and k_e

Assume that the galaxy rotation velocity $\nu(r)$ in Eq. (10) is constant in the neighborhood of radial position r_0 which is beyond essentially all of the baryonic mass of the galaxy. Then, take the derivative of $\nu(r)$ with respect to r and set the result to zero at r_0, where $dM(r)/dr = 0$ at r_0 to obtain

$$0 = -\frac{GM(r_0)}{r_0^2} + \frac{k_f G(M(r_0))^2}{M_b r_0^2} + k_e c H_0, \tag{11}$$

which can be solved for k_e, expressed by

$$k_e = \left(\frac{GM(r_0)}{cH_0 r_0^2}\right)\left(1 - \frac{k_f M(r_0)}{M_b}\right). \tag{12}$$

To obtain a formula for k_f, solve for it using eq. (10) evaluated at $r = r_0$, expressed by

$$k_f = \frac{\left(\nu^2(r_0) - \frac{GM(r_0)}{r_0} - k_e c H_0 r_0\right)}{\left(\frac{M(r_0)}{M_b}\int_0^{r_0} \frac{GM(s)}{s^2}\, ds\right)}, \tag{13}$$

which after substitution for k_e using Eq. (12) reduces to

$$k_f = \frac{\left(v^2(r_0) - \frac{2GM(r_0)}{r_0}\right)}{\left(\frac{M(r_0)}{M_b}\left(\int_0^{r_0} \frac{GM(s)}{s^2}\,ds - \frac{GM(r_0)}{r_0}\right)\right)}. \tag{14}$$

Since the gravitational field of the galaxy extends throughout the entire universe, the coordinate r_0 will be set to the Hubble radius $R_0 = c/H_0$, and this is the location at which the values for the coupling constants will be determined.

4. Fit to spiral galaxy NGC 3198 rotation curve

With the baryonic Tully-Fisher relation (BTFR) [10] determined baryonic mass input of $M_b = 2.464 \times 10^{10} M_\odot$ for NGC 3198, **Tables 1** and **2** show the parameters for the iterative fitting using total rotation energy Eq. (10), gravitation redshift coupling constants Eqs. (12) and (13), and the quadratic equations, described in detail in [11], to determine the baryonic mass. The normalization parameter for the fit was $49.34\ M_\odot\,s^4\,km^{-4}$ which compares well to the default BTFR normalization of $50\,M_\odot\,s^4\,km^{-4}$.

Figure 1 shows the rotation curves for the disk, gas, and gravitational field gravitational and cosmological redshifts and the sum of these fitting to the galaxy rotation velocity. **Figure 2** shows the baryon mass distribution for $M(r)$ which was derived by iteration of the quadratic equation. **Figure 3** shows the masses for the baryons and the field gravitational and cosmological redshifts, the sum of these masses equal to the galaxy mass, and also equal to the galaxy mass determined using the observed rotation velocities and positions.

D	R_f	V_f	M_b BTFR	M_{total}
Mpc	**Kpc**	**km s^{-1}**	**10^{10} M$_\odot$**	**10^{10} M$_\odot$**
13.8	40.1	150	2.464	20.977

Table 1.
SPARC data for galaxy NGC 3198 using the GRST model Eq. (10) with masses from the GRST mass model Eq. (8). D is the distance to the galaxy. R_f is the radial position used for the flat curve velocity. V_f is the flat velocity. M_b BTFR is the baryonic mass used in the fit computed using the flat velocity $V_f = 150$ km s^{-1} and a normalization of 50. M_{total} is the total galaxy mass at R_f.

M_b	k_f	k_e	MAE	A
10^{10} M$_\odot$			**km s^{-1}**	**M$_\odot$ s^4 km^{-4}**
2.464	0.669	9.11×10^{-14}	0.008	49.34

Table 2.
Fit results with SPARC data for galaxy NGC 3198 using the GRST model Eq. (10). The galaxy baryonic mass M_b is the fitted mass at the final flat velocity radial position $R_f = 40.1$ kpc. k_f and k_e are the coupling constants for the gravitational field gravitational and cosmological redshifts, respectively. MAE is the mean absolute error of the rotation curve fit. A is the BTFR normalization value for the fit at R_f.

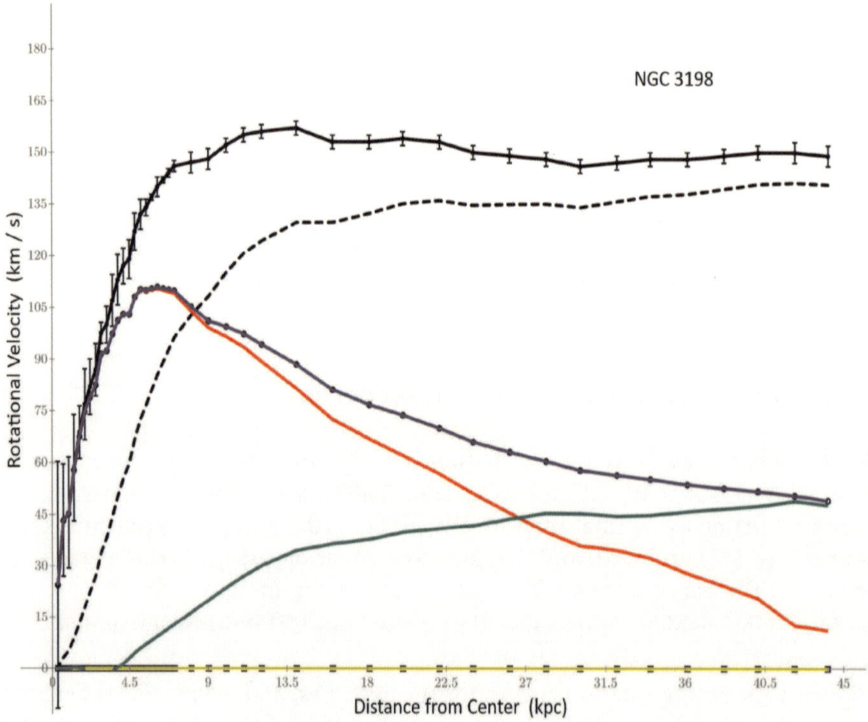

Figure 1.
Fit made to NGC 3198 using SPARC data with velocity profiles for gas, disk, and bulge. The plot shows the rotation velocity (solid black line) and the data points (solid points with error bars.) The curve of black circles is the velocity due to Newtonian theory with baryon mass, GM(r)/r. The black dashed line is the velocity due to the gravitational field gravitational redshift energy loss. The black dashed line with open squares is the velocity due to the gravitational field cosmological redshift energy loss. The solid red line is the difference between the velocities of the fitted baryon mass and the measured gas mass and is assumed to represent the velocity due to the stellar mass. The solid yellow line is the velocity due to the bulge mass (zero since bulge is not present.) The solid green line is the velocity for the gas mass. The solid blue line is the sum of the stellar, bulge, and gas velocities.

5. Gravitational lens including self-interaction of gravity

The gravitational potential $\Phi(r)$ describing the interaction of a photon with a galaxy having total baryonic mass M_b including gravitational self-interacion is defined by

$$\Phi(r) = -\frac{GM_b}{r}\left(1 + k_g + k_c\right), \tag{15}$$

where G is the gravitational constant, M_b is the total baryonic mass of the galaxy, k_g is the gravitational field gravitational redshift coupling, and k_c is the gravitational field cosmological redshift coupling. At radial distance r beyond the galaxy radius R_{gal}, the galaxy baryonic mass M_b along with the coupling coefficients k_g and k_c are all constant. The deviation in the path of the photon as it travels from the distant star, passing a distance b at closest approach to the galaxy and continuing on to the detector at the Earth is given by [12]

$$\nabla_\perp \Phi(r) = \frac{\partial \Phi(r)}{\partial b} = \frac{\partial \Phi(r)}{\partial r}\frac{\partial r}{\partial b}. \tag{16}$$

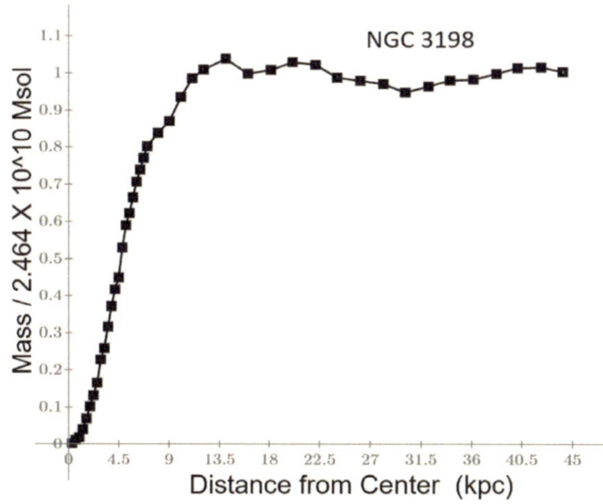

Figure 2.
Baryonic mass distribution for NGC 3198, (Mass / Mass NGC 3198). The baryonic mass of NGC 3198 is computed using the BTFR at the flat galaxy velocity $V_f = 150\ \mathrm{km\ s^{-1}}$ *by the formula* $M_{NGC3198} = 50\ V_f^4 = 2.464 \times 10^{10}\,\mathrm{M_\odot}$. *The black squares are the baryonic mass determined by solving the quadratic equation. The solid black line is the mass determined by the GRST version of the BTFR [11].*

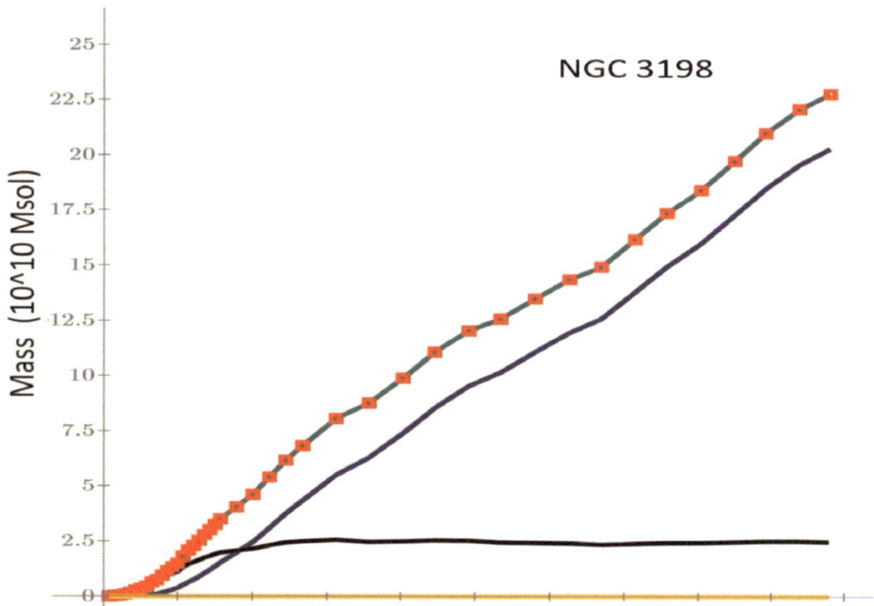

Figure 3.
Fit made to NGC 3198 with SPARC data masses derived from rotation velocity fit. The solid black line is the fitted baryon mass. The solid yellow line is the relativistic mass due to the graviton energy cosmological redshift loss. The solid blue line is the relativistic mass due to the graviton energy gravitational redshift loss. The solid green line is the total galaxy mass $\mathcal{M}(r)$ *of Eq. (8), the sum of baryon mass, graviton gravitational and cosmological redshift relativistic masses. The red squares are the total galaxy mass obtained from the rotation velocity and position data.*

Assume

$$r = \sqrt{b^2 + z^2}, \tag{17}$$

where b is the radial distance from the galaxy center which is beyond the bounds of the galaxy baryonic mass and z is on a line perpendicular to b which extends from the photon source to the observing detector. With this definition of the nearest approach b to the galaxy, the derivative of $\Phi(r)$ is expressed by

$$\begin{aligned} \frac{\partial \Phi(r)}{\partial r} &= \frac{GM_b}{r^2}\left(1 + k_g + k_c\right) = \\ &= \frac{GM_{lens}}{r^2}, \end{aligned} \tag{18}$$

where the lens mass (assumed constant at radial distance b) is defined by

$$M_{lens} = M_b\left(1 + k_g + k_c\right). \tag{19}$$

The derivative of r with respect to b, using Eq. (17), is given by

$$\frac{\partial r}{\partial b} = \frac{b}{r}. \tag{20}$$

Substituting from Eqs. (18) and (20) into Eq. (16) yields

$$\nabla_{\perp}\Phi(r) = \frac{b\,GM_{lens}}{r^3}. \tag{21}$$

Refer to **Figure 4** for an illustration of the ray paths associtated with a gravitational lens system, where it is shown that the total deflection angle of the photon traveling from the source to the observer is given by

$$\theta(b) = \alpha(b) + \beta(b). \tag{22}$$

This total deflection angle $\theta(b)$ is obtained initially by integrating Eq. (21) along the line z, expressed by

$$\begin{aligned} \hat{\alpha}(b) &= \frac{2}{c^2}\int_{-\infty}^{\infty} dz\,\nabla_{\perp}\Phi(r) = \\ &= \frac{2b\,GM_{lens}}{c^2}\int_{-\infty}^{\infty}\frac{dz}{\left(b^2 + z^2\right)^{3/2}} = \\ &= \frac{2GM_{lens}}{c^2 b}\int_{-\infty}^{\infty}\frac{dx}{(1 + x^2)^{3/2}} = \\ &= \frac{4GM_{lens}}{c^2 b}, \end{aligned} \tag{23}$$

where there is a change of variable $x = z/b$. In Eq. (23), the first factor of 2 comes from general relativity theory and the second factor of 2 comes from the integral of $dx\,(1 + x^2)^{-3/2}$ with x going from $-\infty$ to $+\infty$.

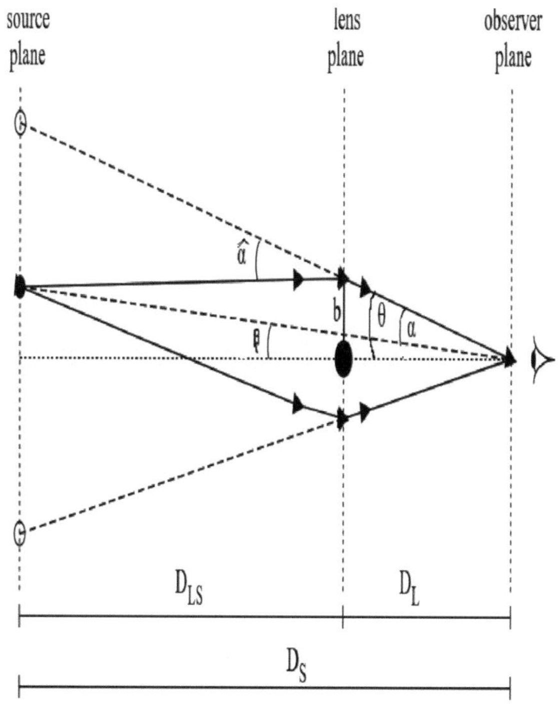

Figure 4.
Geometry of the basic gravitational lens system [13].

Then, from Eq. (23) the angle $\alpha(b)$ is given by

$$\alpha(b) = \left(\frac{D_{LS}}{D_S}\right)\hat{\alpha}(b) = \left(\frac{D_{LS}}{D_S}\right)\frac{4GM_{lens}}{c^2 b}, \tag{24}$$

where D_L is the angular diameter distance from the observing detector to the lens, D_S is the angular diameter distance from the detector to the source, and D_{LS} is the angular diameter distance from the lens to the source. The closest distance of the light path to the lens is the impact parameter b, which is given by

$$b = D_L\theta(b). \tag{25}$$

For a source which is on the center line through the detector and lens, the angle $\beta(b) = 0$, which implies that, combining Eqs. (22, 24), and (25) is

$$\theta(b) = \alpha(b) = \left(\frac{D_{LS}}{D_S}\right)\hat{\alpha}(b) = \left(\frac{D_{LS}}{D_S}\right)\frac{4GM_{lens}}{c^2 D_L\theta(b)}. \tag{26}$$

In this case, where $\beta(b) = 0$, from Eq. (26), the angle $\theta_E = \theta(b)$ of the Einstein ring is given by

$$\theta_E = \sqrt{\left(\frac{D_{LS}}{D_L D_S}\right)\frac{4GM_{lens}}{c^2}}, \tag{27}$$

which is the standard expression. For $\beta(b) \neq 0$, use Eqs. (22, 24, 25), and (27) to derive the quadratic equation in θ

$$\theta^2(b) - \beta(b)\theta(b) - \theta_E^2 = 0, \tag{28}$$

which has the two solutions:

$$\theta(b)_\pm = \left(\frac{1}{2}\right)\left(\beta(b) \pm \sqrt{\beta^2(b) + 4\theta_E^2}\right). \tag{29}$$

6. Einstein ring JWST-ER1r

JWST-ER1 is a gravitational lens system at redshift z = 1.94. It was detected by the James Webb Space Telescope [14] in 2023 [15]. It is a galaxy of estimated stellar mass $M_b = 1.1 \times 10^{11} M_\odot$ (JWST-ER1g). The Einstein ring (JWST-ER1r) subtends an angle of 1.54 arcsec in diameter and is determined to have a radius at the center of the ring of 6.6 kpc. The source of the ring is determined to be a galaxy at redshift $z = 2.98$.

The comoving distance determined by the general gelativity Λ Cold Dark Matter model is given by

$$D_c = \frac{c}{H_0}\int_0^z \frac{du}{\sqrt{\Omega_M (1+u)^3 + \Omega_\Lambda}}, \tag{30}$$

where H_0 is the Hubble constant, Ω_M is the matter density parameter, Ω_Λ is the mass density parameter due to the cosmological constant, and z is the redshift of the source. For this gravitational lens, use $\Omega_M = 0.271$ and $\Omega_\Lambda = 0.729$, with $H_0 = 70$ km s^{-1} mpc^{-1}.

For an intermediate source at redshift z_1, the comoving distance Eq. (30) can be expressed as follows:

$$D_c = D_{c1} + D_{c12}, \tag{31}$$

where the comoving distance

$$D_{c1} = \frac{c}{H_0}\int_0^{z1} \frac{du}{\sqrt{\Omega_M (1+u)^3 + \Omega_\Lambda}}, \tag{32}$$

and the comoving distance

$$D_{c12} = \frac{c}{H_0}\int_{z1}^{z} \frac{du}{\sqrt{\Omega_M (1+u)^3 + \Omega_\Lambda}}. \tag{33}$$

The angular diameter distance DA of the comoving distance D_c at redshift z is defined as follows:

$$DA = \frac{D_c}{1+z}. \tag{34}$$

The angular diameter distance DA_1 of the comoving distance D_{c1} at redshift $z1$ is defined as follows:

$$DA_1 = \frac{D_{c1}}{1+z1}. \quad (35)$$

Combining Eqs. (34) and (35) into Eq. (31), with some manipulations, it yields

$$DA = \frac{(1+z1)DA_1}{1+z} + \frac{D_{c12}}{1+z}, \quad (36)$$

where the angular diameter distance DA_{12} is defined as follows:

$$DA_{12} = \frac{D_{c12}}{1+z}. \quad (37)$$

Substituting Eq. (37) into Eq. (36) and simplifying to get DA_{12} in terms of the other distances, it yields

$$DA_{12} = DA - \left(\frac{1+z1}{1+z}\right)DA_1. \quad (38)$$

7. Derivations for Einstein ring JWST-ER1r

Regarding JWST-ER1, the angular diameter distance D_L of the lensing galaxy at redshift 1.94, from Eq. (30), where $D_L = D_c/(1+z)$ is given by

$$D_L = \left(\frac{1}{1+1.94}\right)\frac{c}{H_0}\int_0^{1.94}\frac{du}{\sqrt{0.271(1+u)^3+0.729}}, \quad (39)$$

where $H_0 = 70.0 \text{ km s}^{-1}\text{mp c}^{-1}$ and gives a value $D_L = 1.7767 \times 10^3\,\text{mpc}$. Likewise, the angular diameter distance D_S of the source at redshift 2.98, from Eq. (30), is given by

$$D_S = \left(\frac{1}{1+2.98}\right)\frac{c}{H_0}\int_0^{2.98}\frac{du}{\sqrt{0.271(1+u)^3+0.729}}, \quad (40)$$

which gives a value $D_S = 1.6400 \times 10^3\,\text{mpc}$. Finally, using Eq. (38), the value of the angular diameter distance D_{LS} from the lens to the source at redshift 2.98 is given by

$$D_{LS} = D_S - \left(\frac{1+1.94}{1+2.98}\right)D_L, \quad (41)$$

which derives a value $D_{LS} = 0.3275 \times 10^3\,\text{mpc}$.

The lens mass M_{lens} of galaxy JWST-ER1g is determined by inverting Eq. (27)

$$
\begin{aligned}
M_{lens} &= \theta_E^2 \left(\frac{D_L D_S c^2}{4 G D_{LS}}\right) = \\
&= \left(0.77 \left(\frac{\pi}{180 \times 3600}\right)\right)^2 \left(\frac{(1.7767 \times 10^3\,\text{mpc})\,(1.6400 \times 10^3\,\text{mpc})\,c^2}{4G\,(0.3275 \times 10^3\,\text{mpc})}\right) = \\
&= 6.476 \times 10^{11}\,\text{M}_\odot.
\end{aligned}
\tag{42}
$$

The impact parameter b of the light path at the lens galaxy JWST-ER1g is given by Eq. (25), which can be determined using the observed ring diameter angle $\theta = 0.77$ arcsec and the distance D_L to the lens, expressed by

$$
\begin{aligned}
b &= D_L\,\theta \\
&= (1.7767 \times 10^3\,\text{mpc})\left(0.77\left(\frac{\pi}{180 \times 3600}\right)\right) \\
&= 6.632\,\text{kpc}.
\end{aligned}
\tag{43}
$$

The half and full luminosity limits for galaxy JWST-ER1 are shown in **Figure 5(a)**. The galaxy circles of effective radius R_e (dashed red circle), full galaxy radius R_{gal} (solid red circle), and Einstein ring (solid black circle) are shown in **Figure 5(b)**. All angular diameter distances have been double checked for accuracy using the cosmological calculator at [19]. **Figure 5(c)** shows the combined mass distributions of the galaxy and ring system, where the dashed blue line is the sum of the stellar mass and dark matter mass using the standard mpdel, which amounts to just 46% of the lens mass at the distance of the impact parameter $b = 6.632$ kpc.

8. About the lensing galaxy JWST-ER1g

For this study, assume that the lensing galaxy is an elliptical with a simulated stellar mass distribution based on the Hernquist profile [16], defined in [18, 20], in the form:

$$
M(r) = M_*^{Sal}\,\alpha_{SPS}^{Sal}\left(\frac{r}{R_b}\right)^2\left(\frac{1 - F(r/R_b)}{(r/R_b)^2 - 1}\right),
\tag{44}
$$

where M_*^{Sal} is the stellar (baryonic) mass, α_{SPS}^{Sal} is the Salpeter Stellar Population Synthesis parameter representing the uncertainty of the initial mass function (IMF), R_b is a reference radius based on an effective radius R_e, and $F(u)$ is defined by

$$
\begin{aligned}
F(u) &= \left(\frac{1}{\sqrt{1-u^2}}\right)\,\text{atanh}\left(\sqrt{1-u^2}\right)\ \text{if } u < 1, \\
&= \left(\frac{1}{\sqrt{u^2-1}}\right)\,\text{atan}\left(\sqrt{u^2-1}\right)\ \text{if } u > 1, \\
&= 1\ \text{if } u = 1.
\end{aligned}
\tag{45}
$$

The values for $\alpha_{SPS}^{Sal} = 0.92^{+0.09}_{-0.08}$ are from stellar population studies [18]. The galaxy stellar mass equals M_b within the full galaxy radius R_{gal}. The standard model dark matter mass profile is provided from [18] for comparative purposes. This is defined as follows:

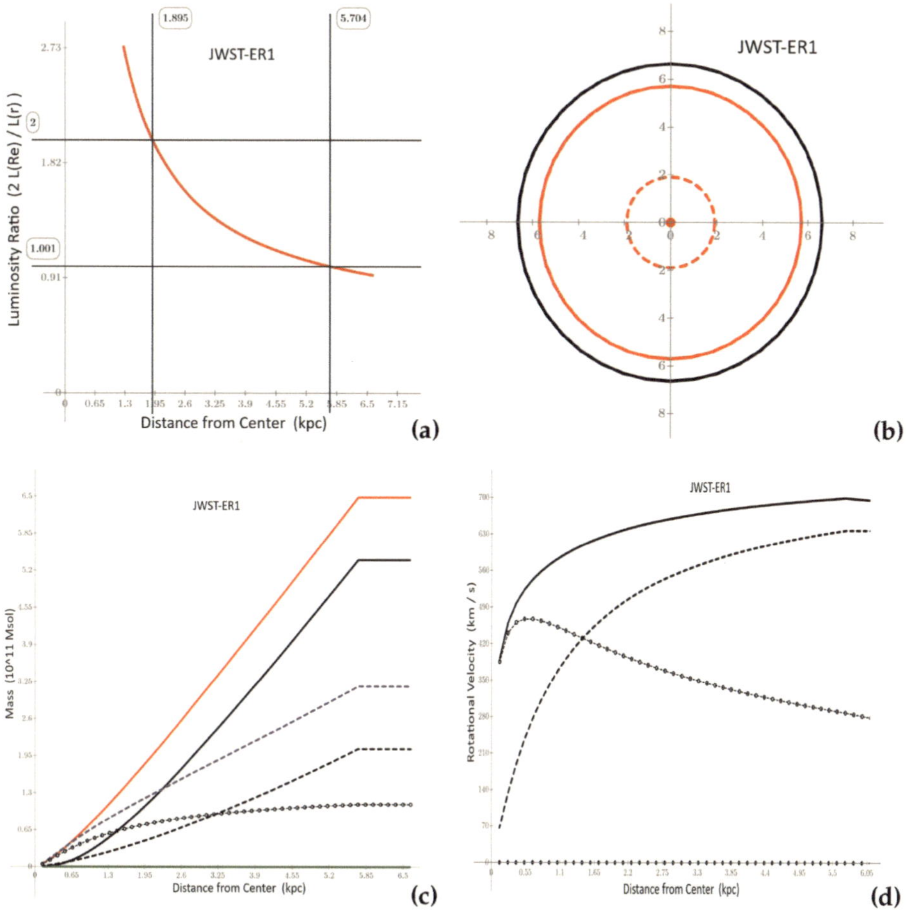

Figure 5.
JWST-ER1. (a) Galaxy luminosity ratio. The solid red line is the curve for $2L(R_e)/L(r)$ *where r is the distance from the galaxy center. At a Sérsic index of* $n = 5.0$ *[15], half of the luminosity is within effective radius* $R_e = 1.895$ kpc *and the full luminosity is within radius* $R_{gal} = 5.495$ kpc. *(b) Galaxy and Einstein ring curves. The solid black circle represents the center of the Einstein ring. The solid red circle represents the bounds of the galaxy and the dashed red line shows the effective half-light radius* R_e. *(c) Stellar, relativistic, and standard dark matter mass distribution curves. The black dashed line with open circles is the stellar mass using the Hernquist profile [16, 17]. The solid black line is the field self-interaction gravitational redshift mass using the Hernquist profile. The solid green line is the gravitational cosmological redshift mass (essentially zero) The solid red line is the sum of the stellar mass and the field self-interaction gravitational and cosmological redshift masses. The dashed black line is the dark matter mass using a simple power law mass distribution as in [18]. The dashed blue line is the sum of the stellar mass and dark matter mass, which amounts to just 46% of the lens mass at the distance of the impact parameter* $b = 6.632$ kpc. *(d) Distribution curves for the rotational velocities using the Hernquist profile [18]. The black dashed line with open circles is the Newtonian velocity curve. The black dashed line is the field self-interaction gravitational redshift boosted velocity curve. The black diamond point line is the gravitational self-interaction cosmological redshift boosted velocity curve. The solid black line is the Newtonian plus self-interaction gravitational and cosmological redshift boosted theoretical circular velocities.*

$$M_{DM}(r) = M_*^{Sal} A_{DM} \left(\frac{r}{R_e} \right)^{3+\gamma_{DM}}, \tag{46}$$

where A_{DM} and γ_{DM} are parameters determined by best fit from population studies. As described in [18], the best fit parameters for $A_{DM} = M_{DM}(r < R_e)/M_*^{Sal} = 0.21 \pm 0.04$ and for $\gamma_{DM} = -1.60^{+0.18}_{-0.13}$.

The circular velocity $\nu(r)$ of objects within the galaxy is given by Eq. (10), with Newtonian energy and gravitational self-interaction redshift energies

$$\nu(r) = \sqrt{\frac{GM(r)}{r} + k_f\left(\frac{M(r)}{M_b}\right)\int_0^r \frac{GM(u)}{u^2}du + k_e cH_0 r}, \quad (47)$$

where k_f is the gravitation coupling constant, k_e is the expansion coupling constant and the baryon mass beyond the galaxy radius, and R_{gal} is the galaxy total baryon mass $M(r)$ where

$$M(r > R_{gal}) = M_b. \quad (48)$$

The total galaxy mass $M_{gal}(r)$ within central distance r, which includes stellar mass, molecular gas mass, and gravitational and cosmological redshift masses, from Eq. (47), is given by,

$$\begin{aligned} M_{gal}(r) = M(r) + k_f Rx(r)\left(\frac{M(r)}{M_b}\right)\int_0^{Rx(r)} \frac{M(u)}{u^2}du + \\ + \frac{k_e cH_0}{G}(Rx(r))^2, \end{aligned} \quad (49)$$

where $Rx(r) = r$ if $r < R_{gal}$, otherwise $Rx(r) = R_{gal}$. Assuming that the impact parameter b is greater than the full galaxy radius R_{gal}, the lens mass, Eq. (42), equals the galaxy total mass at the galaxy radius R_{gal}, that is, $M_{lens} = M_{gal}(R_{gal})$.

From Eq. (49), setting $r = b$, this is expressed by

$$M_{lens} = M_{gal}(b) = M_b + k_f R_{gal}\int_0^{R_{gal}} \frac{M(u)}{u^2}du + \frac{k_e cH_0}{G}r_0^2, \quad (50)$$

where the baryon mass $M(b) = M_b$, the total baryon mass of the galaxy. Eqs. (42) and (50) imply that

$$\begin{aligned} M_{lens} = \theta_E^2\left(\frac{D_L D_S c^2}{4GD_{LS}}\right) = M_{gal}(R_{gal}) = \\ = M_b\left(1 + \frac{k_f R_{gal}}{M_b}\int_0^{R_{gal}} \frac{M(u)}{u^2}du + \frac{k_e cH_0}{GM_b}r_0^2\right). \end{aligned} \quad (51)$$

Assuming that the gravitational cosmological redshift coupling $k_c = k_e cH_0 r_0^2/GM_b$, where $r_0 = R_0 = c/H_0$ is the Hubble radius, then from Eq. (51), an expression for the self-interaction gravitational redshift coupling k_g defined in Eq. (15) is expressed by

$$\begin{aligned} k_g = \left(\frac{k_f R_{gal}}{M_b}\right)\int_0^{R_{gal}} \frac{M(u)}{u^2}du + \\ + \frac{k_e cH_0}{GM_b}R_0^2 - k_c = \\ = \left(\frac{k_f R_{gal}}{M_b}\right)\int_0^{R_{gal}} \frac{M(u)}{u^2}du, \end{aligned} \quad (52)$$

which ensures that, in Eq. (18), $\partial k_g/\partial r = 0$ for $r > R_{gal}$. From Eqs. (51) and (52), it is evident that the galaxy mass increase $\left(k_g M_b + k_e c H_0 R_0^2/GM_b\right)$ due to the energy increase of the galaxy because of the loss in gravitational field energy, due to gravitational and cosmological redshifts, resembles the dark matter and dark energy.

The lensing galaxy JWST-ER1 is at redshift $z_l = 1.94$. The source galaxy which forms the ring is at redshift $z_s = 2.98$. The effective radius of the galaxy is $R_e = 0.22$ arcsec $\times \pi/(180 \times 3600) \times DL = 1.895$ kpc with Sérsic index $n = 5$. The Einstein ring radius is $b = \Theta_E \times DL = 0.77$ arcsec $\times \pi/(180 \times 3600) \times DL = 6.632$ kpc. The mass within the ring radius is $M_{lens} = 6.476 \times 10^{11}\,M_\odot$. The radius parameter $R_b = 0.511 \times R_e = 0.968$ kpc. The stellar mass of the galaxy is estimated to be $M_b = 1.10 \times 10^{11}\,M_\odot$. The stellar mass plus dark matter mass, using a simple power law mass distribution as in [18] shown in Eq. (46), amounts to about 46% of the lens mass. The galaxy size radial distance was found to be $R_{gal} = 5.704$ kpc.

Figure 5(d) shows the gravitational self-interaction redshift theoretical components of the rotation velocities for galaxy JWST- ER1 [15]. The velocity distributions of the Newtonian rotation curve (black dashed line with open circles), the gravitational redshift boost to the velocity curve (black dashed line), the cosmological redshift boost to the velocity curve (black diamond point line), and the sum of these curves (solid black line) form the GRST predicted rotational velocity curve for massive objects rotating in the galaxy.

9. Two lensing systems from the COSMOS field

Two other lens systems with Einstein rings are analyzed, by the same methods as for JWST-ER1, from the COSMOS field [21]. These are COSMOS 5921 + 0638 [22] at a redshift $z_l = 0.551$ using cosmological parameters $H_0 = 70$ km s^{-1} mpc^{-1}, $\Omega_M = 0.3$, and $\Omega_\Lambda = 0.7$ and COSMOS 0018 + 3845 [21] at a redshift $z_l = 0.9755$ using parameters $H_0 = 72$ km s^{-1} mpc^{-1}, $\Omega_M = 0.258$, and $\Omega_\Lambda = 0.742$.

Figure 6(a)-(d) show the results for COSMOS 5921 + 0638. The solid red line on **Figure 6(c)** is the sum of stellar and gravitational redshift masses, and the dashed blue line is for the stellar plus dark matter masses, which amounts to 100% of the lens mass for both methods at the distance of the impact parameter $b = 4.556$ kpc. The mass within the ring radius is $M_{lens} = 1.199 \times 10^{11}\,M_\odot$.

The source galaxy which forms the ring is at redshift $z_s = 3.14$. The effective radius of the galaxy is $R_e = 0.45$ arcsec $\times \pi/(180 \times 3600) \times DL = 2.888$ kpc with Sérsic index $n = 150$. The Einstein ring radius is $b = \Theta_E \times DL = 0.71$ arcsec $\times \pi/(180 \times 3600) \times DL = 4.556$ kpc. The radius parameter $R_b = 0.511 \times R_e = 1.476$ kpc. The stellar mass of the galaxy is estimated to be $M_b = 6.166 \times 10^{10}\,M_\odot$. The stellar mass plus dark matter mass amounts to 100% of the lens mass. The galaxy size radial distance was found to be $R_{gal} = 4.084$ kpc.

It is remarkable that the gravitational self-interaction redshift fit for this galaxy is nearly identical to the standard dark matter model fit. More details on the galaxy are shown in the caption of **Figure 6**.

The results for COSMOS 0018 + 3845 are shown in **Figure 7(a)-(d)**. The dashed blue line on **Figure 7(c)** is the sum of the standard model stellar mass and dark matter mass, which amounts to just 32% of the lens mass at the distance of the impact parameter $b = 10.480$ kpc. The mass within the ring radius is $M_{lens} = 6.307 \times 10^{11}\,M_\odot$.

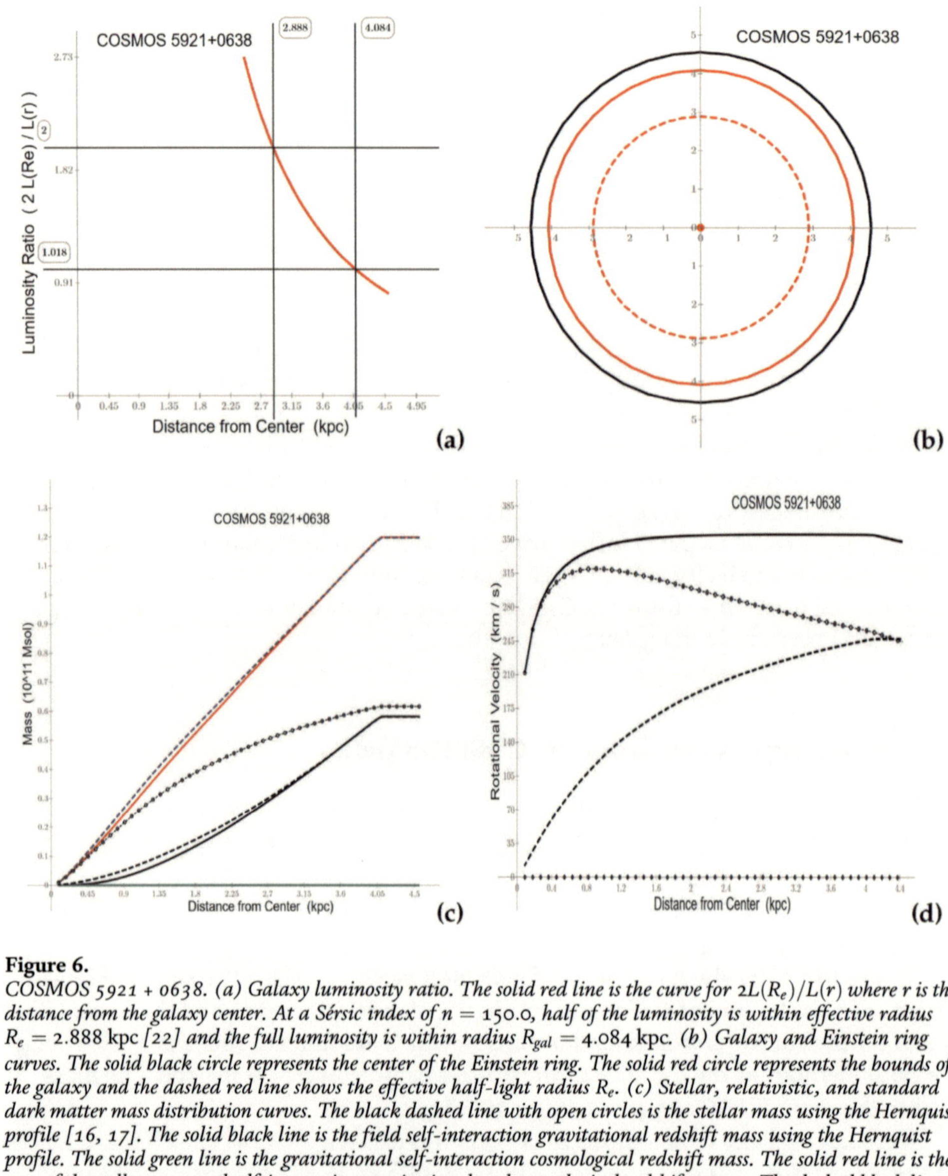

Figure 6.

COSMOS 5921 + 0638. (a) Galaxy luminosity ratio. The solid red line is the curve for $2L(R_e)/L(r)$ *where r is the distance from the galaxy center. At a Sérsic index of* $n = 150.0$*, half of the luminosity is within effective radius* $R_e = 2.888$ kpc *[22] and the full luminosity is within radius* $R_{gal} = 4.084$ kpc. *(b) Galaxy and Einstein ring curves. The solid black circle represents the center of the Einstein ring. The solid red circle represents the bounds of the galaxy and the dashed red line shows the effective half-light radius* R_e*. (c) Stellar, relativistic, and standard dark matter mass distribution curves. The black dashed line with open circles is the stellar mass using the Hernquist profile [16, 17]. The solid black line is the field self-interaction gravitational redshift mass using the Hernquist profile. The solid green line is the gravitational self-interaction cosmological redshift mass. The solid red line is the sum of the stellar mass and self-interaction gravitational and cosmological redshift masses. The dashed black line is the dark matter mass using a simple power law mass distribution as in [18]. The dashed blue line is the sum of the stellar mass and dark matter mass, which amounts to 100% of the lens mass at the distance of the impact parameter* $b = 4.621$ kpc. *(d) Distribution curves for the rotational velocities using the Hernquist profile [18]. The black dashed line with open circles is the Newtonian velocity curve. The black dashed line is the field self-interaction gravitational redshift boosted velocity curve. The black diamond point line is the field self-interaction cosmological redshift boosted velocity curve. The solid black line is the Newtonian plus self-interaction gravitational and cosmological redshift boosted theoretical circular velocity.*

The source galaxy which forms the ring is at redshift $z_s = 3.14$. The effective radius of the galaxy is $R_e = 0.30$ arcsec $\times \pi/(180 \times 3600) \times DL = 2.382$ kpc with Sérsic index $n = 4.8$. The Einstein ring radius is $b = \Theta_E \times DL = 1.32$ arcsec $\times \pi/(180 \times 3600) \times$ $DL = 10.480$ kpc. The radius parameter $R_b = 0.511 \times R_e = 1.217$ kpc. The stellar mass of the galaxy is estimated to be $M_b = 5.373 \times 10^{10}\,M_{\odot}$. The stellar mass plus dark matter mass amounts to about 32% of the lens mass. The galaxy size radial distance was found to be $R_{gal} = 7.860$ kpc. More details on the galaxy are shown in the caption of **Figure 7**.

Tables 3-6 show the parameters used and results obtained for the analysis of all three lenses. **Table 7** shows the dispersion velocities, measured and predicted, for the elliptical lensing galaxies. The GRST predicted dispersion velocity $\sigma_{grst} = \nu_{avg}/\sqrt{3}$, Eq. (55), for JWST-ER1 of 360.5 km s^{-1} compares to within 2% of the singular

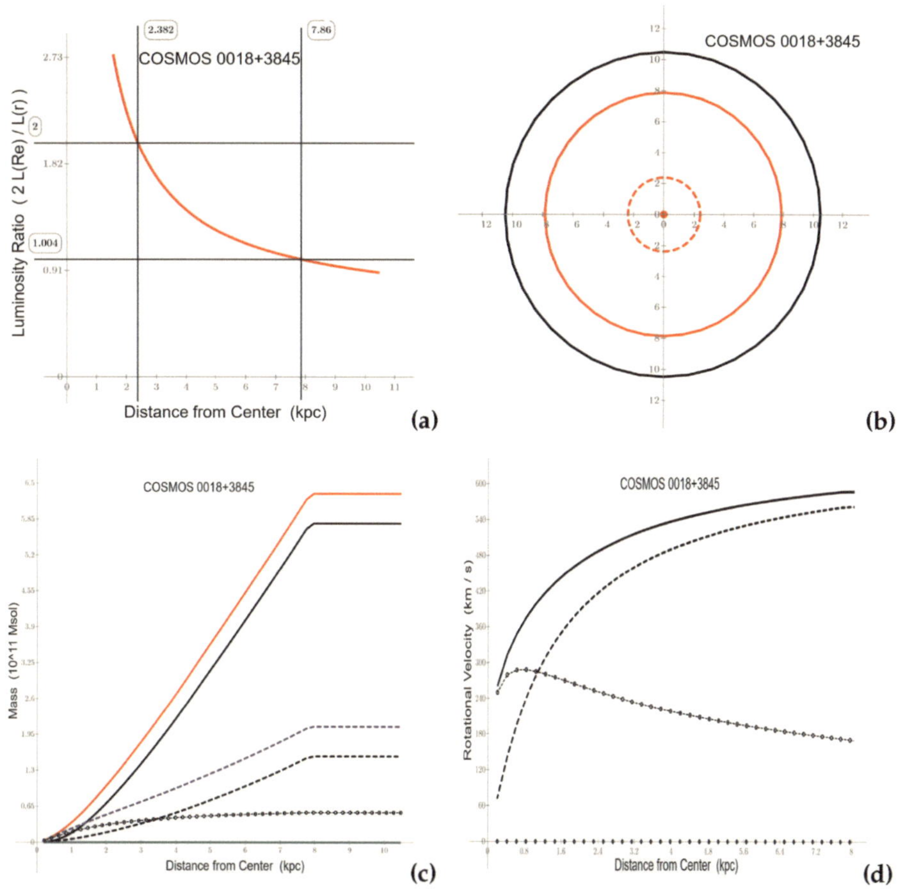

Figure 7.
COSMOS 0018 + 3845. (a) Galaxy luminosity ratio. The solid red line is the curve for $2L(R_e)/L(r)$, where r is the distance from the galaxy center. At a Sérsic index of $n = 4.8$ [21], half of the luminosity is within effective radius $R_e = 2.382$ kpc and the full luminosity is within radius $R_{gal} = 7.860$ kpc. (b) Galaxy and Einstein ring curves. The solid black circle represents the center of the Einstein ring. The solid red circle represents the bounds of the galaxy and the dashed red line shows the effective half-light radius R_e. (c) Stellar, relativistic and standard dark matter mass distribution curves. The black dashed line with open circles is the stellar mass using the Hernquist profile [16, 17]. The solid black line is the field self-interaction gravitational redshift mass using the Hernquist profile. The solid green line is the field self-interaction cosmological redshift mass. The solid red line is the sum of the stellar mass and the self-interaction gravitational and cosmological redshift masses. The dashed black line is the dark matter mass using a simple power law mass distribution as in [18]. The dashed blue line is the sum of the stellar mass and dark matter mass, which amounts to just 32% of the lens mass at the distance of the impact parameter $b = 10.508$ kpc. (d) Distribution curves for the rotational velocities using the Hernquist profile [18]. The black dashed line with open circles is the Newtonian velocity curve. The black dashed line is the field self-interaction gravitational redshift boosted velocity curve. The black diamond point line is the field self-interaction cosmological redshift boosted velocity curve. The solid black line is the Newtonian plus self-interaction gravitational and cosmological redshift boosted theoretical circular velocity.

Galaxy	H_0	Ω_M	Ω_Λ	Z	Z
	km s^{-1} mpc^{-1}			**Lens**	**Source**
JWST-ER1	70	0.271	0.729	1.94	2.98
COSMOS 5921 + 0638	70	0.3	0.7	0.551	3.14
COSMOS 0018 + 3845	72	0.258	0.742	0.9755	3.96

Table 3.
Lens galaxy parameters-Part 1.

Galaxy	DL	DS	DLS	Θ_E	M_{lens}
	10^3 mpc	**10^3 mpc**	**10^3 mpc**	**arcsec**	**10^{11} M$_\odot$**
JWST-ER1	1.777	1.640	0.3275	0.77	6.476
COSMOS 5921 + 0638	1.324	1.567	1.071	0.71	1.199
COSMOS 0018 + 3845	1.638	1.467	0.8152	1.32	6.307

Table 4.
Lens galaxy parameters-Part 2.

Galaxy	M_b	R_e	b	α_{SPS}^{Sal}	A_{DM}	γ_{DM}
	10^{10} M$_\odot$	**kpc**	**kpc**			
JWST-ER1	11.0	1.895	6.632	0.667	0.21	−1.60
COSMOS 5921 + 0638	6.166	2.888	4.556	0.76	0.253	−1.40
COSMOS 0018 + 3845	5.373	2.382	10.480	0.602	0.21	−1.42

Table 5.
Lens galaxy parameters-Part 3.

Galaxy	M_*^{Sal}	R_b	R_{gal}	k_f	k_e	k_g	n
	10^{11} M$_\odot$	**kpc**	**kpc**		**10^{-13}**		
JWST-ER1	2.109	0.968	5.704	0.457	6.67	4.893	5.0
COSMOS 5921 + 0638	1.318	1.476	4.084	0.163	5.767	0.944	150
COSMOS 0018 + 3845	1.120	1.217	7.860	0.925	0.460	10.738	4.8

Table 6.
Lens galaxy parameters-Part 4.

Galaxy	σ	σ_{grst}	σ_{sis}
	km s^{-1}	**km s^{-1}**	**km s^{-1}**
JWST-ER1	n/a	363.8	365.6
COSMOS 5921 + 0638	189.3	196.1	189.8
COSMOS 0018 + 3845	289	290.0	287.0

Table 7.
Lens galaxy velocity dispersions.

isothermal sphere model [23, 24] dispersion σ_{sis} of 365.6 km s^{-1}, with no measurement value available. For the other galaxies the GRST dispersion velocity is within 3% of the measured value.

10. Some derivations

10.1 Coupling constant k_f

Inverting Eq. (52), we get an expression for the galaxy gravitation coupling constant k_f in terms of the lens field self-interaction gravitational redshift coupling k_g,

$$k_f = \frac{k_g M_b}{R_{gal} \int_0^{R_{gal}} \frac{M(u)}{u^2} du}, \tag{53}$$

which is used in fits of the rotation energy, Eq. (10), to the lensing galaxies using the result of the lensing mass and the ring radius. **Table 6**, columns 5 and 6, gives values for the gravitational coupling constant k_f derived from Eq. (53) and the expansion coupling constant k_e from Eq. (12).

10.2 Average rotational velocity ν_{avg}

The average galaxy rotation velocity ν_{avg}, using Eq. (47), is expressed by

$$\begin{aligned} v_{avg} &= \frac{1}{R}\int_0^R \nu(s) ds = \\ &= \frac{1}{R}\int_0^R \left(\sqrt{\frac{GM(s)}{s} + k_f \left(\frac{M(s)}{M_b}\right) \int_0^s \frac{GM(u)}{u^2} du + k_e c H_0 s} \right) ds, \end{aligned} \tag{54}$$

where R is the radius of the galaxy. Define the GRST dispersion velocity σ_{grst} (by)

$$\sigma_{grst} = \frac{\nu_{avg}}{\sqrt{3}}, \tag{55}$$

where division by $\sqrt{3}$ gives the velocity component along any one of the three perpendicular directions.

A common theoretical velocity dispersion parameter is based on a Singular Isothermal Sphere model [23] for a galaxy lens system. This is given by

$$\sigma_{sis} = \sqrt{\frac{c^2 \Theta_E D_S}{4\pi D_{LS}}}. \tag{56}$$

The observed and theoretical velocity dispersions can be compared

$$\sigma \approx \sigma_{grst} \approx \sigma_{sis}. \tag{57}$$

Table 7 lists the observed velocity dispersion and the theoretical velocity dispersions using Eqs. (55) and (56).

10.3 Luminosity ratio $L_{2R_e}(R)$

The Sérsic light intensity profile for a galaxy in terms of the half-light radius is given by

$$I(r) = I_0 e^{7.67} \exp\left(-7.67\left(\frac{r}{R_e}\right)^{\frac{1}{n}}\right), \tag{58}$$

where I_0 is the light intensity at $r = 0$, R_e is the effective (half light) radius, and n is the Sérsic index. $n = 1$ for disk spiral galaxies, and $n > 1$ for elliptical galaxies. The luminosity is the surface integral of the surface brightness at a particular radius R, the integral of $I(r)$ through a surface area $2\pi r dr$. Then, the ratio of twice the half light luminosity over the luminosity at radial distance R, calling it $L_{2R_e}(R)$, is expressed by

$$\begin{aligned} L_{2R_e}(R) &= \frac{2\int_0^{R_e} 2\pi u\, I(u)\, du}{\int_0^{R} 2\pi u\, I(u)\, du}, \\ &= \frac{2\int_0^{R_e} u \exp\left(-7.67\left(\frac{u}{R_e}\right)^{\frac{1}{n}}\right) du}{\int_0^{R} u \exp\left(-7.67\left(\frac{u}{R_e}\right)^{\frac{1}{n}}\right) du}. \end{aligned} \tag{59}$$

This algorithm is used to determine the radial extent R_{gal} of the galaxy, that is, R_{gal} is the radius at which $L_{2R_e}(R_{gal}) = 1$.

11. Conclusions

The fitting of spiral galaxy rotation curves using the GRST of gravitational self-interaction producing gravitational redshift and cosmological redshift in the galaxy field may be perceived as just another explanation for high velocity rotations which flatten out at large radial distances, as these are adequately explained by other dark matter theories, particularly, in principle, in the framework of extended gravity [25], although gravitational self-interaction is a more likely reality than dark matter particles and the gravitational self-interaction theory can explain binary star system decay without the emission of gravitational waves, as described in [26] which was defined in terms of the self-interaction of gravitons.

However, it is in gravitational lensing where this theory may have revealed its significance, because it has been shown how the GRST gravitation self-interaction redshift theory can explain the lens mass when the standard model dark matter theory cannot, as in the cases of the ring systems in JWST-ER1 and COSMOS 0018 + 3845. The fundamental property about gravitational self-interaction in galaxies is that the total galaxy mass $M_{gal}(r) \propto r$ at large distance, since the gravitational and cosmological redshifts continue to act at larger distances r, adding more relativistic mass to the system. This can be shown as follows.

Assume that most of the baryonic mass M_b is within a radius $R_b < R_{gal}$. Then, for $r > R_b$, expand Eq. (51) into the form,

$$M_{gal}(r > R_b) \approx M_b + k_f r\left(\int_0^{R_b} \frac{M(u)}{u^2} du + M_b \int_{R_b}^{r} \frac{du}{u^2}\right) + \frac{k_e c H_0 r^2}{G} \approx$$
$$\approx M_b + k_f r\left(\frac{M_b}{R_a} + \frac{M_b}{R_b} - \frac{M_b}{r}\right) + \frac{k_e c H_0 r^2}{G} \approx \tag{60}$$
$$\approx (1 - k_f) M_b + Br + \frac{k_e c H_0 r^2}{G}$$

where $\frac{M_b}{R_a} = \int_0^{R_b} \frac{M(u)}{u^2} du$ for some distance R_a and $B = k_f M_b \left(\frac{1}{R_a} + \frac{1}{R_b}\right)$. It is assumed that $0 \le k_f \le 1$. All three of the lens systems in this study displayed this approximately $M_{gal}(r) \propto r$ relation, and there is not any appreciable r^2 behavior on account of the small values for the cosmological coupling constant $k_e < 10^{-12}$ for these galaxies. These results are due to determining the values for k_f at the edge of the galaxy and k_e at the edge of the universe $R_0 = c/H0$.

A new result in this research is the ability to use the surface brightness distribution $I(r)$ to obtain the galaxy radius R_{gal} from the half to full luminosity ratio function $L_{2R_e}(R)$, which has been crucial in determining the overall mass and velocity distributions.

Another new feature in this research is the determination of the galaxy average rotation velocity v_{avg} which is a simple average of the integral of the rotation velocity over the total radial galaxy size. This result makes a good match, to within a few percent, to the measured velocity dispersion for the galaxy. This has been supportive in justifying the derived galaxy rotation velocity distribution.

The discovery of the gravitational self-interaction cosmological redshift of the field energy cannot be overemphasized, as it makes galaxy dynamics a microcosm of the universe dynamics, possessing both dark matter and dark energy profiles.

The GRST self-interaction of the gravitational field describes how the mass of a galaxy increases relativistically due to the gravitational and cosmological redshifts of the field itself. This is a non-electromagnetic effect; therefore, it is not measureable by photometric instruments. These effects increase the relativistic mass of the galaxy system which boosts the rotation velocity of the orbiting masses and also increases the strength of the gravitational lensing abilities. In this regard, these attributes satisfy the basic properties of dark matter and dark energy, being non-photometric and observable only by gravitational effects.

Conflict of interest

The author declares no conflict of interest.

Author details

Firmin Oliveira
East Asian Observatory/James Clerk Maxwell Submillimetre Telescope, Hilo, Hawaii, USA

*Address all correspondence to: firmjay@hotmail.com

References

[1] Oliveira FJ. How the redshift of gravitons explains dark matter and dark energy. Journal of Modern Physics. 2022; **13**(11):1348-1368. DOI: 10.4236/jmp.2022.1311084

[2] Lelli F, McGaugh SS, Schombert JM. SPARC (Spitzer Photometry & Accurate Rotation Curves) database [Internet]. 2016 [cited 2025 Nov 25]. Available from: http://astroweb.cwru.edu/SPARC/

[3] Lelli F, McGaugh SS, Schombert JM. SPARC: Mass models for 175 disk galaxies with Spi tzer photometry and accurate rotation curves. Astronomy Journal. 2016;**152**(6):157-170. DOI: 10.3847/0004-6256/152/6/157

[4] Einstein A. The Foundation of the General Theory of Relativity. New York: Dover Publications, Inc; 1952

[5] Misiaszek M, Rossi N. Direct detection of dark matter: A critical review. Symmetry. 2024;**16**(2):id. 201. DOI: 10.3390/sym16020201

[6] Meng Y, Wang Z, Tao Y, Abdukerim A, Bo Z, Chen W, et al. Dark matter search results from the PandaX-4T commissioning run. Physical Review Letters. 2021;**127**:261802. DOI: 10.1103/PhysRevLett.127.261802

[7] Sabulsky DO, Dutta I, Hinds EA, Elder B, Burrage C, Copeland EJ. Experiment to detect dark energy forces using atom interferometry. Physical Review Letters. 2019;**123**(6):061102. DOI: 10.1103/PhysRevLett.123.061102

[8] Oliveira FJ. Evaluating the effects of graviton redshift upon spiral galaxy rotation curves, surface brightness magnitudes and gravitational lensing. JHEPGC. 2024;**10**(3):967-985. DOI: 10.4236/jhepgc.2024.103059

[9] Deur A. Implications of graviton-graviton interaction to dark matter. 2009. Available from: https://arxiv.org/pdf/astro-ph/0901.4005

[10] McGaugh SS. The baryonic Tully-fisher relation of galaxies with extended rotation curves and the stellar mass of rotating galaxies. The Astrophysical Journal. 2005;**632**(2):859-871. DOI: 10.1086/432968

[11] Oliveira FJ. New insights into the action of gravitons in spiral galaxies. JHEPGC. 2023;**9**(4):968-983. DOI: 10.4236/jhepgc.2023.94072

[12] Pritchard J. Lecture Notes: Gravitational Lensing. 2009. Available from: https://lweb.cfa.harvard.edu/~dfabricant/huchra/ay202/lectures/lecture12.pdf

[13] By Krishnavedala - Own work, CC0. Available from: https://commons.wikimedia.org/w/index.php?curid=20298731

[14] STScI. Webb Telescope [Internet]. Baltimore, MD: Space Telescope Science Institute; [cited 2025 Nov 25]. Available from: https://webbtelescope.org/home

[15] van Dokkum P, Brammer G, Wang B, Leja J, Conroy C. A massive compact quiescent galaxy at z=2 with a complete Einstein ring in JWST imaging. Nature Astronomy. 2023;**8**:119-125. DOI: 10.1038/s41550-023-02103-9

[16] Hernquist L. An analytical model for spherical galaxies and bulges. The Astrophysical Journal. 1990;**356**:359. DOI: 10.1086/168845

[17] Keeton CR. A Catalog of Mass Models for Gravitational Lensing. 2001. Available from: arXiv:astro-ph/0102341

[18] Oguri M, Rusu CE, Falco EE. The stellar and dark matter distributions in elliptical galaxies from the ensemble of strong gravitational lenses. Monthly Notices of the Royal Astronomical Society. 2014;**439**(3):2494-2504. DOI: 10.1093/mnras/stu106

[19] Wright EL. A cosmology calculator for the world wide web. The Publications of the Astronomical Society of the Pacific. 2006;**118**(850):1711-1715. DOI: 10.1086/510102

[20] Bruzual G, Charlot S. Stellar population synthesis at the resolution of 2003. Monthly Notices of the Royal Astronomical Society. 2003;**344**: 1000-1028. DOI: 10.1046/j.1365-8711.2003.06897.x

[21] Faure C, Anguita T, Alloin D, Bundy K, Finoguenov A, Leauthaud A, et al. On the evolution of environmental and mass properties of strong lens galaxies in COSMOS[å]. Astronomy and Astrophysics. 2011;**529**:A72. DOI: 10.1051/0004-6361/200913498

[22] Aguita T, Faure C, Kneib J-P, Wambsganss J, Knobel C, Koekemoer AM, et al. COSMOS 5921 +0638: Characterization and analysis of a new strong gravitationally lensed AGN[å]. Astronomy and Astrophysics. 2021; **507**(1):35-46. DOI: 10.1051/0004-6361/200912091

[23] Keeton CR. A Catalog of Mass Models for Gravitational Lensing. 2002. Available from: https://arxiv.org/abs/astro-ph/0102341

[24] Mozumdar P, Fassnacht CD, Treu T, Spiniello C, Shajib AJ. New lensing galaxy redshift and velocity dispersion measurements from keck spectroscopy of eight lensed quasar systems. Astronomy & Astrophysics. 2023;**672**:A20. DOI: 10.1051/0004-6361/202245082

[25] Corda C. Interferometric detection of gravitational waves: The definitive test for general relativity. International Journal of Modern Physics D. 2009;**18**: 2275-2282. DOI: 10.1142/S0218271809015904

[26] Oliveira FJ. Binary star system decay by graviton interaction. Journal of High Energy Physics, Gravitation and Cosmology. 2022;**8**:317-329. DOI: 10.4236/jhepgc.2022.82026

Chapter 9

The Gravitational Analog of Maxwell's Equations: A Solution to the Vacuum-Energy and Cosmological-Constant Problems within Conventional General Relativity

Richard J. Cook

Abstract

The Einstein field equation of general relativity suffers from two long-standing and closely related problems: the vacuum energy problem and the cosmological constant problem. Here we point out there is a set of equations for the curvature tensor in conventional general relativity, the 'curvature equations' or (CEs), that are strikingly similar to the Maxwell equations of electrodynamics, and these equations do not suffer from the vacuum energy problem or the cosmological constant problem. According to the CEs, the vacuum energy does not effect the curvature of spacetime, any solution of Einstein's equation is also a solution of the CEs, and the cosmological constant is an integration constant unrelated to the vacuum energy of quantum fields. If the CEs are viewed as the field equations of general relativity, the vacuum-energy and cosmological-constant problems are resolved within conventional general relativity.

Keywords: general relativity, vacuum energy problem, cosmological constant problem, cosmological constant, relativistic cosmology

1. Introduction

Ever since Dirac's development of quantum electrodynamics in 1927, with its $(1/2)\hbar\omega$ ground-state energy for each mode of the electromagnetic field, it was evident that the vacuum could have a very large energy density [1]. The energy-momentum tensor of the quantum vacuum is necessarily of the form

$$T^{\mu\nu}_{qv} = -\rho_{qv} c^2 g^{\mu\nu}, \tag{1}$$

[for a metric with signature $(-, +, +, +)$] so it is the same in each local inertial frame and is invariant under local Lorentz transformations. Here ρ_{qv} is the mass density equivalent to the vacuum energy density. Estimates of the vacuum energy place the equivalent mass density in a range from $\rho_{qv} \sim 2 \times 10^{18}$ kg m^{-3}$(8 \times 10^{-3}\,\mathrm{GeV}^4)$, for quantum electrodynamics alone with a cutoff at the proton energy [2] to $\rho_{qv} \sim 5 \times 10^{91}$ kg m^{-3}$(2 \times 10^{71}\,\mathrm{GeV}^4)$, if the Planck scale is chosen as cutoff [3].

1.1 The vacuum energy problem

Such large values for the vacuum energy must not be included in the source term of Einstein's equation

$$G^{\mu\nu} = \frac{8\pi G}{c^4} T^{\mu\nu}, \tag{2}$$

if that equation is to give results in agreement with observation, and we do not know why the vacuum energy does not contribute to the curvature of spacetime in the same way as any other form of energy. For example, for a vacuum density ρ_{qv} in the above range, the Einstein equation predicts the universe, in its present state of expansion, would expand and then collapse to a "big crunch" in less than a microsecond! And we cannot simply set the vacuum energy to zero, because the measured Casimir effect seems to provide direct evidence for the existence of such a vacuum energy, [4, 5].

Therefore, we have both theoretical and experimental evidence that the vacuum energy of quantum fields exits and is large, and yet we cannot use this energy in Einstein's equation without making predictions that are vastly at variance with observation. This has been called the "Vacuum Energy Citastrophy" [6], and the question to be answered may be stated as follows:

The Vacuum Energy Problem: Why does the vacuum energy of quantum fields not act as a source of curvature in Einstein's equation in the same manner as any other form of energy?

1.2 The cosmological constant problem

A very different view of the vacuum energy developed after the discovery of the accelerating expansion of the universe by Perlmutter and Riess [7–10]. According to this view, the cosmological term in Einstein's equation is brought across the equal sign and interpreted as the *total* energy-momentum tensor of the quantum vacuum,

$$T_{\Lambda}^{\mu\nu} = -\frac{\Lambda c^4}{8\pi G} g^{\mu\nu} = -\rho_{\Lambda} c^2 g^{\mu\nu}. \tag{3}$$

Then, with the cosmological constant Λ chosen to account for the accelerating expansion of the universe, the effective mass density of the vacuum $\rho_{\Lambda} = \Lambda c^2/8\pi G\,[= 6 \times 10^{-27}$ kg m^{-3}$(2.5 \times 10^{-47}\mathrm{GeV}^4)]$ is found to be many orders of magnitude smaller than the above estimates for the quantum vacuum energy. This is the "cosmological constant problem," [3, 7] which again we express as a question to be answered:

The Cosmological Constant Problem: If the cosmological term in Einstein's equation represents the vacuum energy of quantum fields, why is the observed energy density

$\rho_\Lambda c^2 = \Lambda c^4/8\pi G$ so much smaller than estimates of the vacuum energy density $\rho_{qv}c^2$ of quantum fields?

This has famously been called the largest discrepancy between theory and experiment in the history of physics. There is extensive literature describing attempts to answer this question, ranging from modifications of general relativity to alternative approaches to calculating the vacuum energy: [11–20]. But none has yet received general acceptance.

Point of Interpretation: The identification of the cosmological term in Einstein's equation with the vacuum energy of quantum fields is an hypothesis, and the 50 to 120 orders of magnitude difference between estimates of the vacuum energy density ρ_{qv} and the density ρ_Λ that follows from the measured accelerating expansion of the universe, could just as well be understood as strong evidence falsifying this hypothesis. The cosmological term is identified with vacuum energy merely because it has the same mathematical form as the vacuum energy-momentum tensor (a constant multiple of the metric tensor). But this is a weak argument. Coulomb's Law and Newton's law of gravitation have the same mathematical form, but this does not mean Coulomb's law is a gravitational effect.

The vacuum energy problem and the cosmological constant problem are problems with the Einstein field equation of general relativity.

1.3 Energy-momentum tensors

The energy-momentum tensors $T^{\mu\nu}_{qv}$ and $T^{\mu\nu}_{\Lambda}$ are defined in Eqs. (1) and (3), with one or the other zero depending on one's point of view. We also define $\hat{T}^{\mu\nu}$, with a hat "∧", to represent the energy and momentum of excitations of quantum fields above the vacuum state (real rather than virtual particles) or classical fields. Then the total energy-momentum tensor is

$$T^{\mu\nu} = \hat{T}^{\mu\nu} + T^{\mu\nu}_{qv} + T^{\mu\nu}_{\Lambda}, \tag{4}$$

and Einstein's equation, including the cosmological term and perhaps vacuum energy, reads

$$G^{\mu\nu} = \frac{8\pi G}{c^4}\left(\hat{T}^{\mu\nu} + T^{\mu\nu}_{qv} + T^{\mu\nu}_{\Lambda}\right). \tag{5}$$

Then, by metric compatibility ($\nabla_\alpha g^{\mu\nu} = 0$), and because $T^{\mu\nu}_{qv}$ and $T^{\mu\nu}_{\Lambda}$ are constant multiples of the metric tensor $g^{\mu\nu}$, we have $\nabla_\nu T^{\mu\nu}_{qv} = 0$ and $\nabla_\nu T^{\mu\nu}_{\Lambda} = 0$; and using $\nabla_\nu G^{\mu\nu} = 0$, the divergence of Eq. (5) gives

$$\nabla_\nu \hat{T}^{\mu\nu} = 0. \tag{6}$$

That is to say, only the above vacuum energy-momentum tensor $\hat{T}^{\mu\nu}$ plays a role in local energy and momentum conservation.

2. The curvature equations

Here we suggest there is a set of equations for the curvature tensor in conventional general relativity, the 'curvature equations' (or CEs)

$$\nabla_\omega R_{\alpha\beta\mu\nu} + \nabla_\nu R_{\alpha\beta\omega\mu} + \nabla_\mu R_{\alpha\beta\nu\omega} = 0, \tag{7}$$

$$\nabla_\nu R_{\alpha\beta}{}^{\mu\nu} = -4\pi J_{\alpha\beta}{}^{\mu}, \tag{8}$$

that may better serve as the field equations of general relativity in place of the Einstein Eq. (2). The CEs are strikingly similar to the Maxwell equations of electrodynamics,

$$\nabla_\omega F_{\mu\nu} + \nabla_\nu F_{\omega\mu} + \nabla_\mu F_{\nu\omega} = 0, \tag{9}$$

$$\nabla_\nu F^{\mu\nu} = -4\pi J^{\mu}. \tag{10}$$

In fact, if we follow Cartan and view the curvature tensor $R_{\alpha\beta}{}^{\mu\nu}$ and $J_{\alpha\beta}{}^{\mu}$ as tensor-valued 2-forms $\mathscr{R}^{\mu\nu}$ and $\mathcal{J}^{\mu}$, then the curvature Eqs. (7) and (8) become formally identical to Maxwell's equations:

$$\nabla_\omega \mathscr{R}_{\mu\nu} + \nabla_\nu \mathscr{R}_{\omega\mu} + \nabla_\mu \mathscr{R}_{\nu\omega} = 0, \tag{11}$$

$$\nabla_\nu \mathscr{R}^{\mu\nu} = -4\pi \mathcal{J}^{\mu}. \tag{12}$$

The formal similarity of the curvature equations to Maxwell's equations has been noted by others [21–24].

It is reasonable to suggest that the CEs replace the Einstein equation because: (1) the CEs are true and valid equations in conventional general relativity (no new or speculative physics); (2) any solution of Einstein's equations is also a solution of the CEs, and hence any successful prediction of the Einstein equation could just as well be counted as evidence supporting the curvature equations; and (3) the CEs do not suffer from the vacuum energy problem or the cosmological constant problem, but allow us to understand why the vacuum energy does not influence the curvature of spacetime and why the cosmological term does not represent the quantum vacuum energy.

Normally a suggestion to change the field equation of a successful theory would be a bold step indeed. But here this is not the case. To replace the Einstein equation by the CEs is merely a change of viewpoint as to which of valid equations within Einstein's theory is to be viewed as more fundamental. This step in no way constitutes a new theory of gravitation. It is Einstein's theory with the mildest possible reinterpretation of some of the equations. Moreover, a change to the CEs is a change to a pattern for a physical law whose form has already been battle tested in Electrodynamics and whose internal consistency is assured.

Finally, as we shall see, the CEs have a first integral in the form of an Einstein equation, without the vacuum energy in its source term, and with a cosmological constant unrelated to the vacuum energy of quantum fields. These are the properties that solve the vacuum-energy and cosmological-constant problems.

3. Derivation of the curvature equations

The first curvature equation (7) is the Bianchi identity whose validity in general relativity is not in question. The second curvature equation (8) is derived as follows: Begin with the once contracted Bianchi identity

$$\nabla_\nu R_{\alpha\beta}{}^{\mu\nu} = \nabla_\beta R^\mu{}_\alpha - \nabla_\alpha R^\mu{}_\beta \tag{13}$$

and use the Einstein equation in the form

$$R^\mu{}_\alpha = \frac{8\pi G}{c^4}\left(T^\mu{}_\alpha - \frac{1}{2}\delta^\mu{}_\alpha T\right), \tag{14}$$

to obtain the second curvature Eq. (8) with source term

$$J_{\alpha\beta}{}^\mu = \frac{2G}{c^4}\left(\nabla_\alpha \overline{T}^\mu{}_\beta - \nabla_\beta \overline{T}^\mu{}_\alpha\right), \tag{15}$$

where

$$\overline{T}^\mu{}_\alpha = T^\mu{}_\alpha - \frac{1}{2}\delta^\mu{}_\alpha T \tag{16}$$

is the Hilbert conjugate of $T^\mu{}_\alpha$, with $T = T^\mu{}_\mu$.

Using Eqs. (13), (15), and (16), the second CE (8) can also be written in the form

$$\nabla_\alpha\left[R^\mu{}_\beta - \frac{8\pi G}{c^4}\left(T^\mu{}_\beta - \frac{1}{2}\delta^\mu{}_\beta T\right)\right] - \nabla_\beta\left[R^\mu{}_\alpha - \frac{8\pi G}{c^4}\left(T^\mu{}_\alpha - \frac{1}{2}\delta^\mu{}_\alpha T\right)\right] = 0. \tag{17}$$

From this we see that, if we have a solution of the Einstein equation (if the quantities in square brackets vanish), then that is also a solution of the curvature equations.

4. Curvature equations as solution of the vacuum-energy problem

The solution to the vacuum-energy problem provided by the curvature equations is very simple indeed. For the energy-momentum tensor $T_{qv}{}^\mu{}_v = -\rho_{qv}c^2\delta^\mu{}_v$ of the quantum vacuum, the Hilbert conjugate $\overline{T}_{qv}{}^\mu{}_v = \rho_{qv}c^2\delta^\mu{}_v$ is also a constant multiple of the metric tensor $g^\mu{}_v = \delta^\mu{}_v$, and by metric compatibility the covariant derivatives in the source term (15) acting on $\overline{T}_{qv}{}^\mu{}_v$ vanish. Therefore, according to the curvature equations the vacuum energy of quantum fields has no effect on the curvature of spacetime, *and this is what is observed in practice.* This also means that any equations derived solely from the curvature equations will contain no information about the vacuum energy. We break this out as.

> *Result I: Contributions to the energy-momentum tensor in the form of a vacuum energy, $T_{qv}{}^\mu{}_v = -\rho_{qv}c^2 g^\mu{}_v = -\rho_{qv}c^2\delta^\mu{}_v$, are annihilated by the covariant derivatives in the source term of the curvature equations and have no effect on the curvature of spacetime. Thus no information about the vacuum energy is contained in any relation derived solely from the curvature equations.*

We view this as a simple and elegant solution to the vacuum-energy problem, which eliminates the vacuum energy catastrophe, but only if the curvature equations are viewed as the field equations of general relativity.

The solution to the vacuum-energy problem in Result I creates another apparent problem. If the cosmological term is viewed as part of the energy-momentum tensor, it also is a constant multiple of the metric tensor and would be nulled by the covariant derivatives in the source term (15). This makes it appear we would have no way to account for the accelerating expansion of the universe. But, as shown in the following section, this is not the case, and the cosmological term resurfaces as part of the first integral of the CEs.

5. A first integral of the curvature equations

With the vacuum-energy parts of the energy-momentum tensor $T^{\mu}{}_{v}$ annihilated by the covariant derivatives in Eq. (17), this equation becomes

$$\nabla_{\alpha}\left[R^{\mu}{}_{\beta}-\frac{8\pi G}{c^4}\left(\hat{T}^{\mu}{}_{\beta}-\frac{1}{2}\delta^{\mu}{}_{\beta}\hat{T}\right)\right]-\nabla_{\beta}\left[R^{\mu}{}_{\alpha}-\frac{8\pi G}{c^4}\left(\hat{T}^{\mu}{}_{\alpha}-\frac{1}{2}\delta^{\mu}{}_{\alpha}\hat{T}\right)\right]=0, \tag{18}$$

and $T^{\mu}{}_{\alpha}$ has been replaced by $\hat{T}^{\mu}{}_{\alpha}$. So, if the Einstein equation without vacuum energy and without cosmological term is satisfied (if the quantities in square brackets is Eq. (18) vanish), then the CEs are also satisfied. This is the formal content of the previous section.

But the curvature equations have a somewhat wider set of solutions than the Einstein equation

$$R^{\mu}{}_{\alpha}=\frac{8\pi G}{c^4}\left(\hat{T}^{\mu}{}_{\alpha}-\frac{1}{2}\delta^{\mu}{}_{\alpha}\hat{T}\right). \tag{19}$$

If the Einstein equation is not satisfied, then the quantity

$$Z^{\mu}{}_{\alpha}\equiv R^{\mu}{}_{\alpha}-\frac{8\pi G}{c^4}\left(\hat{T}^{\mu}{}_{\alpha}-\frac{1}{2}\delta^{\mu}{}_{\alpha}\hat{T}\right) \tag{20}$$

in Eq. (18) does not vanish, and the curvature Eq. (18) now reads

$$\nabla_{\alpha}Z^{\mu}{}_{\beta}-\nabla_{\beta}Z^{\mu}{}_{\alpha}=0. \tag{21}$$

Using the Hilbert conjugate pairs

$$X^{\mu}{}_{\alpha}=Z^{\mu}{}_{\alpha}-\frac{1}{2}\delta^{\mu}{}_{\alpha}Z, \quad Z^{\mu}{}_{\alpha}=X^{\mu}{}_{\alpha}-\frac{1}{2}\delta^{\mu}{}_{\alpha}X, \tag{22}$$

and

$$G^{\mu}{}_{\alpha}=R^{\mu}{}_{\alpha}-\frac{1}{2}\delta^{\mu}{}_{\alpha}R, \quad R^{\mu}{}_{\alpha}=G^{\mu}{}_{\alpha}-\frac{1}{2}\delta^{\mu}{}_{\alpha}G^{\sigma}{}_{\sigma}, \tag{23}$$

and taking the Hilbert conjugate of both sides of Eq. (20), this equation becomes the Einstein equation

$$G^{\mu}{}_{\alpha}=\frac{8\pi G}{c^4}\hat{T}^{\mu}{}_{\alpha}+X^{\mu}{}_{\alpha}. \tag{24}$$

The interpretation of Eq. (24) is clear. The curvature equations are differential equations containing derivatives of the metric one order higher than those in Eq. (24). Therefore (24) is a first integral of the curvature equations, and the $X^{\mu}{}_{\alpha}$ are somewhat arbitrary integration functions determined by the initial conditions for the universe. We call $X^{\mu}{}_{\alpha}$ the 'integration tensor' for the curvature equations, and this tensor must satisfy Eq. (21), here written in terms of $X^{\mu}{}_{\alpha}$ as

$$\nabla_{\alpha}\left(X^{\mu}{}_{\beta} - \frac{1}{2}\delta^{\mu}{}_{\beta}X\right) - \nabla_{\beta}\left(X^{\mu}{}_{\alpha} - \frac{1}{2}\delta^{\mu}{}_{\alpha}X\right) = 0. \tag{25}$$

Taking the divergence of Eq. (24) and using $\nabla_{\nu}G^{\nu}{}_{\alpha} = 0$ and $\nabla_{\nu}\hat{T}^{\nu}{}_{\alpha} = 0$, we have $\nabla_{\nu}X^{\nu}{}_{\alpha} = 0$. Then contracting on indices β and μ in Eq. (25), we have $\nabla_{\alpha}X = \partial_{\alpha}X = 0$ or

$$X = X^{\mu}{}_{\mu} = \text{constant}, \tag{26}$$

and this simplifies condition (25) to read $\nabla_{\alpha}X^{\mu}{}_{\beta} - \nabla_{\beta}X^{\mu}{}_{\alpha} = 0$. We summarize these findings in the following result:

Result II: The Einstein equation

$$G^{\mu}{}_{\alpha} = \frac{8\pi G}{c^4}\hat{T}^{\mu}{}_{\alpha} + X^{\mu}{}_{\alpha}, \tag{27}$$

without vacuum energy in its source term $\hat{T}^{\mu}{}_{\alpha}$, is a first integral of the curvature equations if the integration tensor $X^{\mu}{}_{\alpha}$ satisfies conditions

$$\nabla_{\nu}X^{\nu}{}_{\alpha} = 0, \quad \text{and} \quad \nabla_{\alpha}X^{\mu}{}_{\beta} - \nabla_{\beta}X^{\mu}{}_{\alpha} = 0. \tag{28}$$

Solutions of this Einstein equation are solutions of the curvature equation when these conditions are satisfied.

The one solution of conditions (28) that is the same in all spacetimes is an integration tensor in the form of a cosmological term, $X^{\mu}{}_{\alpha} = -\Lambda\delta^{\mu}{}_{\alpha} = (1/4)X\delta^{\mu}{}_{\alpha}$, with cosmological constant $\Lambda = -(1/4)X$. Because X is constant, all of the covariant derivatives in conditions (28) give vanishing results (by metric compatibility), and these conditions are satisfied. According to the curvature equations, X and Λ are integration constants having no necessary relationship to the vacuum energy of quantum fields, and this solves the cosmological constant problem. In fact, an interpretation of the cosmological term as representing a vacuum energy seems to be inconsistent with the CEs for the following reason: A cosmological term $X^{\mu}{}_{\alpha} = -\Lambda\delta^{\mu}{}_{\alpha}$ in the Einstein Eq. (27) clearly contributes to the curvature of spacetime, whereas, if the cosmological term in interpreted as the energy-momentum tensor (3), the source term in the CEs annihilates this part of the energy-momentum tensor, and $T^{\mu\nu}_{\Lambda}$ does not even appear in the CEs to influence the curvature. This inconsistency seems to rule out an interpretation of the cosmological term as a vacuum energy-momentum tensor. We collect these observations as:

Result III: The conditions (28) on the integration tensor admit a solution in the form of a cosmological term for the Einstein Eq. (27),

$$X^{\mu}{}_{\alpha} = -\Lambda\delta^{\mu}{}_{\alpha} = \frac{1}{4}X\delta^{\mu}{}_{\alpha}, \tag{29}$$

with cosmological constant $\Lambda = -(1/4)X$. According to the curvature equations, X and Λ are integration constants unrelated to the vacuum energy of quantum fields. Hence there is no expectation that the supposed energy equivalent of the cosmological term $\rho_{\Lambda}c^2 = \Lambda c^4/8\pi G$ should have the magnitude of a quantum vacuum energy. This solves the cosmological constant problem simply because the cosmological term does not represent a vacuum energy.

6. Evolution of the integration tensor

The conditions (28) on the integration tensor have solutions other than the cosmological term (29). In this section we show that deviations from the time-independent cosmological term $X^{\mu}{}_{\nu} = -\Lambda\delta^{\mu}{}_{\nu}$ decay to zero as the universe expands.

In a Friedmann-Robertson-Walker universe with flat spatial section, the line element

$$ds^2 = -c^2dt^2 + a^2(t)\left(dx^2 + dy^2 + dz^2\right) \tag{30}$$

has non-vanishing Christoffel symbols

$$\Gamma^{0}{}_{ij} = \frac{a\dot{a}}{c}\delta_{ij}, \quad \text{and} \quad \Gamma^{k}{}_{0i} = \Gamma^{k}{}_{i0} = \frac{\dot{a}}{ca}\delta^{k}{}_{i}. \tag{31}$$

Using these to expand the first of Eq. (28) we obtain

$$\frac{1}{c}\frac{\partial X^{0}{}_{0}}{\partial t} = -\frac{\partial X^{k}{}_{0}}{\partial x^k} - \frac{\dot{a}}{ca}\left(3X^{0}{}_{0} - X^{k}{}_{k}\right), \quad (\text{k summed, k} = 1,2,3) \tag{32}$$

$$\frac{1}{c}\frac{\partial X^{0}{}_{i}}{\partial t} = -\frac{\partial X^{k}{}_{i}}{\partial x^k} - \frac{3\dot{a}}{ca}X^{0}{}_{i}, \quad (\text{i} = 1,2,3) \tag{33}$$

and the second of Eq. (28) gives

$$\frac{1}{c}\frac{\partial X^{0}_{i}}{\partial t} = \frac{\partial X^{0}_{0}}{\partial x^i} - \frac{\dot{a}}{ca}X^{0}_{i}, \ (\text{i} = 1,2,3) \tag{34}$$

$$\frac{\partial X^{0}_{j}}{\partial x^i} = \frac{\partial X^{0}_{i}}{\partial x^j} \ (\text{i, j} = 1,2,3) \tag{35}$$

$$\frac{1}{c}\frac{\partial X^{k}_{i}}{\partial t} = \frac{\partial X^{k}_{0}}{\partial x^i} - \frac{\dot{a}}{ca}\left(X^{k}_{i} - \delta^{k}_{i}X^{0}_{0}\right), \ (\text{k, i} = 1,2,3) \tag{36}$$

$$\frac{\partial X^{k}_{j}}{\partial x^i} = \frac{\partial X^{k}_{i}}{\partial x^j} - \frac{\dot{a}}{ca}\left(\delta^{k}_{i}X^{0}_{j} - \delta^{k}_{j}X^{0}_{i}\right). \ (\text{k, i, j} = 1,2,3) \tag{37}$$

And by spatial contractions in Eqs. (36) and (37) we obtain

$$\frac{1}{c}\frac{\partial X^{k}_{k}}{\partial t} = \frac{\partial X^{k}_{0}}{\partial x^k} - \frac{\dot{a}}{ca}\left(X^{k}_{k} - 3X^{0}_{0}\right), \tag{38}$$

$$\frac{\partial X^k_j}{\partial x^k} = \frac{\partial X^k_k}{\partial x^j} - \frac{2\dot{a}}{ca} X^0_j. \quad (\mathrm{j} = 1{,}2{,}3) \tag{39}$$

The deviations of $X^\mu{}_v$ away from the cosmological term $X^\mu{}_v = -\Lambda\delta^\mu{}_v = \frac{1}{4}X\delta^\mu{}_v$ are

$$\Delta X^0_0 = X^0_0 - \frac{1}{4}X, \tag{40}$$

$$\Delta X^0_i = X^0_i, \tag{41}$$

$$\Delta X^i_0 = X^i_0, \tag{42}$$

$$\Delta X^i_j = X^i_j, \quad (\text{for i} \neq \text{j}) \tag{43}$$

$$\Delta X^{\bar{i}}_{\bar{i}} = X^{\bar{i}}_{\bar{i}} - \frac{1}{4}X, \quad (\bar{\text{i}}\text{ not summed}) \tag{44}$$

$$\Delta X^k_k = X^k_k - \frac{3}{4}X. \quad (\text{k summed, k} = 1{,}2{,}3) \tag{45}$$

Then from $X = X^0_0 + X^k_k = \text{constant}$ we have $\Delta X^0_0 = -\Delta X^k_k$, and using this, the equations for deviations containing time derivatives are found to be

$$\frac{1}{c}\frac{\partial \Delta X^0_0}{\partial t} = -\frac{4\dot{a}}{ca}\Delta X^0_0 - \frac{\partial \Delta X^k_0}{\partial x^k}, \tag{46}$$

$$\frac{1}{c}\frac{\partial \Delta X^0_i}{\Delta t} = -\frac{3\dot{a}}{ca}\Delta X^0_i - \frac{\partial \Delta X^k_i}{\partial x^k}, \tag{47}$$

$$\frac{1}{c}\frac{\partial \Delta X^0_i}{\partial t} = -\frac{\dot{a}}{ca}\Delta X^0_i + \frac{\partial \Delta X^0_0}{\partial x^i}, \tag{48}$$

$$\frac{1}{c}\frac{\partial \Delta X^k_i}{\partial t} = -\frac{\dot{a}}{ca}\left(\Delta X^k_i + \delta^k_i \Delta X^0_0\right) + \frac{\partial \Delta X^k_0}{\partial x^i}, \tag{49}$$

$$\frac{1}{c}\frac{\partial \Delta X^k_k}{\partial t} = -\frac{4\dot{a}}{ca}\Delta X^k_k + \frac{\partial \Delta X^k_0}{\partial x^k}. \tag{50}$$

$$\frac{1}{c}\frac{\partial \Delta X^{\bar{i}}_{\bar{i}}}{\partial t} = -\frac{\dot{a}}{ca}\left(\Delta X^{\bar{i}}_{\bar{i}} - \Delta X^0_0\right) + \frac{\partial \Delta X^{\bar{i}}_0}{\partial x^{\bar{i}}}, \tag{51}$$

where the bar on index $\bar{i}$ indicates that index is not summed.

6.1 A homogeneous and isotropic integration tensor

If the integration tensor, like other sources of gravitation, obeys the cosmological principle and is position-independent and invariant under spatial rotations $(\Delta X^0_i = \Delta X^i{}_0 = 0$ and $\Delta X^i{}_j = (1/3)\Delta X^k_k(t)\delta^i_j)$, then Eqs. (46)–(51) reduce to

$$\frac{d\Delta X^0_0}{dt} = -\frac{4\dot{a}}{a}\Delta X^0_0, \tag{52}$$

$$\frac{d\Delta X^{\bar{i}}_{\bar{i}}}{dt} = -\frac{\dot{a}}{a}\left(\Delta X^{\bar{i}}_{\bar{i}} - \Delta X^0_0\right), \tag{53}$$

with solution

$$\Delta X^0_0(t) = \frac{\Delta X^0_0(t_0)}{a^4(t)}, \tag{54}$$

$$\Delta X^{\bar{i}}_{\bar{i}}(t) = -\frac{1}{3}\frac{\Delta X^0_0(t_0)}{a^4(t)}. \tag{55}$$

Hence the deviations $\Delta X^0_0(t)$ and $\Delta X^{\bar{i}}_{\bar{i}}(t)$ tend to zero as the universe expands.

If at some early time t_i, say at recombination, the deviations from the cosmological term are of the same order of magnitude as $\Lambda = -(1/4)X$, then at the present time t_0 the deviations,

$$\Delta X^0_0(t_0) = a^4(t_i)\Delta X^0_0(t_i), \tag{56}$$

$$\Delta X^{i}_{\bar{i}}(t_0) = a^4(t_i)\Delta X^{\bar{i}}_{\bar{i}}(t_i), \tag{57}$$

are very much smaller than the cosmological term, and only the cosmological term is significant at late times.

6.2 A homogeneous anisotropic solution

For a position-independent but anisotropic integration tensor, Eqs. (46)–(51) yield

$$\Delta X^0_i = 0, \quad \Delta X^i_0 = 0, \tag{58}$$

where $\Delta X^k{}_0 = -\Delta X^0_k/a^2$ was used, and equations

$$\frac{d\Delta X^0_0}{dt} = -\frac{4\dot{a}}{a}\Delta X^0_0, \tag{59}$$

$$\frac{d\Delta X^k_i}{dt} = -\frac{\dot{a}}{a}\Delta X^k_i, \quad \text{(for k} \neq \text{i)} \tag{60}$$

$$\frac{d\Delta X^{\bar{i}}_{\bar{i}}}{dt} = -\frac{\dot{a}}{a}\left(\Delta X^{\bar{i}}_{\bar{i}} - \Delta X^0_0\right), \tag{61}$$

with solutions

$$\Delta X^0_0(t) = \frac{\Delta X^0_0(t_0)}{a^4(t)}, \tag{62}$$

$$\Delta X^k_i(t) = \frac{\Delta X^k_i(t_0)}{a(t)}, \quad \text{(for k} \neq \text{i)} \tag{63}$$

$$\Delta X^{\bar{i}}_{\bar{i}}(t) = \left[\Delta X^{\bar{i}}_{\bar{i}}(t_0) + \frac{1}{3}\Delta X^0_0(t_0)\right]\frac{1}{a(t)} - \frac{1}{3}\frac{\Delta X^0_0(t_0)}{a^4(t)}. \tag{64}$$

Here the deviations from the cosmological term decay with factors $1/a(t)$ and $1/a^4(t)$ as the universe expands.

This solution is not entirely self-consistent. An anisotropic integration tensor in the Einstein Eq. (27) will cause the metric to differ from the isotropic line element (30) used in the calculation. However, if the anisotropic deviation from the cosmological term is of the same order of magnitude as the latter in the early universe, it is dominated by matter and radiation in the Einstein equation at early times and likely has decayed by the time the cosmological term is dominant, making the line element (30) a good approximation at all times.

6.3 General short-term decay of integration tensor

We do not have a general solution of Eqs. (46)–(51), but the behavior of the solution can be inferred from a solution over a short period of time during which a and the Hubble parameter $\dot{a}/a$ can be treated as constant. In this case, the solution $\Delta X^{\mu}{}_{\nu}(t_i + t)$ for a short duration t starting at time t_i can be expanded as a Fourier integral, i.e., as a linear combination of modes of the form

$$G(t_i + t, \mathbf{x}) = exp\left[i\omega(t_i + t) - i\mathbf{k}\cdot\mathbf{x}\right]. \tag{65}$$

Then the algebraic equations resulting from Eqs. (46)–(51) have non-trivial solutions only if the dispersion relation

$$\left(i\omega + \frac{4\dot{a}}{a}\right)\left(i\omega + \frac{\dot{a}}{a}\right) + \frac{c^2k^2}{a^2} = 0 \tag{66}$$

is satisfied. This is a quadratic equation for $i\omega$ with solution

$$i\omega = -\frac{5}{2}\frac{\dot{a}}{a} \pm \frac{3}{2}\sqrt{\left(\frac{\dot{a}}{a}\right)^2 - \frac{4}{9}\frac{c^2k^2}{a^2}}. \tag{67}$$

The real part of this solution is negative for all spatial scales k of the mode (65), suggesting an exponential decay of all Fourier components. But as this solution is only good over the short duration t during which a and $\dot{a}/a$ are essentially constant, solution (65) is only good to first order in t, and a first order expansion of (65) contains all relevant information:

$$G(t_i + t, \mathbf{x}) = G(t_i, \mathbf{x})\,(1 + i\omega t). \tag{68}$$

This may be compared to the first order expansions of $1/a(t)$ and $1/a^4(t)$ obtained in the previous section,

$$\frac{1}{a(t_i + t)} = \frac{1}{a(t_i)}\left[1 - \frac{\dot{a}}{a}t\right], \tag{69}$$

$$\frac{1}{a^4(t_i + t)} = \frac{1}{a^4(t_i)}\left[1 - \frac{4\dot{a}}{a}t\right]. \tag{70}$$

For the homogeneous case $k = 0$, Eq. (67) gives the same fractional amounts of decay per unit time, $i\omega = -\dot{a}/a;\ -4\dot{a}/a$, as in Eqs. (69) and (70), as it should. And for smaller scales (larger k), Eq. (67) gives decay rates between the two values $-\dot{a}/a$ and $-4\dot{a}/a$ with a final value $-(5/2)\dot{a}/a$ when the radical in Eq. (67) becomes imaginary.

We summarize these results as:

Result VI: The deviations $\Delta X^{\mu}{}_{v}$ of the integration tensor away from a cosmological term decay to zero as $1/a(t)$ or faster as the Universe expands. Thus the cosmological term becomes dominant at late times.

The more slowly decaying parts of the deviations $\Delta X^{\mu}{}_{v}(t)$ decay slower than pressureless matter $[\sim 1/a^3(t)]$, and might be important for the evolution of the universe when present as a source of gravitation in the Einstein Eq. (27). This is a topic for future research.

7. A universe with vacuum energy, matter, and cosmological constant

A concrete example should clarify any remaining obscure points. Consider a homogeneous universe containing pressureless matter of density $\rho_m(t)$ described by energy-momentum tensor $T_m{}^{\mu}{}_{v}$ with only one non-vanishing component

$$\hat{T}_{m\ 0}^{0} = -\rho_m(t)c^2 = -\rho_{cr}c^2\frac{\Omega_m}{a^3(t)}, \tag{71}$$

and a large vacuum energy of quantum fields,

$$T_{qv}{}^{\mu}_{v} = -\rho_{qv}c^2\delta^{\mu}{}_{v}, \tag{72}$$

where $\rho_{cr} = 3H_0^2/8\pi G$ is the critical density, H_0 is the Hubble constant, and $\Omega_m = \rho_m(t_0)/\rho_{cr}$ is the density parameter for matter.

Then, when the total energy-momentum tensor, $T^{\mu}{}_{v} = \hat{T}^{\mu}{}_{m\ v} + T^{\mu}_{qv\ v}$ is used in the source term of the curvature equation

$$\nabla_{v}R_{\alpha\beta}{}^{\mu v} = \frac{8\pi G}{c^4}\left[\nabla_{\beta}\left(T^{\mu}{}_{\alpha}\right) - \nabla_{\alpha}\left(T^{\mu}{}_{\beta}\right)\right] = \frac{8\pi G}{c^4}\left[\nabla_{\beta}\left(\hat{T}^{\mu}_{m\ \alpha}\right) - \nabla_{\alpha}\left(\hat{T}^{\mu}_{m\ \beta}\right)\right], \tag{73}$$

the vacuum energy is nulled by the covariant derivatives in the source term, and only the matter tensor $\hat{T}^{\mu}_{m\ \beta}$ remain as the source of curvature, as indicated in (73). Therefore no remnant of the vacuum energy remains in the curvature equations or in any equation derived solely from the CEs.

Using energy-momentum tensor (71) and the Friedmann-Robertson-Walker line element (30), the non-vanishing components of Eq. (73) are all found to be of the form

$$\frac{\dddot{a}}{a} + \frac{\dot{a}\ddot{a}}{a^2} - 2\left(\frac{\dot{a}}{a}\right)^3 = -\frac{3}{2}H_0^2\frac{\Omega_m}{a^3}\left(\frac{\dot{a}}{a}\right). \tag{74}$$

If we assume the integration tensor $X^{\mu}{}_{v}(t)$ is the cosmological term (29), the first integral of the curvature Eq. (73) is

$$G^{\mu}_{v} = \frac{8\pi G}{c^4}\hat{T}^{\mu}_{m\ v} - \Lambda\delta^{\mu}_{v}. \tag{75}$$

We emphasize that the vacuum energy (72) does not appear in this Einstein equation because the first integral of the curvature equations is derived from the curvature equations which do not contain the vacuum energy. This is why one must ignore the vacuum energy when applying Einstein's equation in practice.

The $\mu = 0, v = 0$ component of Eq. (75) is the Friedmann equation

$$\left(\frac{\dot{a}}{a}\right)^2 = H_0^2\left(\frac{\Omega_m}{a^3} + \Omega_\Lambda\right), \tag{76}$$

where $\Omega_\Lambda = \Lambda c^2/3H_0^2$.

Now the integration tensor $X^\mu{}_v = -\Lambda\delta^\mu{}_v$ in Eq. (75), and the density parameter Ω_Λ in the Friedmann Eq. (76) certainly effect the curvature tensor. But the right hand side of Eq. (73), and the right hand side of Eq. (74) do not contain these parameters. Therefore these parameters must cancel out and not appear on the left sides of these equations. To check this, we write Eq. (76) as $\dot{a}^2 = H_0^2(\Omega_\Lambda a^2 + \Omega_m/a)$ and take derivatives to obtain

$$\ddot{a} = H_0^2\left(\Omega_\Lambda a - \frac{\Omega_m}{2a^2}\right), \tag{77}$$

and

$$\dddot{a} = \frac{\dot{a}^3}{a^2}. \tag{78}$$

Then substitution of these relations into Eq. (74) shows that Ω_Λ does indeed cancel out on the left side, and this equation is satisfied. This is simply a verification that Eqs. (75) and (76) are indeed first integrals of the curvature equations with arbitrary integration constants Λ or Ω_Λ. This example approximates the Universe in which we live during the period after radiation is dominant.

8. Quantized curvature equations?

In this section we ask what is, at this point, a fairly obvious question: With the curvature equations so similar to Maxwell's equations, can the curvature equations be quantized in a manner analogous to Maxwell's equations to give a satisfactory quantum theory of gravitation? In this section we do not answer this question, but outline a number of features of the curvature equations that differ from previous attempts to quantize general relativity, and may be helpful in the quest to achieve a quantum theory of gravitation.

The gravitational analogy with Maxwell's equations extends further than we have thus far indicated, [25]. For instance, the analogue of the vector potential A_μ, which determines the electromagnetic field tensor

$$F_{\mu\nu} = \partial_\mu A_\nu - \partial_\nu A_\mu, \tag{79}$$

is the set of affine connection coefficients $\Gamma^\alpha_{\beta\mu}$ (Christoffel symbols), which determines the curvature tensor

$$R^\alpha{}_{\beta\mu\nu} = \partial_\mu\Gamma^\alpha_{\beta\nu} - \partial_\nu\Gamma^\alpha_{\beta\mu} + \Gamma^\alpha_{\tau\mu}\Gamma^\tau_{\beta\nu} - \Gamma^\alpha_{\tau\nu}\Gamma^\tau_{\beta\mu}. \tag{80}$$

Note the similarity of the first two terms in (80) and the terms in (79) [the non-linear terms in (80) presumably being associated with the fact that the (pseudo) energy of the gravitational field is itself a source of gravitational field].

The source-free Maxwell Eq. (9) is satisfied by an electromagnetic field tensor derived from A_μ as in Eq. (79); and the source-free CE (7), the Bianchi identity, is satisfied by a curvature tensor derived from $\Gamma^\alpha_{\beta\mu}$ as in Eq. (80).

The second Maxwell Eq. (10) is derived from a variational principle. For arbitrary but small variations δA_μ of the vector potential vanishing on the boundary $\partial\Sigma$ of the 4-volume Σ, the action for the electromagnetic field,

$$S_{EM} = \frac{-1}{16\pi}\int_\Sigma \left(F^{\mu\nu}F_{\mu\nu} + 16\pi J^\nu A_\nu\right)\sqrt{-g}\,d^4x, \tag{81}$$

is stationary ($\delta S_{EM} = 0$), if and only if $F^{\mu\nu}$ satisfies the second Maxwell Eq. (10). Similarly, for variations $\delta\Gamma^\alpha_{\beta\mu}$ of the connection coefficients vanishing on the boundary $\partial\Sigma$ of the 4-volume Σ, the action for the gravitational field

$$S_G = \frac{-1}{16\pi}\int_\Sigma \left(R^{\alpha\beta\mu\nu}R_{\alpha\beta\mu\nu} + 16\pi J^{\alpha\beta}_\mu \Gamma^\mu_{\alpha\beta}\right)\sqrt{-g}\,d^4x. \tag{82}$$

is stationary ($\delta S_G = 0$), if and only if $R_{\alpha\beta}{}^{\mu\nu}$ satisfies the second curvature Eq. (8). The key relation in this derivation is the Palatini identity

$$\delta R^\alpha_{\beta\mu\nu} = \nabla_\mu\left(\delta\Gamma^\alpha_{\beta\nu}\right) - \nabla_\nu\left(\delta\Gamma^\alpha_{\beta\mu}\right), \tag{83}$$

which is exactly analogous to the relation

$$\delta F_{\mu\nu} = \nabla_\mu(\delta A_\nu) - \nabla_\nu(\delta A_\mu) \tag{84}$$

in electrodynamics.

These considerations suggest that the proper field variables for the gravitational field are the Christoffel symbols $\Gamma^\alpha_{\beta\mu}$, and not the metric coeficients $g_{\alpha\beta}$. The above question then becomes whether the action (82), properly extended to include the fields of massive particles that are the sources of gravitational field, can be made the basis of a workable quantum theory of gravitation. This result would seem to be significant, because most of the failed attempts to quantize general relativity assume the metric components $g_{\mu\nu}$ are the proper field variables for quantizing general relativity. Whether this is a productive direction for research remains to be determined.

9. Conclusion

This chapter asks the question: Has Mother Nature used the same basic pattern for her laws of gravitation and electrodynamics? To replace the Einstein equation by the Maxwell-like curvature equations as the field equations of general relativity in no way constitutes a new theory of gravitation. Except for a possible addition of deviations from the cosmological term in Einstein's Eq. (27), not a single equation of conventional general relativity has been altered by this replacement. We merely suggest that a different set of valid equations in Einstein's theory could better serve as the field equations of that theory, because these equations do not suffer from the vacuum-energy or cosmological-constant problems.

We list below the various insights gained by viewing the curvature equations as the field equations of general relativity:

1. Any solution of Einstein's equation is also a solution of the curvature equations. Hence, any successful prediction of Einstein's equation could just as well be counted as evidence supporting the curvature equations.

2. Because the source term of the CEs annihilates any part of the energy-momentum tensor in the form of a vacuum energy, the curvature of spacetime is unaffected by the vacuum energy of quantum fields, as is observed to be the case.

3. For the reason immediately above, there is no reason to question the very large estimates for the vacuum energy of quantum fields made by luminaries such as Feynman [2] and Weinberg [3]. Whether the vacuum energy is large or small, it does not influence the spacetime curvature.

4. An Einstein equation with an additive integration tensor is a first integral of the curvature equations, and the integration tensor naturally evolves to a cosmological term as the universe expands. The cosmological constant is an integration constant unrelated to any vacuum energy. Thus the small observed value of the cosmological constant is not inconsistent with the large estimates for the vacuum energy of quantum fields. This solves the cosmological constant problem.

5. The Einstein equation, which is the first integral of the CEs, does not contain the vacuum energy of quantum fields in its source term because the Einstein equation is derived solely from the curvature equations and the later do not contain this vacuum energy. This solves the vacuum-energy problem for Einstein's equation.

6. The curvature equations identify the nature of "dark energy" as the integration tensor in the first integral of these equations. Thus "dark energy" is not a material substance, but rather is a manifestation of the integration functions in the first integral of the CEs, and these are determined by the initial conditions for the universe. Hence the term "dark energy" would seem to be a misnomer. In addition, the form of "dark energy" can differ from the simple cosmological term, particularly in the early universe.

7. The CEs suggest two directions for future research: (1) Do the deviations of the integration tensor from the cosmological term contribute significantly to the evolution of the universe?, and (2) Can the Maxwell-like CEs be quantized by analogy with Maxwell's equations using the Christoffel symbols as field variables to give a satisfactory quantum theory of gravitation?

We believe the above results make a strong case for viewing the curvature equations as the proper field equations of general relativity, with the Einstein equation emerging as the first integral of the CEs, and no doubt continuing its role as the working equation of the theory, but no longer suffering from the vacuum-energy or cosmological-constant problems, and all within conventional general relativity, without introducing any new or speculative physics.

Acknowledgements

I wish to thank Edward Teller (deceased) for encouraging this line of research and Hans C. Ohanian (deceased) for helpful comments on a very early version of this work (both many years ago). Deep appreciation also to Ronald J. Adler, Robert Wagoner, and Alexander S. Silbergleit for extended discussions of these ideas (again some years ago). And sincere thanks to Chanda Cook for editorial assistance.

Author details

Richard J. Cook
Department of Physics, U.S. Air Force Academy, CO, USA

*Address all correspondence to: richard.cook43@gmail.com

References

[1] Dirac PAM. The quantum theory of the emission and absorption of radiation. Proceedings of the Royal Society of London A. 1927;**114**(767):243-265. DOI: 10.1098/rspa.1927.0039. Bibcode: 1927RSPSA.114..243D

[2] Feynman RP, Hibbs AR. Quantum Mechanics and Path Integrals. New York: McGraw-Hill; 1965. pp. 244-246

[3] Weinberg S. The cosmological constant problem. Reviews of Modern Physics. 1989;**61**(1):1-23

[4] Casimir HBG. On the attraction between two perfectly conducting plates. Proceedings of the Section of Sciences, Koninklijke Nederlandsche Akademie van Wetenschappen. 1948; **51**(7):793-795

[5] Sparnaay MJ. Attractive forces between flat plates. Nature. 1957; **180**(4581):334-335. DOI: 10.1038/180334b0

[6] Adler RJ, Brendon C, Jacob OC. Vacuum catastrophe: An elementary exposition of the cosmological constant problem. American Journal of Physics. 1995;**63**(7):620-626. DOI: 10.1119/1.17850

[7] Peebles PJE, Ratra B. The cosmological constant and dark energy. Reviews of Modern Physics. 2003;**75**(2): 559. DOI: 10.1103/RevModPhys.75.559

[8] Perlmutter S et al. Measurement of omega and lambda from 42 Hight-redshift supernovae. Astrophysical Journal. 1999;**517**:565-586. DOI: 10.48550/arXiv.astro-ph/9812133

[9] Riess AG et al. Observational evidence from supernovae for an accelerating universe and a cosmological constant. Astronomical Journal. 1998; **116**:1009-1039. DOI: 10.1086/300499

[10] Aghanim N, Akrami Y, Ashdown M, et al., Planck Collaboration. Planck 2018 results. VI. Cosmological parameters. 2018;**640**:A6. DOI: 10.1051/0004-6361/201833910

[11] Philip B, Akrami Y, Ademek J, et al. Beyond ?CDM: Problems, solutions, and the road ahead. Physics of the Dark Universe. 2016;**12**:56-99. DOI: 10.1016/j.dark.2016.02.001

[12] Wang Q, Zhen Z, Unruh WG. How the huge energy of quantum vacuum gravitates to drive the slow accelerating expansion of the universe. Physical Review D. 2017;**95**(10):103504. DOI: 10.1103/PhysRevD.95.103504

[13] Ellis GFR. The trace-free Einstein equations and inflation. General Relativity and Gravitation. 2014;**46**: 1619. DOI: 10.1007/s10714-013-1619-5

[14] Percacci R. Unimodular quantum gravity and the cosmological constant. Foundations of Physics. 2018;**48**(10): 1364-1379. DOI: 10.1007/s10701-018-0189-5

[15] Montesinos M, Gonzalez D. Black holes with scalar hair: Extending from and beyond the Schwarzschild solution. Physical Review D. 2023;**108**(12): 124013. DOI: 10.1103/PhysRevD.108.124013

[16] Brodsky SJ, Roberts CD, Shrock R, Tandy PC. New perspectives on the quark condensate. Physical Review C. 2010;**82**(2):022201. DOI: 10.1103/PhysRevC.82.022201

[17] Luongo O, Muccino M. Speeding up the universe using dust with pressure.

Physical Review D. 2018;**98**(10):2-3. DOI: 10.1103/PhysRevD.98.103520

[18] Cohen A, Kaplan DB, Nelson A. Effective field theory, black holes, and the cosmological constant. Physical Review Letters. 1999;**82**(25):4971-4974. DOI: 10.1103/PhysRevLett.82.4971

[19] Blinov N, Draper P. Densities of states and the CKN bound. Physical Review. 2021;**104**:076024. DOI: 10.48550

[20] Linde A. New directions in the pursuit of Majorana fermions in solid state systems. Reports on Progress in Physics. 2017;**8**(2):022001. DOI: 10.1088/1361-6633/aa50e4

[21] Yang CN. Integral formalism for gauge fields. Physical Review Letters. 1974;**33**:445. DOI: 10.1103/PhysRevLett.33.445

[22] Szczyrba V. Stephenson-Kilmister-Yang theory of gravity and its dynamics. Physical Review D. 1987;**36**:351

[23] Gronwald F, Hehl FW. On the gauge aspects of gravity. arXiv:gr-qc/9602013v1. DOI: 10.48550/arXiv.gr-qc/9602013

[24] Dolan P, Kim CW. The wave equation for the Lanczos potential. Proceedings of the Royal Society London. 1994;**A447**:557. and references contained therein

[25] Cook RJ. The gravitational-electromagnetic analogy: A possible solution to the vacuum-energy and dark-energy problems. arXiv:0810,4495v2[gr-qc]. 2009. DOI: 10.48550/arXiv.0810.4495

Section 3

Examining the Cosmological Constant

Chapter 10

Atomic-Clock Detection of CNO-Based Dark Matter

Peter D. Morley and Douglas J. Buettner

Abstract

Laboratory detection of Dark Matter is an active area of research. Instead of detecting the Dark Matter particles directly in underground experiments, an alternative line of research is to detect a Dark Matter macroscopic effect from their clustering. Previous research has shown that the condensation of cosmological neutrinos into condensed neutrino objects (CNO) is a viable Dark Matter candidate [1]. The CNO would be the largest objects in the Universe (with radii millions of light years in size and characteristic masses$\sim 10^{15} M_{\odot}$) [2]. The CNO would have entire galaxy clusters embedded within them. Weak lensing data supports the existence of CNO with galaxy clusters inside [3]. One of these predictions is that the Local Group of Galaxies is embedded in a CNO [4]. We show in this paper that this Local Group CNO may be observationally detected with a measurable gravitational time delay between two synchronized spacecraft atomic clocks that are in an opposition solar orbit of one astronomical unit (au).

Keywords: Dark Matter, Local Group, condensed neutrino object, gravitational time delay, atomic clock

1. Introduction

When Vera Rubin found that galaxy rotational data implied missing matter, she guessed that the missing matter was in the halo of galaxies. We suggest that her guess was wrong—possibly constrained by her belief that the effect on a galaxy's rotation had to be a local galactic phenomena. In fact, Zwicky originally observed the issue using the Coma Galaxy Cluster, that is now the poster child for Dark Matter [5]. In Coma, the fastest bound galaxy is GMP = 3176, which has a relative speed of 3000 km/s compared to the cluster's center-of-mass [6]. This means there is a gigantic gravitational potential well that the entire Coma Galaxy Cluster is embedded in. The individual galaxies are undergoing Kepler orbital motion inside this well and the fastest galaxy is near or at the gravitational well's center. Further, weak lensing data shows that an entire galaxy cluster is embedded in Dark Matter (**Figure 1**), [3].

Despite billions of dollars being spent to find the Dark Matter particle, null results continue to pile up. In [1] we provided a list of the leading Dark Matter candidates with papers and news articles regarding the null status of experiments designed to confirm those potential candidates as the source of the Dark Matter phenomena. We can now add to these null results the recent results from the LUX-ZEPLIN (LZ)

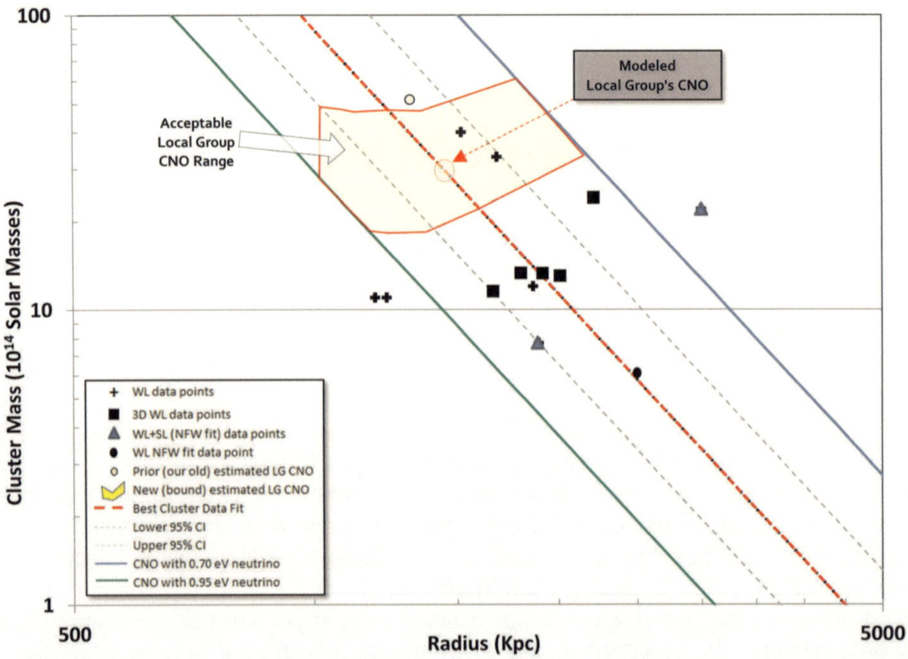

Figure 1.
Weak lensing data associated with galaxy clusters, taken from reference [3]. This data is observational data of the Dark Mass (in units of 10^{14} $M_\odot$*) embedding particular (listed) galaxy clusters vs. the estimated Dark Matter radius (in Kpc). The takeaway is that observational weak lensing data requires Dark Matter to congregate into* 10^{15} $M_\odot$ *class objects having millions of light years in radius.*

Quantity	Mathematical expression
Reduced neutrino momentum x	$x = \frac{p}{m_\nu c}$
Equation of State mass density ρ	$\rho = \frac{6\pi m_\nu^4 c^3}{h^3}\left\{2x(1+x^2)^{3/2}\right.$
	$\left.-x\sqrt{1+x^2} - \sinh^{-1}x\right\}$
Equation of State pressure P	$P = \frac{2\pi m_\nu^4 c^5}{h^3}\{3\sinh^{-1}x$
	$+x(2x^2-3)\sqrt{1+x^2}\}$
Instantaneous relativistic radiation loss P_M	$P_M = \frac{2}{3c^3}\left[\gamma^8(\boldsymbol{\beta}\cdot\dot{\boldsymbol{\beta}})^2\dot{\boldsymbol{\mu}}^2 + \cdots\right]$

Table 1.
Main physics of the CNO objects.

experiment [7], updated results from XENON1T [8], null results from a dark photon Dark Matter search [9] and the additional critical review arguments found in [10].

In **Table 1**, we list the main physics of the condensation of the cosmological neutrinos: the equation of state [2] and the relativistic radiation loss [11] due to its small magnetic dipole flipping $\dot{\boldsymbol{\mu}}^2$. Here m_ν is the neutrino mass scale and the assumption is made of equal neutrino and anti-neutino flavors present. The expansion of spacetime, in conjunction with their radiation losses, causes the condensation of cosmological neutrinos. Gamma (γ) and $\boldsymbol{\beta}$ are the familiar special relativity functions.

CNO	x(0)	mass in $M_{\odot}$	radius r_{CNO} in kpc
Local Group	0.015	3.2024×10^{15}	1834
Coma	0.010	1.744×10^{15}	2244

Table 2.
Comparison of the Coma CNO to the local group CNO for m_ν = 0.75 eV/c² (Note the signature for degenerate Dark Matter).

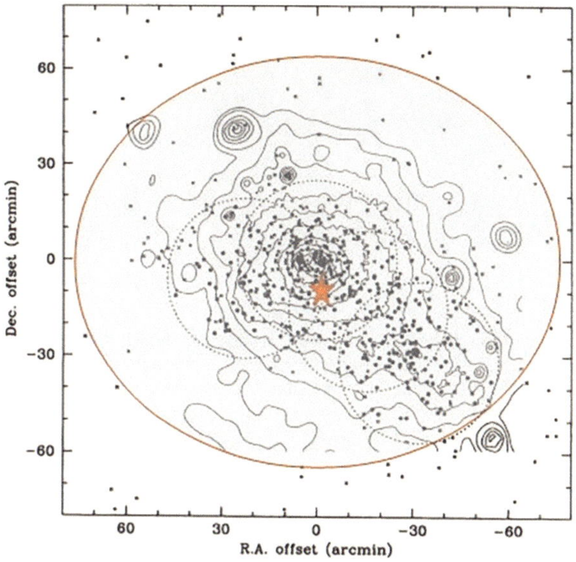

Figure 2.
Coma galaxy cluster embedded in a gravitational field [12], where the "black dots" are the locations of galaxies in the Coma Cluster and with the fastest cluster member identified (located with the red star) GMP = 3176 (reprinted with permission). The circular line is the estimated CNO extent, which is invisible because it does not scatter background light.

The Coma Cluster Dark Matter has been analyzed in Ref. [12], which showed that the Coma Cluster Galaxy dynamics imply a CNO having a radius around~ 2300 kpc and a mass~ 1.8×10^{15} $M_{\odot}$, **Table 2**. The Coma Galaxy Cluster is shown with correct dimensions in **Figure 2** with the CNO radius drawn in and the fastest galaxy GMP = 3176 identified. The value of the neutrino mass scale enters into the physical size of the CNO object and its total mass as $1/m_\nu^2$ so a 0.5 eV/c^2 scale has a 2.25 increase in mass and size compared to a $m_\nu = 0.75$ eV/c^2 mass scale.

Additional observational data is available showing that Rubin's guess of individual galaxy halos is in error, comes from galaxy rotation data itself [13, 14]. The Milky Way (MW) does not have a unique rotation curve, as shown here in **Figure 3**. Compare this to a spiral galaxy that has a unique rotation curve in **Figure 4**. Both M33 and MW are members of the Local Group of galaxies, so an explanation must be made as to why M33 has a unique rotation curve, but MW does not. This asymmetric rotation curve issue has also been discussed in Refs. [13, 14]. The physical explanation is that both galaxies are in a very large gravitational potential field, but their spin vectors have different orientations within that external field, **Figures 5** and **6**.

What we have learned is that the gravitational potentials of CNO Dark Matter have radial distances of millions of light years and their total mass is in the 10^{15} $M_{\odot}$ class [1, 2].

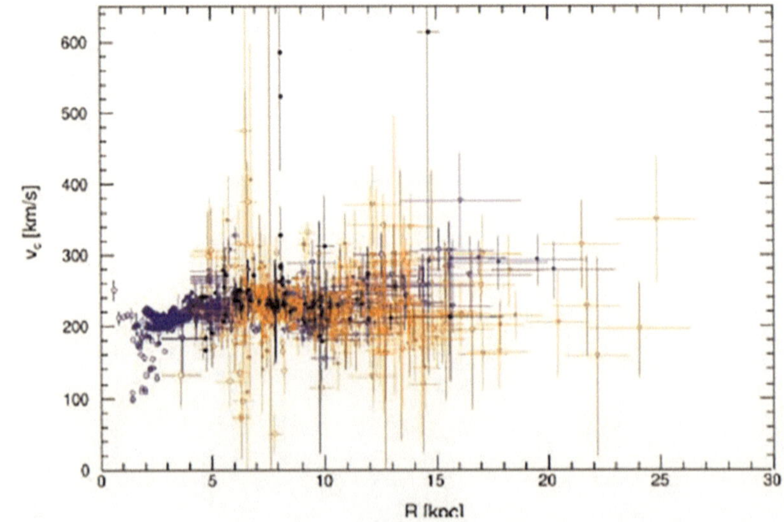

Figure 3.
MW Circular velocities of three tracers as derived from gas kinematics (blue), star kinematics (orange) and masers (black); figure is from reference [15] reused with permission. This observational data shows the same asymmetric rotation curve issue as discussed in Refs. [13, 14]. Thus, the MW does not have a unique rotation curve.

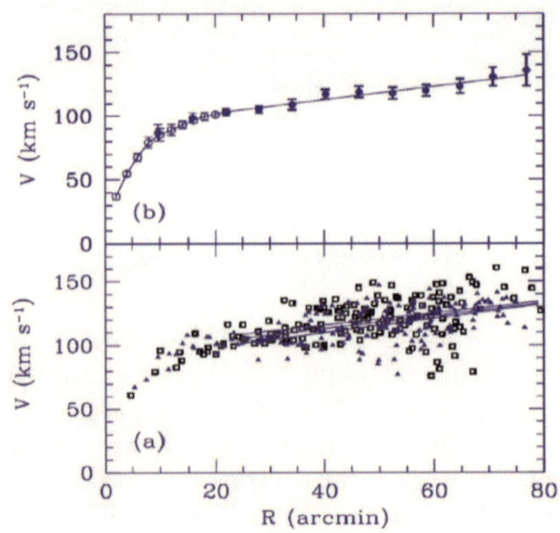

Figure 4.
M33 data [16] (used with permission): The filled triangles refer to the Northern half of the galaxy and the open squares refer to the Southern half. M33 has no apparent azimuthal angle dependence.

Futhermore, the Local Group of Galaxies is embedded in the same type of object as the Coma Galaxy Cluster [4, 12]. The tentative conclusion is that the cosmological neutrinos created in the Big Bang have condensed. We discuss in this paper a method to see if the Local Group CNO can be directly detected using its gravitational well.

Dark Matter as Condensed Neutrino Objects (CNO) was reviewed in Ref. [1]. The mechanism of condensation of the almost unlimited number of neutrinos created in the Big Bang would be rapid magnetic cooling [11]. The most important physics result of magnetic cooling is that it does not depend on the very small neutrino magnetic

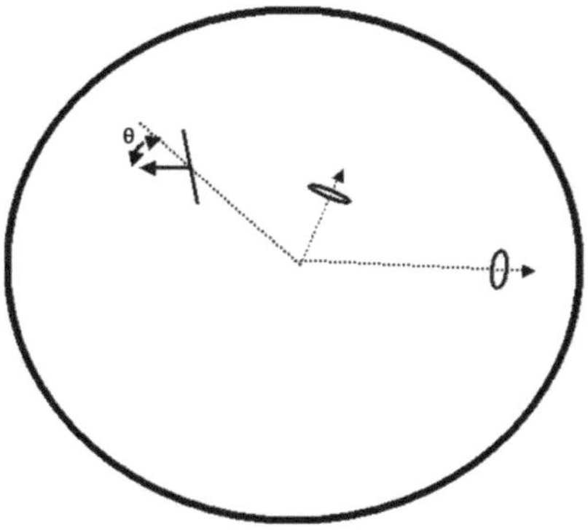

Figure 5.
*M33 (central galaxy) has a spin vector along the radial of the external field, while M31 (far right) is also along a radial. Both have an equilipotential gravity field, for a constant radial distance from their spin axis, giving rise to velocity speeds independent of azimuth angle: a unique rotation curve. MW, however has a tilt angle with respect to the external radial and so has no unique rotation curve, as demonstrated in **Figure 6**. The sphere is the Local CNO (not drawn to scale).*

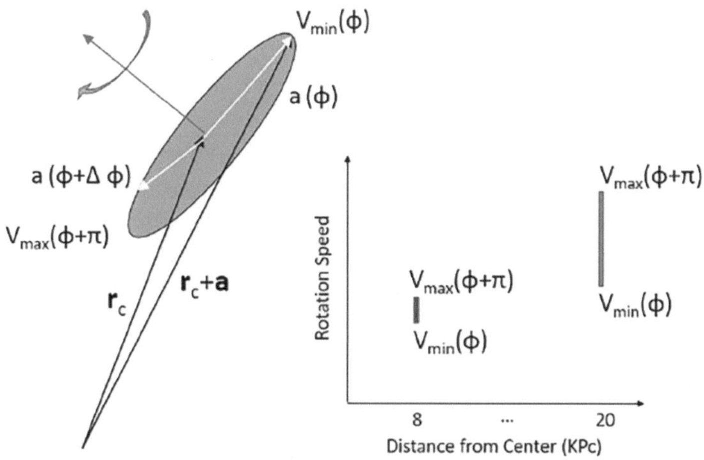

Figure 6.
Explanation why MW has azimuthal angle dependence, preventing it from having a unique rotation curve. The tilt with respect to the Local CNO radial means that a 360 degree rotation about the MW spin axis has varying Local CNO gravitational potential, so stars are going up a gravity hill and coming down a gravity hill.

dipole moment, only its time variation (flipping rate), with a very large relativistic factor enhancement.

The idea of neutrinos forming bound states is a very old idea and has been reviewed in Ref. [2]. All of these old papers have the wrong neutrino equation of state, e.g. [17] which has only one neutrino flavor. The correct neutrino equation of state was given in Ref. [2]. The individual neutrino flavor masses are irrelevant for the equation of state (EOS) and a common neutrino scale mass is all that is required e.g. m_ν = 0.75 eV/c^2. The only remaining parameter for the neutrino EOS is the boundary condition at the center of the CNO $x(0) = p_F/m_\nu c$, where p_F is the Fermi momentum and c is the speed of light. From **Table 2**, we see that CNO all have very non-relativistic neutrino degeneracy. Even with the correct neutrino equation of state, one must identify astronomical data that uniquely identifies CNO as the Dark Matter. This

CNO signature was identified in Ref. [18] using weak lensing data: degenerate matter has the signature that smaller radius objects have higher mass. This is true for neutron stars (neutron degeneracy), White Dwarfs (electron degeneracy) and CNO (neutrino degeneracy). In **Table 2** we demonstrate the CNO signature.

We now list important consequences of CNO as the Dark Matter:

1. Since CNO are degenerate matter, Navarro–Frenk–White (NFW) Dark Matter profiles are incorrect, while Einasto Density Profile fits provide the correct density profiles for Dark Matter [3].

2. The ΛCDM model prediction of the Hubble Constant gets updated, because the neutrinos are no longer in thermal equilibrium with baryonic matter. Instead, magnetic degrees of freedom would replace the neutrino degrees of freedom. Neutrinos would rapidly cool, falling out of equilibrium prior to the observation of the CMB afterglow. This is due to the magnetic cooling of dipole moments in the Early Universe, because the radiated power $P = \frac{2}{3c^3}\left[\gamma^8\left(\vec{\beta}\cdot\dot{\vec{\beta}}\right)\left(\dot{\vec{\mu}}\right)^2 + \cdots\right]$ has a large relativistic γ factor and the small magnetic dipole moment $\vec{\mu}$ does not enter in the leading term; instead, its spin-flipping $\left(\dot{\vec{\mu}}\right)^2$ enters, which can be as large as infinity.

3. Our proposed thermal equilibrium point of departure from the ΛCDM model would not significantly effect the cosmological proton-neutron or deuterium and helium ratios where these abundances are set shortly after neutrino decoupling. While, it is interesting to note that neutrinos are a decay product of ^{7}Be in the creation of primordial ^{7}Li [19], our proposed magnetic-cooling of primordial neutrinos should not directly contribute to the 'lithium problem' [20].

4. The mass scale for neutrinos cannot be arbitrarily small, but has a lower bound [1, 12]. This is a testable prediction for KATRIN [21] to confirm.

5. All degenerate matter, White Dwarfs, Neutron Stars and CNO have a finite mass that can withstand gravity and masses beyond their last stable mass, will collapse. For CNO, this would result in the largest Black Holes in the Universe: This hypothetical Black Hole from neutrino degrees of freedom would approach 2×10^{17} solar masses [22].

6. The hot intra-cluster gas seen in galaxy clusters has nothing whatsoever to do with thermal equilibrium, but is the result of individual galaxy winds that are present as each galaxy undergoes orbital rotation inside the Dark Matter gravitational potential [23]. The expelled galaxy winds themselves undergo orbital rotation inside the Dark Matter potential.

Figure 7 is a three dimensional Mathematica plot with significant galaxies for a local group CNO [4] with neutrino mass scale 0.75 eV/c^2 and boundary condition x (0) = 0.015. In order to determine if we are in fact actually embedded in a CNO, we will model this CNO's gravity well for atomic-clocks in a juxtaposition orbit and in the next section will describe this proposed experiment and confirm that these atomic

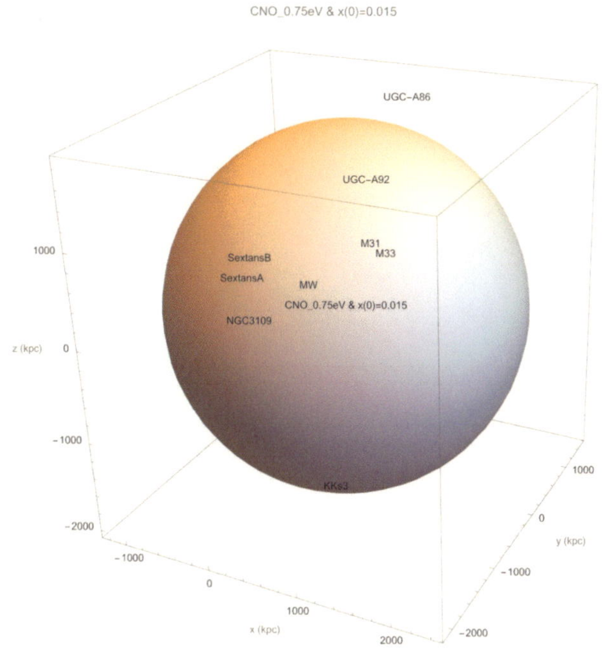

Figure 7.
Local Group CNO with neutrino mass scale 0.75 eV/c² and boundary condition x(0) = 0.015. Some galaxy members of the Local Group are plotted.

clocks would have errors from our local gravitational noise sources below the modeled CNO gravity well signal for our orientation with respect to the CNO center.

To compute the gravitational potential associated with the Local Group CNO, we need the Dark Matter mass enclosed within a radius r. This is given by solving the equation of hydrostatic equilibrium reference [2].

$$M_{CNO}(s) = 5.712 \times 10^{17}\ M_{\odot} I(s), \quad s = r/r_{CNO} \tag{1}$$

where the 10th order polynomial $I(s)$ is given in **Figure 8** and the polynomial expansion is given in Eq. (2) below.

$$\begin{aligned} I(s) = {} & 0.0000144313s^2 + 0.0332517s^3 + 0.00203896s^4 - 0.0729033s^5 \\ & + 0.000801943s^6 + 0.110044s^7 - 0.086918s^8 \\ & + 0.0160271s^9 + 0.00325112s^{10}, \quad s = r/r_{CNO} \end{aligned} \tag{2}$$

2. Physics of atomic-clock detection of CNO dark matter

The CNO embedding the Local Group of galaxies will have gravitational redshift effects [24]. Unfortunately, these are too small to be seen against the background of peculiar galaxy velocities. However, we may be able to measure the gravitational effects of the CNO Dark Matter potential well that the Milky Way resides in by using atomic clocks to detect the gravitational time delay. We show that a CNO signal can be identified by measuring the difference between two atomic clocks, to remove noise sources, with an optimized orbit configuration.

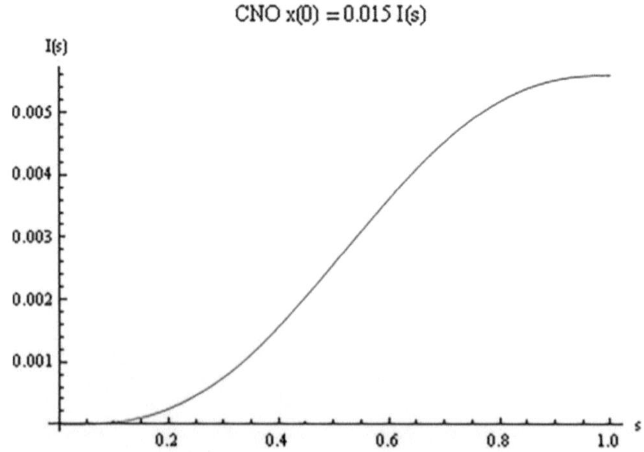

Figure 8.
A 10th order polynomial fit of the scaled Local Group CNO Dark Matter mass $I(s)$ assuming $x(0) = 0.015$ and $m_\nu = 0.75\ eV/c^2$. The Y-axis has no dimensions (see Eq. 1). The x-axis is the variable s.

2.1 CNO signal

In our proposed experiment, two satellite atomic clocks, which had previouly been synchronized, will differ in time due to gravitational time dilation, if the clocks are in different gravitational potentials. This effect is well understood for the Earth's gravitational potential as the GPS constellation satellites must take this effect into account for their signals to allow ground based observers to accurately determine their position [25, 26]. We consider two spacecraft with atomic clocks in an opposition orbit around the Sun and measure their time difference at points P_1 and P_2, where the line from the CNO center to the Sun intersects their 1 au orbit, **Figure 9**. The clock at P_2 is in a deeper CNO gravitational potential well, compared to the clock at P_1 and therefore runs slower. If these clocks have a measurable time difference, then atomic-clock based physics detection of Dark Matter will have been achieved, assuming this signal is not swamped by systemic or significant random noise sources. We now calculate the time difference levels from signal and systemic noise sources for our proposed experiment. The times recorded at points P_1 and P_2 are

$$t_1 = t_s\sqrt{1 - \frac{2GM_1}{r_1c^2}} \tag{3}$$

$$t_2 = t_s\sqrt{1 - \frac{2GM_2}{r_2c^2}} \tag{4}$$

where M_i and r_i are the enclosed CNO mass and radius at these points, i = 1, 2 and t_s is the synchronous time. The measured time delay coefficient Z is

$$Z = \sqrt{1 - \frac{2GM_1}{r_1c^2}} - \sqrt{1 - \frac{2GM_2}{r_2c^2}} \tag{5}$$

To evaluate Eq. (5), we use notation

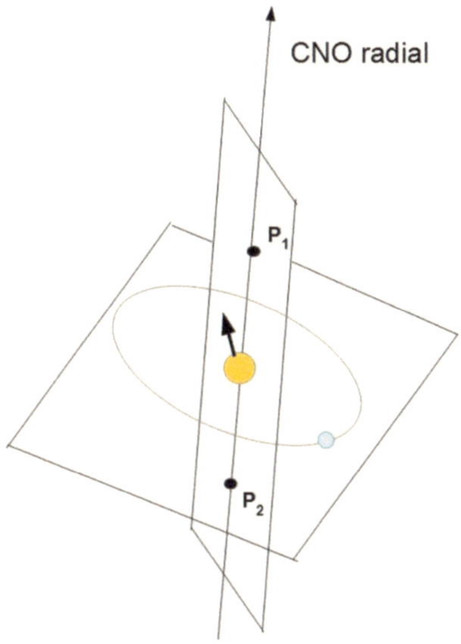

Figure 9.
Two atomic clocks situated along the CNO radial to the Sun at points P_1 and P_2, each in an opposition orbit of 1 au from the Sun (yellow). The Sun's spin vector (black arrow) is shown and the nominal orbit of Earth (blue) is shown. Note that P1 and P2 are just outside of the orbital plane for the Earth (the rectangular plane); the geometry of which with Keplerian parameters are provided in the Orbital Geometry section of the paper. Note also that the CNO line intersecting the 1 au radius from the Sun does not lie in Earth's orbit.

$$2\frac{GM_1}{r_1c^2} \equiv f(r+\Delta r_1) \tag{6}$$

and use the Taylor expansion

$$f(r+\Delta r_1) = f(r) + \left.\frac{df}{dr}\right|_r \Delta r_1 \tag{7}$$

Because $\Delta r_1 = -\Delta r_2 = \Delta r$, we have

$$Z = \frac{1}{\sqrt{1-f(r)}}\left.\frac{df}{dr}\right|_r \Delta r \tag{8}$$

where Δr = 1 au. Since$f = 2GM(r)/rc^2$, Eq. (8) reduces to

$$Z = \frac{\Delta r}{\sqrt{1-f(r)}}\left\{2\frac{G}{rc^2}\left.\frac{dM}{dr}\right|_r - 2\frac{GM(r)}{r^2c^2}\right\} \tag{9}$$

To evaluate Eq. (9), we use the Local Group CNO solution [4], which has the location of the Milky Way/Sun/Earth in CNO coordinate system $s_{MW} = 0.36859$. Using $dM/dr = (dM/ds)(ds/dr)$ and

$$\left.\frac{dM}{ds}\right|_{.36859} = 4.922908 \times 10^{15} M_\odot \tag{10}$$

The following expressions were obtained (r = radius location of Sun in CNO coordinate system)

$$\sqrt{1-2\frac{GM(r)}{rc^2}} = 0.999948 \tag{11}$$

$$\frac{\Delta r}{r_{CNO}}\frac{2G}{rc^2}\frac{dM}{ds} = 1.842\times 10^{-12} \tag{12}$$

$$-2\frac{GM}{r^2c^2}\Delta r = -7.45296\times 10^{-16} \tag{13}$$

Giving the final time delay for 1 au spacecraft in orbit

$$Z = 1.842\times 10^{-12} \tag{14}$$

The atomic clock time delay Eq. (14) is within the time resolution of satellite atomic clocks$\sim 10^{-18}$ fractional frequency difference from gravitational effects (see for example [27]).

2.2 Sun J2 noise

It is well known that the Sun's gravitational potential can be expanded in spherical harmonics [28].

$$\Phi_{Sun} = -\frac{GM_{\odot}}{r}\left\{1-\sum_n J_n\left(R_{eq}/r\right)^n PL_n(\cos(\phi))\right\} \tag{15}$$

where R_{eq} is the Sun's semimajor axis, ϕ is the colatitude, the PL_n are Legendre polynomials and the J_n are the measured zonal spherical coefficients (quadrupole and octopole moments of the Sun) [28].

$$J_2 = -(6.13\pm 2.52)\times 10^{-7} \tag{16}$$

$$J_4 = (2.8\pm 2.1)\times 10^{-12} \tag{17}$$

The octopole moment J_4 is too small and the J_2 contribution cancels in the difference at P_1 and P_2 because ϕ is the colatitude and $PL_2(\cos(\phi))$ is even.

2.3 Milky Way noise

To calculate the Milky Way potential, we require an accurate model of its stellar densities. For our approach, the Dark Matter contribution is just the Local Group CNO, so it does not contribute to the Milky Way stellar noise. We use reference [29] to give us the Milky Way potential at points P_1 and P_2 with the data in **Table 3**, The Milky Way potential in this model [29] is a sum of a disk (cylindrical coordinates R, z) $\Phi_{disk}(R,z)$ and sphere model (spherical coordinates r) $\Phi_{sphere}(r)$.

$$\Phi_{disk}(R,z) = \frac{-GM_{disk}}{\sqrt{R^2+\left(a+\sqrt{z^2+b^2}\right)^2}} \tag{18}$$

Parameter	Value	Reference
Spherical radius from center	8.178 kpc	[30]
Distance above disk	20 pc (z_0)	[31]

Table 3.
Sun's parameters in the Milky Way model calculation.

Parameter	Value
M_{disk}	1×10^{11} M$_\odot$
a	6.5 kpc
b	.26 kpc
M_{sphere}	3.4×10^{10} M$_\odot$
C	0.7 kpc

Table 4.
Milky Way parameters.

$$\Phi_{sphere}(r) = \frac{-GM_{sphere}}{r + C} \tag{19}$$

The total potential energy at P_1 with (epsilon in au/kpc) is

$$\Phi_{P_1}(\epsilon) = \Phi_{P_1 \ disk}(\epsilon) + \Phi_{P_1 \ sphere}(\epsilon) \tag{20}$$

and at P_2

$$\Phi_{P_2}(\epsilon) = \Phi_{P_2 \ disk}(\epsilon) + \Phi_{P_2 \ sphere}(\epsilon) \tag{21}$$

The Milky Way noise Z_{MW} (is)

$$Z_{MW} = \sqrt{1 - \frac{2}{c^2}\Phi_{P_1}(\epsilon)} - \sqrt{1 - \frac{2}{c^2}\Phi_{P_2}(\epsilon)} \tag{22}$$

The parameters [29] are in **Table 4**. The critical observation is that the Milky Way has a tilt angle (θ) in the CNO frame, **Figure 10**. This adds a little complexity to the geometry, see **Figure 10**.

From this figure, we obtain (using notation ϵ = au/kpc) the following list of coordinates.

1. P_1 has z-coordinate $z_0 + (au)\cos(\theta)$.

2. P_2 has z-coordinate $z_0 - (au)\cos(\theta)$.

3. The Sun has cylindrical coordinate $\rho = \sqrt{\left((8.178)^2 - (.02)^2\right.}$ in kpc. We use notation $\rho = \tilde{\rho}$ kpc.

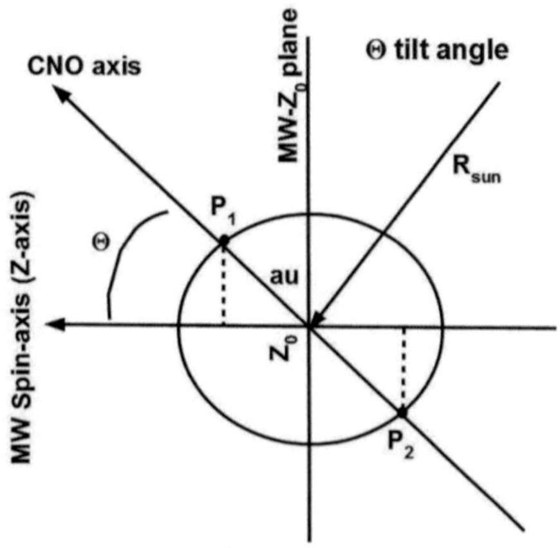

Figure 10.
Milky Way coordinates of P_1 and P_2.

4. P_1 has cylindrical coordinate $\rho - (au)\sin(\theta)$.

5. P_2 has cylindrical coordinate $\rho + (au)\sin(\theta)$.

6. P_1 has spherical coordinate $\text{kpc}\sqrt{(\tilde{\rho} - \epsilon\sin(\theta))^2 + (0.02 + \epsilon\cos(\theta))^2}$.

7. P_2 has spherical coordinate $\text{kpc}\sqrt{(\tilde{\rho} + \epsilon\sin(\theta))^2 + (0.02 - \epsilon\cos(\theta))^2}$.

The tilt angle from reference [4] is $\theta = 47.221°$, so $\sin(\theta) = 0.73397$, $\cos(\theta) = 0.67917$. Doing the disk contribution

$$-\frac{GM_{disk}}{kpc} = -4.3009\ldots \times 10^{11}\ m^2/s^2 \equiv -B \tag{23}$$

The disk contribution at P_1 (is)

$$\Phi_{P_1\ disk}(\epsilon) = \frac{-B}{\sqrt{(\tilde{\rho} - \epsilon\sin(\theta))^2 + \left(6.5 + \sqrt{(.02 + \epsilon\cos(\theta))^2 + (.26)^2}\right)^2}} \tag{24}$$

We use Mathematica to obtain the ϵ expansion of Eq. (24)

$$\Phi_{P_1\ disk}(\epsilon) = -B\{0.0942444 + 0.00472974\epsilon - 0.00484651\epsilon^2 + O(\epsilon^3)\} \tag{25}$$

The value of the disk potential for point P_2 just has $\epsilon \to -\epsilon$

$$\Phi_{P_2\ disk}(\epsilon) = -B\{0.0942444 - 0.00472974\epsilon - 0.00484651\epsilon^2 + O(\epsilon^3)\} \tag{26}$$

We now do the spherical potential

$$\frac{-GM_{sphere}}{kpc} = -1.4623\ldots \times 10^{11} m^2/s^2 \equiv -D \tag{27}$$

Then

$$\Phi_{P_1\ sphere}(\epsilon) = \frac{-D}{.7 + \sqrt{(\tilde{\rho} - \epsilon \sin(\theta))^2 + (.02 + \epsilon \cos(\theta))^2}} \tag{28}$$

and

$$\Phi_{P_2\ sphere}(\epsilon) = \frac{-D}{.7 + \sqrt{(\tilde{\rho} + \epsilon \sin(\theta))^2 + (.02 - \epsilon \cos(\theta))^2}} \tag{29}$$

Mathematica gives the ϵ expansion

$$\Phi_{P_1\ sphere}(\epsilon) = -D\{0.112638 + .00929112\epsilon + 0.000406691\epsilon^2 + O(\epsilon^3)\} \tag{30}$$

The value of the spherical potential for point P_2 just has $\epsilon \to -\epsilon$

$$\Phi_{P_1\ sphere}(\epsilon) = -D\{0.112638 - .00929112\epsilon + 0.000406691\epsilon^2 + O(\epsilon^3)\} \tag{31}$$

Adding the disk and sphere contributions together gives

$$\frac{-2}{c^2}\Phi_{P_1}(\epsilon) = 1.269 \times 10^{-6} + 7.5 \times 10^{-8}\epsilon \tag{32}$$

$$\frac{-2}{c^2}\Phi_{P_2}(\epsilon) = 1.269 \times 10^{-6} - 7.5 \times 10^{-8}\epsilon \tag{33}$$

Doing the expansion

$$\sqrt{1 - \frac{2}{c^2}\Phi_{P_1}(\epsilon)} = 1 + \frac{1}{2}\left(\frac{-2\Phi_{P_1}(\epsilon)}{c^2}\right) + \ldots \tag{34}$$

Then

$$Z_{MW} = 7.5 \times 10^{-8}\epsilon \tag{35}$$

With ϵ =au/kpc = 5×10^{-9}, we finally get

$$Z_{MW} = 3.75 \times 10^{-16} \tag{36}$$

This value is four orders of magnitude below our threshold of $< 10^{-12}$.

2.4 Earth noise

The satellites are placed in an orbit such that they are always within the line of sight of the Earth (see the orbital geometry discussion) and always remaining at approximately the same Earth-Satellite distance. Hence, the equation for the Earth's noise $Z_\oplus$ (is)

$$Z_{\oplus} = \sqrt{1 - \frac{2GM_{\oplus}}{\zeta_1 c^2}} - \sqrt{1 - \frac{2GM_{\oplus}}{\zeta_2 c^2}} \tag{37}$$

where the distances from Earth to points P_1 and P_2 are parameterized as ζ_1 in au, ζ_2 in au, with ζ_1 and $\zeta_2 \sim 1$, and the value is always positive based on which spacecraft is at P_1.

$$Z_{\oplus} = 2.96 \times 10^{-14}\left(\frac{1}{\zeta_2} - \frac{1}{\zeta_1}\right) \tag{38}$$

This is evaluated in the section Orbital Geometry.

2.5 Venus, comets, asteroids and other noise sources

Satellites in an opposition orbit around the sun allows smaller objects not in close proximity to either spacecraft to be at least partially canceled by differencing the two clocks. With their observed gravitational potential falling off as m_i/r_j, where m_i is the mass of the object, and r_j each spacecraft's distance to that object. Prior to the actual launch of such a mission, a more detailed orbit analysis should be performed for all known Earth crossing objects. We show in this section an assessment for Venus with our orbital configuration. Eq. (39), below is similar to Eq. (38) for the Earth, however, it uses the mass for Venus and each satellite's distance from it.

$$Z_{venus} = 2.417 \times 10^{-14}\left(\frac{1}{\zeta_2} - \frac{1}{\zeta_1}\right) \tag{39}$$

This is also evaluated in the section Orbital Geometry.

3. Orbital geometry

We are using an opposition orbit around the sun, such that both spacecraft will always be approximately equidistant from the Earth, and both in view from our orbital vantage point when one of the spacecraft is closest to the CNO center of mass. For our simulation, we used Aerospace Corporation's (Aerospace) Satellite Orbital Analysis Program (SOAP) to extract ephemeris information for both spacecraft for specific dates in the future when these spacecraft would be in alignment with the CNO center. **Table 5** provides the distance values to the Earth dependent on the spacecraft closest to the predicted CNO center for a full orbit.[1]

For the Earth noise: Evaluating Eq. (38) using the values in **Table 5** when SC_1 is at P_1 and SC_2 is at P_2, both converted into au from units of kilometers ζ_1 = 1.5936 and ζ_2 = 1.18558, resulting in $Z_{\oplus} = 6.3923 \times 10^{-15}$. When SC_2 is at P_1 and SC_1 is at P_2 gives ζ_1 = 1.581328 and ζ_2 = 1.2461, resulting in $Z_{\oplus} = 5.03567 \times 10^{-15}$. Both of these values are about three orders of magnitude below our threshold of $<10^{-12}$.

[1] Aerospace's SOAP does not provide ephemeris data for the Earth in orbit around the sun, hence we used the Jet Propulsion Laboratory's Horizons Web Application (https://ssd.jpl.nasa.gov/horizons/app.html#/) for the location of the Earth in orbit around the sun on these dates and times.

Date time in UTC	SC_# at $P_\#$ distance to Earth	Distance (km)
2026/11/09 16:00:00.0	SC_1 at P_1 nearest to CNO	177360340.6
2026/11/09 16:00:00.0	SC_2 at P_2 furthest from CNO	238399888.3
2027/05/11 04:00:00.0	SC_2 at P_1 nearest to CNO	186413072.8
2027/05/11 04:00:00.0	SC_1 at P_2 furthest from CNO	236562780.1

Table 5.
Spacecraft Earth distance values from the simulated orbit.

Date time UTC	SC_# at $P_\#$ distance to Venus	Distance (km)
2026/11/09 16:00:00.0	SC_1 at P_1 nearest to CNO	96623790.03
2026/11/09 16:00:00.0	SC_2 at P_2 furthest from CNO	241764182.2
2027/05/11 04:00:00.0	SC_2 at P_1 nearest to CNO	182734399.4
2027/05/11 04:00:00.0	SC_1 at P_2 furthest from CNO	186090325.7

Table 6.
Spacecraft Venus distance values from the simulated orbit.

For the Venus noise: Evaluating Eq. (39) using the values in **Table 6** when SC_1 is at P_1 and SC_2 is at P_2, both converted into au from units in kilometers ζ_1 = 1.6156 and ζ_2 = 0.6458, resulting in $Z_{venus} = 2.24661 \times 10^{-14}$. When SC_2 is at P_1 and SC_1 is at P_2 gives ζ_1 = 1.24394 and ζ_2 = 1.221508, resulting in $Z_{venus} = 3.5682 \times 10^{-16}$. These values are two and four orders of magnitude below our threshold of $<10^{-12}$ respectively. The case with Venus nearest to SC_1 when it is nearest to the center of the CNO at P_1 at about two orders of magnitude less than our noise level is our predicted worst case noise scenario for this orbital geometry.

4. Conclusion

The idea that an atomic-clock can be used to detect Dark Matter is not new [32–36]. However, all of these published concepts are looking for particle Dark Matter bound to the sun. They do not account for our hypothesis that Dark Matter is a condensed neutrino object (CNO). Our proposed opposition orbit around the sun experiment not only provides a systematic method for supporting the CNO-based Dark Matter hypothesis, it would also support or refute the local halo conjecture. We postulate however, that even though it may be possible for a hypothetical Dark Matter halo to have the same mass distribution in the same direction as our predicted CNO location, we find this extremely unlikely based on our use of other Local Group spiral galaxy spin vectors to find the CNO's center [4].

We solved these equations for spacecraft in a circular opposition orbit around the sun for the predicted CNO location, finding that this orbital geometry should see the gravitational time dilation effect. Further work could consider multiple orbit averaging or eccentric orbits such as with both spacecraft's aphelion in the direction of the CNO center, or potentially having one aphelion in the direction of the CNO, and the other's aphelion in the opposite direction. Potentially, this may improve the signal to noise ratio from our solar system's planets and other masses. We should mention that placing atomic clocks on spacecraft not in Earth orbit is a desirable secondary payload on missions to other planets, as they would provide unique data from their solar orbital trajectories.

The calculations presented here are for a neutrino mass scale of 0.75 eV/c^2 and a Local Group CNO value x(0) = 0.015. If KATRIN finds a smaller mass signal or if x(0) is enlarged to 0.02, the gravitational time delay would be larger than the value calculated here. Direct atomic-clock based physics detection of Dark Matter is possible as long as the KATRIN experiment has a signal.

A. Appendix

A.1 Estimation of the relativistic correction value

General Relativity has numerous additive correction factors. The relativistic correction factor ζ refers to velocity-dependent correction factors [37].

$$\zeta \sim \frac{GM}{c^2 r} \sim \frac{v^2}{c^2} \tag{40}$$

For our case

$$\zeta \sim \left[v^2(P1) - v^2(P2)\right]/c^2 \tag{41}$$

We'll show that a reasonable evaluation of this shows it is negligible. Going into the calculation: we already know that General Relativity corrections are only important for compact objects, such as neutron stars and Black Holes. CNO are not compact objects. Assuming a 4 giga-year period (author unpublished - typical value of galaxies rotating in a CNO):

$$V = \frac{2\pi r_{MN}}{4 \times 10^9 \text{years}} \tag{42}$$

where r_{MW} is the Milky Way radius in the CNO, r_{MW} = 676 kpc. Approximating the P1, P2 positions relative to r_{MW} as P2 $\sim r_{MW}$ + P2, P1 $\sim r_{MW}$ - P1, with P1 = P2 = 1 AU, then

$$V^2(P2) \simeq \left[2\pi(r_{MW} + P2)/4 \times 10^9 \text{years}\right]^2 \tag{43}$$

$$V^2(P1) \simeq \left[2\pi(r_{MW} - P1)/4 \times 10^9 \text{years}\right]^2 \tag{44}$$

Then

$$V^2(P2) - V^2(P1) \sim 31\ m^2/s^2 \tag{45}$$

Finally:

$$\left[V^2(P2) - V^2(P1)\right]/c^2 \sim 3 \times 10^{-16} \tag{46}$$

Author details

Peter D. Morley[1*] and Douglas J. Buettner[2]

1 Blue Ridge Scientific LLC, Front Royal, VA, USA

2 Department of Mechanical Engineering, University of Utah, Salt Lake City, UT, USA

*Address all correspondence to: peter3@uchicago.edu

References

[1] Buettner DJ, Morley PD. Neutrinos as the cold dark matter: A critical review, KATRIN and new research directions. Physics of the Dark Universe. 2022. Article 101149

[2] Morley PD, Buettner DJ. SHM of galaxies embedded within condensed neutrino matter. International Journal of Modern Physics D. 2015;**24**: 1550004

[3] Morley P, Buettner D. Weak lensing data and condensed neutrino objects. Universe. 2017;**3**:81

[4] Morley PD, Buettner DJ. Dark matter in the local group of galaxies. International Journal of Modern Physics D. 2017;**26**:1750069

[5] Zwicky F. On the masses of nebulae and of clusters of nebulae. The Astrophysical Journal. 1937;**86**(10):217. DOI: 10.1086/143864

[6] Colless M, Dunn AM. Structure and dynamics of the coma cluster. The Astrophysical Journal. 1996;**458**:435-454

[7] Aalbers J et al. Dark Matter Search Results from 4.2 Tonne-Years of Exposure of the LUX-ZEPLIN (LZ) Experiment. arXiv:2410.17036v2

[8] Aprile E et al. Search for coherent elastic scattering of solar 8B neutrinos in the XENON1T dark matter experiment. Physical Review Letters. 2021;**126**: 091301. DOI: 10.1103/physrevlett.126.091301

[9] Knirck S et al. (BREAD collaboration) first results from a broadband search for dark photon dark matter in the 44 to 52 μeV range with a coaxial dish antenna. Physical Review Letters. 2024;**132**: 131004. DOI: 10.1103/PhysRevLett.132.131004

[10] Misiaszek M, Rossi N. Direct detection of dark matter: A critical review. Symmetry. 2024;**16**(2):201. DOI: 10.3390/sym16020201

[11] Morley PD, Buettner DJ. Instantaneous power radiated from magnetic dipole moments. Astroparticle Physics. 2015;**62**:7-11

[12] Morley PD. Prediction of the Neutrino Mass Scale Using Coma Galaxy Cluster Data. Symmetry, June 2020. DOI: 10.3390/sym12061049

[13] Fridman AM et al. The orientation parameters and rotation curves of 15 spiral galaxies. A&A. 2005;**430**:67-81. DOI: 10.1051/0004-6361:200400087

[14] Jog CJ. Large-scale asymmetry of rotation curves in lopsided spiral galaxies. A&A. 2002;**391**:471-479. DOI: 10.1051/0004-6361:20020832

[15] Patoa M, Iocco F. Galkin: A new compilation of milky way rotation curve data. SoftwareX. 2017;**6**:54-62. DOI: 10.1016/j.softx.2016.12.006

[16] Corbelli E, Salucci P. Monthly Notices of the Royal Astronomical Society. The extended rotation curve and the dark matter halo of M33. 2000;**311**: 441. DOI: 10.1046/j.1365-8711.2000.03075.x

[17] Markov MA. On possible existence of neutrino superstars. Physics Letters. 1964;**10**:122

[18] Morley PD, Buettner DJ. A dark matter signature for condensed neutrinos. International Journal

of Modern Physics D. 2016;**25**: 1650089

[19] Khatri R, Sunyaev RA. Time of primordial ^{7}Be conversion into ^{7}Li, energy release and doublet of narrow cosmological neutrino lines. Astronomy Letters. 2011;**37**:367-373. DOI: 10.1134/S1063773711060041

[20] Fields BD. The primordial lithium problem. Annual Review of Nuclear and Particle Science. 2011;**61**:47-68. DOI: 10.1146/annurev-nucl-102010-130445

[21] KATRIN. Available from: https://www.katrin.kit.edu

[22] Morley PD. New class of black holes. Physics of the Dark Universe. 2020;**30**: 100643

[23] Morley PD. Non-Keplerian Orbits in Dark Matter, Dark-Matter-Recent Observations and Theoretical advances. Intechopen Books. 2001. DOI: 10.5772/intechopen.99243

[24] Morley PD. Dark matter gravitational spectral shifts in the local group. Astrophysics and Space Science. 2020;**365**(3):57

[25] Love AW. GPS, atomic clocks and relativity. IEEE Potentials. 1994;**13**(2): 11-15. DOI: 10.1109/45.283881

[26] Ashby N. Relativity in the global positioning system. Living Reviews in Relativity. 2003;**6**:1. DOI: 10.12942/lrr-2003-1

[27] Mehlstäubler TE et al. Atomic clocks for geodesy. Reports on Progress in Physics. April 2018;**81**. DOI: 10.1088/1361-6633/aab409

[28] Rozelot JP, Godier S, Lefebvre S. On the theory of the oblateness of the sun. Solar Physics. 2001;**198**:223-240. DOI: 10.1023/A:1005238718479

[29] Price-Whelan AM et al. Inferring the gravitational potential of the milky way with a few precisely measured stars. ApJ. 2014;**794**. DOI: 10.1088/0004-637X/794/1/4

[30] Particle Data group. Available from: http://pdg.lbl.gov

[31] Anderson LD et al. A galactic plane defined by the milky way HII region distribution. ApJ. 2019;**871**. DOI: 10.3847/1538-4357/aaf571

[32] Sherrill N et al. Analysis of atomic-clock data to constrain variations of fundamental constants. New Journal of Physics. 2023;**25**:093012. DOI: 10.1088/1367-2630/aceff6

[33] Tsai YD, Eby J, Safronova MS. Direct detection of ultralight dark matter bound to the sun with space quantum sensors. Nature Astronomy. 2023;**7**:113-121. DOI: 10.1038/s41550-022-01833-6

[34] Derevianko A, Pospelov M. Hunting for topological dark matter with atomic clocks. Nature Physics. 2014;**10**:933-936. DOI: 10.1038/nphys3137

[35] Roberts BM et al. Search for domain wall dark matter with atomic clocks on board global positioning system satellites. Nature Communications. 2017;**8**:1195. DOI: 10.1038/s41467-017-01440-4

[36] Kennedy CJ et al. Precision metrology meets cosmology: Improved constraints on ultralight dark matter from atom-cavity frequency comparisons. Physical Review Letters. 2020;**125**:20130. DOI: 10.1103/PhysRevLett.125.201302

[37] Benitez F, Gallardo T. The relativistic factor in orbital dynamics of point masses. Celestial Mechanics and Dynamical Astronomy. 2008;**101**: 289-307. DOI: 10.1007/s10569-008-9146-5

Chapter 11

The Problems of Modern Cosmology, the Fine Tuning of Constants and Parameters, and the Notion of Time

Yves Gaspar

Abstract

The Standard Hot Big Bang theory is successful from the observational point of view, yet a quite substantial list of unresolved problems still exists, resisting the endeavour of researchers who try to tackle these issues. Recent treatises claim that cosmology, and fundamental physics, currently faces a crisis. The initial conditions existing close to the Big Bang singularity, the value of the fundamental parameters of the Standard Model and of the constants of fundamental physics, as well as the form of the laws of nature remain without a satisfactory explanation. In this chapter, we will first review the proposal of Lee Smolin and Roberto Manabeira Unger, which essentially motivates a re-thinking regarding the notion of time, which would be considered as fundamental, irreducible and real. This suggested agenda for cosmology also considers the mathematical description of the world to be powerful indeed, but also that mathematics has immense limitations—it is claimed that mathematics cannot deal with phenomenal particularity as well as with time. We will subsequently present theoretical research done by the author, mostly at a heuristic level, which examines why fundamental aspects of mathematics would have difficulties when dealing with time and with the specificity of natural phenomena.

Keywords: time, fine tuning, limits of mathematics, initial conditions, laws of nature

1. Introduction

The three main parts of theoretical physics that form the central body of modern cosmology are general relativity, the Standard Model of particle physics and the underlying structure of quantum field theory, as well as thermodynamics and statistical physics.

One could say that the central theory at the heart of cosmology is the so-called *Standard Hot Big Bang theory*, which implies that the observable universe has evolved from a very dense and hot (energetic) initial phase, which underwent cosmological expansion to produce the actual universe. The Einstein Field Equations (EFE) governs cosmic evolution, which are the central equations of general relativity.

As is well known, this theory certainly has four important observational confirmations: the expansion of the universe, the existence and properties of the cosmic microwave background radiation (the CMBR), the relative abundances of light elements including isotopes in the observable universe, and the number counts of distant radio sources compared to nearer ones.

However, this cosmological theory also raises at least 10 questions or problems that remain unsolved, or which are still debated to this day. Among these are the flatness, the horizon and the smoothness problems. The flatness problem consists in the fact that the spatial sections on large distance scales of the observable universe are very close to flatness, while the EFE imply that the flat solutions are unstable and are the most improbable models.

The horizon problem implies that very distant regions in the actual universe have almost the same CMBR temperature. Yet, the Big Bang did not provide sufficient time for such regions to establish causal contact and thermalize in order to reach almost the same temperature.

The smoothness problem corresponds to the fact a precise and adequate amplitude of density fluctuations is needed in order to act as seeds for structure formation (to form galaxies, stars, ...): if the fluctuations are too small, then not enough cosmic structure forms, and if they are too large, then too much over dense structures would form, such as black holes. One would need a correct mechanism in order to produce in the early universe adequate primordial density fluctuations.

Further problems correspond to the initial singularity problem and the nature of a quantum theory of gravity, the matter/antimatter asymmetry problem, and the topological defect production (such as magnetic monopoles) after Grand Unified Theories (GUT) phase transitions (symmetry breaking) in the very early universe.

In addition, the *general initial conditions problem* and the *general fine tuning problem* of constants of physics and of cosmological parameters are still debated, as well as the dark matter and dark energy problems. Cosmic evolution strongly depends on the general initial conditions existing in the initial phase of the universe—for instance, if the geometry of space–time was too chaotic or inhomogeneous at early stages, then cosmic expansion would not have produced our flat and relatively smooth universe. The fine-tuning problem of constants of physics and parameters of cosmology entails that even very tiny variations of these numbers could imply that the observable universe today would be very different. It seems that, with respect to these numbers, the universe is *structurally unstable* and that their values turn out to be critical—such numbers correspond to the speed of light in vacuum, the electric charge of the electron, the ratio of the electron to proton mass and the fine structure constant. Yet, there exists today no well-established theory or explanation of the fine-tuned values of these quantities and of their ratios: we know their values through measurement, not through computation from a well-defined theoretical framework.

In modern cosmology, one strategy to try to tackle some of these issues corresponds to the so-called *inflationary paradigm*, which is often invoked in order to address the flatness, the smoothness and the topological defect problems. Cosmological inflation means that the universe in its very early stages underwent a very short period of rapid, accelerated cosmological expansion. This would entail that the actual universe has emerged from a much smaller region, enabling all of its parts to thermalize and solve the horizon problem. Moreover, due to this phase of accelerated expansion, the eventual topological defects would get diluted such that their local density would be virtually zero. Furthermore, an inflationary phase of expansion would turn the instability of the spatially flat models into a stable future attractor solution, rendering the flat solution the most probable one, thereby solving the flatness problem. Also, results found by Garry Gibbons and Stephen Hawking provide an adequate primordial density fluctuation spectrum as being generated by quantum fluctuations in the so-called de Sitter space–time during the inflationary phase, thus providing a solution to the smoothness problem.

However, inflationary cosmology not only does not provide a solution to the other problems of the Standard Hot Big Bang model, but also introduces new problems of its own.

First, in the original model proposed by Alan Guth in 1981 (called "old inflation"), the driver of the inflationary accelerated cosmic expansion is the potential energy density of the so-called Higgs field. However, it turned out that the universe that emerges from such an inflationary phase would not correspond to the observed one, unless some parameters were fine tuned. Subsequently, new models have been proposed having a different shape of the potential energy, such as new inflation, extended inflation, hybrid inflation and assisted inflation. In fact, a *proliferation of possible models occurred*—indeed, there are more than 100 different models.

Second, if the initial conditions characterising the primordial space–time are too inhomogeneous and chaotic, and if the "potential terms" do not dominate over the kinetic terms and over the spatial gradients of the field, then it is possible that inflation does not even start. The so-called chaotic inflationary model and the eternal inflation model try to overcome these issues linked to particular initial conditions: for some regions of space–time, inflation can start and generate a suitable macroscopic domain resembling the observable universe, whereas in other regions inflation does not start or produces a domain, which is different from our universe. An associated major problem of these inflationary scenarios is that the identity of the quantum field, whose potential energy density would be the driver of the accelerated expansion, is still unknown—it is generically called "the inflation field".

In such a context, ideas such as the *multiverse* have been developed, essentially suggesting that our macroscopic universe is one amidst a huge (possibly infinite) number of other causally disconnected universes, and that our universe happens to be characterised by the observed fine-tuned constants and parameters, entailing the possibility to develop carbon-based life forms. The possible values of the fundamental constants and parameters would be distributed essentially randomly over the "multiverse landscape".

At the heart of the questions regarding the nature of the universe in its very early stages, the so-called *quantum theory of gravity* would be helpful to address the singularity problem. Also, it would help understand the general initial conditions with which the universe started its evolution towards a remarkably homogeneous and isotropic stage at late times, far from the initial Big Bang singularity. It is remarkable that although general relativity and quantum physics are each on their own very well-tested theories till now, when one tries to unify them in a single theoretical framework, a great number of difficulties emerge such that there is actually no certainty on which proposal is moving in the right direction. Jay Armas has published 37 interviews of leading researchers in the field of quantum gravity in his book "Conversations on quantum gravity" [1], all of the experts have different approaches. It turns out that there is no general consensus regarding which approach is the correct one, and after some 50 years of research in this field no new fundamental leading principle or idea seems to have been proposed in order to motivate theoretically a given approach to quantum gravity. Current research essentially extends, modifies or simply uses known techniques from quantum physics and general relativity, but no novel idea has yet emerged.

Popular proposals are loop quantum gravity, Euclidean quantum gravity and superstring theory or M-theory. Considering the latter, in a way which is similar to the case of inflationary cosmology, one has a *proliferation of possible universe models*—of the order of 10^500. This substantially leaves the general initial conditions problem and the fine tuning problem unsolved, apart from stating that the fundamental constants and parameters are distributed randomly over such a "string theory landscape". This situation is rather peculiar, since the history of physics generally shows a tendency to *unification and simplification* of the theoretical models—think of the unification between "terrestrial and celestial mechanics" with Newton, thermodynamics and the kinetic model of matter, electromagnetism and optics, mass and energy in special relativity. Instead, in recent times, one has a proliferation of possible explanatory solutions, or a *complexification of the models*—this is reminiscent of the ancient Ptolemaic universe, where one has cycles and epicycles upon cycles in order to explain the observed motion of the planets on the celestial sphere—the heliocentric Copernican model has provided a significant simplification.

In the subsequent sections, we will consider another possible road towards a solution to this "crisis of modern cosmology", recently suggested in the work of Lee Smolin and Roberto Mangabeira Unger [2]. They argue in their book "The singular universe and the reality of time" that superstring theory or the multiverse/landscape model could well turn out not to be a correct solution to the crisis of modern cosmology. A new methodology, similar to natural philosophy, could provide new heuristic ways to penetrate deeper into the structure of theoretical physics, thereby outlining new possible paths towards the solution of the problems inherent to modern cosmology. In particular, with respect to the fine tuning problem and the general initial conditions problem, the models studied in Ref. [2] consider that there exists only one universe. New paradigms should be invoked which regard the *nature of physical laws* and the *concept of time*, as well as the way mathematics is used as a language to describe scientifically the natural world and to express the inherent physical laws.

In the next section, we will outline what the critical aspects actually are of the current paradigms that are used to describe the world, and subsequently, we will develop ideas suggested in Ref. [2] which entail that mathematics has great power but also immense limitations, providing examples showing how fundamental mathematics has difficulties when dealing with time and with phenomenal particularity.

2. Critical aspects of current theoretical models

In their work [2], Unger and Smolin explain which shortcomings appear to be present within the theoretical and mathematical formulation of models that aim to describe the world. We will list these critical aspects and subsequently analyse some particular features of the problems inherent to the use of mathematics to formulate theoretical models, which ought to provide a better understanding of nature and of the properties of the observable universe, developing examples showing how fundamental mathematics have difficulties when dealing with time and with phenomenal particularity.

- A first aspect of fundamental physics theories corresponds to *reductionism*, and to a wider extend to *methodological individualism* in science: one assumes that the properties of a system can be explained using the properties of its constituents or parts. The laws of thermodynamics explain the macroscopic thermal properties

DOI: http://dx.doi.org/10.5772/intechopen.1009433

of physical systems, and these are interpreted in terms of statistical and mechanistic classical Newtonian physics applied to particles, atoms or molecules. The behaviour of these atoms and molecules depends, however, on their microscopic properties, ultimately explained by the characteristics of elementary particles such as the quark and the electron in the Standard model. The following question naturally arises, namely: where do the elementary properties of these particles, like mass, electric charge and the colour charge of quarks within quantum chromodynamics, come from? The Standard model of particle physics does not explain these parameters. Superstring theory or M-theory tries to offer an explanation going beyond the Standard model. In these theories, elementary particles correspond essentially to vibration modes in a higher dimensional space–time of extended objects such as one-dimensional strings—the whole known particle spectrum should be produced in this way, but a very special kind of topology of compactified higher dimensions, such as the Calabi-Yau manifolds, is needed in order to produce the observed particle species. Compactification entails that the extra space dimensions are "curled up" on a very small scale, since they ought not to be observable at ordinary energy levels, but only at the so-called Planck scale. However, the problem remains to explain why this extra-dimensional space would have exactly this very special topology. How does hyperspace gets its topology? There is at present no convincing compactification mechanism, which produces these topologies, and, most importantly, which guarantees their stability. Why would only three space dimensions be extended and macroscopic, while the other six extra space dimensions of the theory remain curled up on a very small scale? Again, instead of simplification and unification, one seems to have complexification.

- As opposed to reductionism, *relationalism* is also part of modern theoretical physics, such as relativity theory and Yang-Mills quantum field theory. However, one ought to keep in mind that this approach too has possible problems. First, relationalism implies a fundamental dynamical and changeable character in the network of interconnections, yet relativity and other relational models presuppose immutable laws. Second, relationalism entails the existence of individual entities that form the subject of the relations—as explained in Ref. [2], not every property can be a relation, there ought to be intrinsic properties, which the relations relate.

- According to Unger and Smolin [2], another aspect corresponds to the *assumption of the Newtonian paradigm*, which entails that any physical system can be described by a fixed set of unchangeable laws and that there is a distinction between laws and initial conditions. While varying the initial conditions, and using a fixed set of laws, one can in principle predict the behaviour of the system in various circumstances. Theoretical physics models usually boil down to a fixed set of differential-type equations, which have infinity of solutions, so, a particular choice of initial conditions or boundary conditions needs to be specified. Also, any set of equations will always be effectively characterised by a certain number of fixed parameters or constants. In the case of cosmology, this duality between laws and initial conditions leads to the general initial conditions problem. An approach due to Hartle and Hawking in Euclidean quantum gravity, called the no-boundary proposal, argues that in the quantum regime at the initial stage of the evolution of the universe, one does not need boundary conditions:

there would be essentially no boundary. On the other hand, the approach suggested in Ref. [2] argues that time would be real and would affect everything, including the laws of physics themselves, and that one would not have a fixed immutable set of equations describing the dynamics of the universe. One then has no distinction between initial conditions and laws.

- A further characteristic mentioned in Ref. [2] corresponds to *"false universality"* present essentially in the Newtonian paradigm. It consists in assuming that locally determined laws of physics are valid for the entire universe, in any circumstance and at any energy level. This assumption is still part of general relativity and even of quantum physics. Moreover, general relativity and quantum physics are supposed to hold also in other "universes" in the multiverse/string landscape scenario. Smolin in Ref. [2] argues that close to black hole singularities, or close to the initial Big Bang singularity, the laws of physics could undergo change—events "without precedents" in this case would occur and would produce genuine novelty inherent to time and true change.

- Another critical aspect of modern theories corresponds to what one could call *"universal anachronism"*, which assumes that physical laws pertinent only to a small part of the history of the universe hold throughout the whole cosmic history. Physical laws in the very early universe might have been different from the laws we witness in our large and low-energy universe. High-energy environments (or other environments, which are different from the ones we observe in the local universe) might stimulate degrees of freedom, which are otherwise not excited in the dynamics of a low energy region, which would exhibit regular, stable and thus fixed physical laws.

- Smolin and Unger argue that mathematics has great power to describe the natural world, but it is also important to realise that it has immense limitations—they propose a "selective realism of mathematics". Mathematics would have difficulties when dealing with time and change, and with phenomenal particularity. Mathematics only captures selected parts of the dynamics of natural systems. It is claimed in Ref. [2] that no mathematical structure can be isomorphic to the history of the universe.

- An interesting concept, which is mentioned in recent treatises such as in Ref. [2], but also in Ref. [3], corresponds the so-called *"qualia"*, that is to the properties of systems that are observed through conscious perception. This refers to an old philosophical problem which has been, and which still is, very much debated. In order to illustrate this point briefly, consider the perception of green colour. As argued by Roderich Tumulka in Ref. [3], *no mathematical structure or computation can explain how green colour appears* to the conscious observer. Of course, green colour has its wavelength and its refractive properties; it is an electromagnetic wave and one knows most of the mechanisms that generate electro-chemical impulses in the eye, which travel to particular areas of the brain, and cognitive neurosciences provide a lot of information on these processes. But the knowledge of electromagnetism and optics, of neuroscience and cognitive neuroscience is not capable to explain to someone *how green colour appears* to the conscious observer—the only way would be to *directly experience* green colour. Someone who knows all of physics and cognitive neuroscience, but which has never experienced green colour, will not be able to deduce how green actually would appear to his consciousness. Max Tegmark has argued that the universe might actually be a purely mathematical entity [4]—the universe not only would be well described by mathematics, it actually *is* mathematics—but the existence of qualia and of conscious perception certainly poses an important challenge to such an idea. Smolin in Ref. [2] argues that qualia are not just "illusions", but ought to be considered as novel physical phenomena, which might correspond to events without precedents.

3. The limits of mathematics and time

Is it possible to illustrate why and how mathematics has difficulties when dealing with phenomenal particularity, time and change? One way would be to consider a fundamental part of mathematics, that is, set theory, which forms the basis of most of mathematical concepts, including numbers.

A standard axiomatic theory in pure mathematics corresponds to the so-called *Zermelo-Fraenkel set theory*, with the addition of the axiom of choice (ZFC set theory). Before entering into set theory, let us consider some more conceptual or philosophical aspects regarding the notion of time and change. The American philosopher Irving Copi has highlighted one of these: consider some entity A. If A is subject to change, it will become a distinct entity B. Thus, we end up with two distinct entities, and the following question arises: how one can speak of the change of one single entity? In other words, *is there an identity over change*? In order to deal with this paradox of time and change, we will heuristically discuss an alternative set theory, which differs from the standard ZFC model. The essential idea is that sets are "collections" of elements, but, in addition, they also have *information content*. In a sense, this alternative set theory is analogous to non-Euclidean geometry, because one of the fundamental axioms of ZFC theory is not valid in this new model. Our approach at this stage is non-formal and heuristic, and will need further research in order to develop rigorous formulations of the models.

The starting point is the concept of *information and sets*. The difference or *duality between information and meta-information* plays an essential role. In order to illustrate this aspect, consider as a concrete example a book. The title, the author and the whole content of the book represents *information*. On the other hand, the colour of the cover represents *meta-information* with respect to the information—it cannot be deduced, in any way, from the information content. The location of the book in space also represents *meta-information*: the book could be on a table, or on a shelf of some library. Now, as far as sets are concerned, consider a set S containing three books. This set contains the information, which describes in detail the content of the three books. However, considering meta-information, one can construct a set S′ that contains exactly the same three books, but which has *different meta-information* content: for instance, in S the books could be aligned on the same shelf, while in the set S′ the books could be on the vertices of an equilateral triangle. The elements of the sets S and S′ are the same, but these sets are different. This means that the so-called *extensionality axiom of ZFC theory*, which states that two sets are equal if their elements are the same, is not valid. In the real physical world, the extensionality axiom is in fact not always true because of the meta-information content characterising location in space, or the relative ordering of elements in space; that is, sets of things can have an information content with respect to its elements as well as meta-information that is essentially *relational*.

As is explained by George Tourlakis in Ref. [5], sets are considered in ZFC theory regardless of their "inner structure" or of "intention", of how the sets came about, that is formal or axiomatic mathematics aims at *representing* reality within an artificial but formal and precise language. In this representation, as is argued in Ref. [5], there is *always* something lost, partly due to the decisions we make about what features of reality are essential. This artificiality, or abstraction, represents in this case an instance showing that mathematics does not capture *phenomenal particularity*.

One can try to generalise ZFC theory by violating the extensionality axiom, and by considering sets as defined by their elements, as well as by a meta-information content, in order to attempt to derive a richer mathematical description, which would offer possibilities to include some phenomenal particularity. Moreover, this setting is suitable for our idea about time and change. If a set S is described by its elements, its information content as well as by the meta-information content, then the identity during change corresponds to the elements of the set S. In addition, the set S at some time t and the set S′ at time t' have the same elements and in this part the extensionality axiom would hold through change, guaranteeing identity. But *meta-information* can be added or can change, and this entails that change really can occur and that the sets S and S′ are not the same. Since meta-information is independent from the information regarding the elements, change would indeed introduce genuine novelty.

From a heuristic (not formal), but slightly more technical perspective, the following departure from ZFC theory could be formulated. A set S can be described by its elements e(i) and its information content I(S) regarding its elements:

$$S=\{e(i),I(S)\} \tag{1}$$

For *any* such set S, meta-information, Im(S), can always be added yielding a new set

$$S^{'}=\{e(i),I(S)+Im(S)\}. \tag{2}$$

In this context, the empty set plays a crucial role. Consider, for instance, two distinguishable sets S and S′, characterised by *different* elements:

$$S=\{e(i),I(S)\}, \tag{3}$$

$$S^{'}=\{f(i),I(S^{'})\} \tag{4}$$

These two sets can *share meta-information*, which can be of *relational type* between the elements e(i) and f(i), so that their intersection contains no elements, but contains shared meta-information. This intersection would correspond to an "empty" set Φ with no elements, but with a meta-information content:

$$S\cap S^{'}=\Phi\{Im(S,S^{'})\}, \tag{5}$$

where Im(S,S′) represents the meta-information between S and S′, such as the "distance" between the elements of S and S′, or the difference in "colour" between the elements, or any difference that distinguishes the elements in some suitable *configuration space*.

This idea of considering an empty set having no elements but carrying meta-information is suitable with respect to the above explained idea of change: a system can change not only because new elements are added, but because new meta-information content is added—that is, while preserving identity (elements), change can occur (meta-information). Indeed, consider the union set between a set S and the empty set Φ, that is S U Φ. No new elements are added, but this union set has an information content given by

$$I(S\,U\,\Phi) = I(S) + I(\Phi) - I(A \cap \Phi) = I(S) + Im(S) > I(S), \tag{6}$$

since

$$I[S\{e(i), I(S)\} \cap \Phi\{Im(S)] = 0. \tag{7}$$

This last relation means that the intersection between any set S and the empty set Φ cannot be the empty set Φ, unlike in ordinary ZFC set theory, such that S∩Φ differs from Φ, and carries no elements nor information. In this alternative theory, the empty set carries non-trivial meta-information and thus cannot be part of any non-empty set S.

This is paradoxical from ordinary ZFC set theory. Indeed, suppose, by absurdum, that the empty set Φ is not part of any non-empty set S. But then there must exist some x, not belonging to S, and which is element of Φ, which leads to a contradiction. In the alternative theory, non-empty sets S and the empty set Φ cannot be viewed as existing "at the same level", in a way which is analogous to the difference between the real part and the imaginary part of complex numbers. It is as if S would lie on a "real" axis, while Φ would lie along the imaginary axis of some abstract Wessel plane. With respect to the notion of time, set Φ is never "simultaneous" to set S—it "lies somehow in the future of S". One can thus see how a simple idea of change easily leads to paradoxes or contradictions in fundamental mathematics if tools from elementary set theory are used.

These is another area in elementary set theory, which would suggest, in similar way, a notion of time, namely the paradoxes of set theory were first discovered Georg Cantor. The use of the power set theorem and of Cantor's so-called diagonal argument shows that given some non-empty set, it is always possible to construct a larger set, which contains the original set as a subset. Consider a set S containing three elements A, B and C. It is possible to construct the set S′ of all the subsets of S (as explained above, we do not include the empty set in S′, and we also take into account the order between the elements):

$$S' = \{\{A, B, C\}, A, B, C, AB, AC, BC, BA, CA, CB\}, \tag{8}$$

which contains S as a subset.

Now, if one considers the "set of all sets"—or the universe set—one has a paradox. Indeed, if the universe set can be specified, then given Ω, it is always possible to construct the set of all the subsets of Ω, which is a larger set containing Ω as a subset.

This power set theorem works even in the case of infinite sets, as has been shown by G. Cantor, and gives rise to the possibility to formulate a theory of transfinite numbers. This suggests a notion of time related to the above-mentioned ideas, and entails that change is always possible, and that *no set can be considered as a completed, fixed whole*. Whenever a possible universe set Ω is given, it is always possible to construct *a different* set Ω', not included in Ω. The different subsets of Ω' contain the original elements listed in Ω but are associated in different ways and in distinct combinations—this represents *relational* information and corresponds actually to a form of *meta-information*.

These set theoretic paradoxes, which in a sense represent limits of mathematics and can be dealt with by choosing appropriate axioms or definitions of what a set ought to be. One can distinguish between sets and classes, and introduce axiomatically the notion of proper class, which, by definition cannot be part any other set.

However, according to our viewpoint these limits of mathematics actually point towards a notion of time. No set can be conceived as a *completed immutable whole*—there is no timeless "block universe set"—the possibility of relational or meta-information always opens up new possibilities.

In the next section, we try establish heuristically links between these theoretical reflections and the physical world—in particular with features of quantum physics.

4. Links with the physical world, quantum physics

In Ref. [6], we discussed the philosophical concept referred to as the *nomological machine* idea of philosopher Nancy Cartwright. Essentially, the viewpoints highlighted by Smolin and Unger in their work entail that the natural or physical states form the laws, rather than the states being subject to immutable laws.

Moreover, *causality* is considered as a primitive concept in Ref. [2]. Usually, laws are considered as fundamental, whereas causal relations are instances of the laws. However, according to our approach, laws are particular instances of causal relationships or interconnections. The natural or physical systems can be considered as nomological machines, which in situation-specific circumstances can produce or exhibit regular law-like dynamics, whereas in other circumstances, some unpredictable, uncomputable or changeable dynamics occurs.

In Ref. [6], the idea has been discussed which considers the standard cosmological model as a nomological machine. Another system was considered too, which corresponds to *a set of mutually reflecting spheres*, which produces in each sphere sequences of iterated reflections within reflections, ad infinitum. An observer can, within each sphere, perceive sequences of nested spheres, which can generate regular, recursive patterns, but also irregular and unpredictable sequences. The recursive patterns would correspond to law-like behaviour, while chaotic sequences would correspond to non-recursive causal interconnections, which can entail the possibility of changing or mutable laws.

One can consider the physical example of a set of four perfectly reflecting spheres a, b, A and B. Light rays will be reflected back and forth between the spheres. *This would correspond to a primitive causal connection*. In this way, an observer can perceive a sequence of nested spheres within each sphere, which can be infinite in principle. For instance, the observer could perceive in the sphere b either the reflection of b (itself), of a, of B and A. This leads to four "two lettered words," bb, ba, bB and bA. Within each of these secondary "spheres", subsequent reflections can be observed:

bbb, bba, bbB, bbA or bab, baa, baB, baA or bBb, bBa, bBB, bBA or bAb, bAa, bAB and bAA. If the reflection process proceeds, then the observable sequences reach infinite length and correspond to "infinite words", like bBaBbbaBAAbaA...

In Ref. [7], D. Mumford, C. Series and D. Wright explore such mathematical objects by using essentially *Möbius transformations in the complex Wessel plane and on the Riemann sphere*. These are the most general transformations that map circles into circles, and one of the cases corresponds to four circles in the complex plane a, b, A and B, which are mapped one into each other by the iteration of a Möbius transformation. This iteration generates the successive reflections which we mentioned earlier, and which generate infinite sequences of nested disks. The "words" that are generated by the transformations are represented differently in Ref. [7]: the sequence of letters represent the sequences of iterated maps. The transformation named "a" maps the outside of disk A into the inside of disk a, while the transformation "b" maps the outside of disk B into the inside of disk b etc. Two lettered words are in this case aB, ab, ba, bb, aa, bA, Ab, AB, BA, BB, Ba and AA. These two lettered words can also be generated if one would create all the possible "binary" subsets of the set {a, b, A, B}, allowing the same element to be repeated and taking ordering into account. Only four combinations are not present in the reflections studies in [7], namely Aa, aA, Bb and Bb. Thus, the mappings between circles (or spheres) generate a new set of 12 elements (words) that are part of the possible subsets of the set S = {a, b, A, B}.

When united with the set {a, b, A, B}, one obtains a larger set S′ which contains the original set as a subset and contains 16 elements. Sequences of three lettered words can also be generated by iterated maps, aaa, aaB, aab, aba, abb, abA, baB, baa, bab, bba, bbb, bbA, bAb, bAA, etc. All these new words are part of the set of all possible three lettered subsets of S′. If the reflection process is iterated ad infinitum, one obtains infinitely long words that correspond to the infinite successions of nested disks, and whose ultimate intersection corresponds to a limit point—all the possible limit points generate the *limit set*. As explained in Ref. [7], the number of limit points is actually *uncountable*. Computable sequences form a countable set, and thus, it follows that *most of the sequences are uncomputable*.

Turning our attention to *quantum physics* and to the orthodox Copenhagen interpretation in particular, the following property must hold, namely: *any system quantum mechanics applies to must be part of a larger system that contains the system as subsystem*. Any system ought to be potentially observable "from the outside". This leads to *a paradox when one would try to apply quantum mechanics to the universe—how could an act of measurement be performed on the universe*? This is reminiscent of the paradoxes discussed in the previous sections in the context of elementary set theory: the universe set Ω cannot be conceived as a completed whole. In same way, one can assume that it is not possible to conceive "the universe" as a completed whole. In addition, if one would adopt the fundamental axioms suggested by Unger in Ref. [2], then the reflecting sphere model provides a way to generate in any case a new and larger set, which contains the original set as a subset, thus making sure that quantum mechanics can always be applied to any set or system. In a sense, the universe, having constituents or parts with reflect each other or which interconnect with each other, would "observe itself from the inside".

Another interesting feature of quantum mechanics, which suggests interesting viewpoints, corresponds to the *"No-Cloning Theorem"*. This theorem states that it is not possible to clone a general quantum state through unitary processes, see Ref. [8]. If this "quantum xeroxing" would be possible, then one could perform the measurement of two conjugate variables, one on each copy, thereby violating the Heisenberg uncertainty principle.

Now, within our alternative set theory, suppose that it would be possible to clone any set S, such that the set S and its copy are distinguishable in some suitable configuration space: this means that S and its copy *share relational meta-information*. Then, we have that, since S ∩ S = S:

$$S \cap S = S\{\mathrm{Im}(S,S)\}, \tag{9}$$

where Im(S,S) represents the meta-information that distinguishes between S and its copy S.

However, in this case we contradict the fact that, as stated earlier in Section 3, the intersection between two distinguishable sets is *the empty set* $\Phi\{Im(S,S')\}$. Moreover, it not possible that a set S determines by itself its own meta-information content, which is relational in nature. Thus, in some suitable configuration space, it is not possible to observe a set S and an identical copy. One can observe a set S and an identical copy only at different "instants of time", but not "simultaneously".

We hope that these heuristic arguments can be developed further through future research, in order to find a way towards a unifying framework, which would contain general relativity as well as quantum physics. If time is real and fundamental, then everything, including the laws of physics and the associated fundamental constants, might be subject to change. There have been already proposals that study possible variations of the constants, such as the fine structure constant, or varying speed of light theories were the speed of light changes through cosmic evolution. All of the properties of the observable universe might be the result of a unique dynamical process, which would ultimately affect everything.

Acknowledgements

I thank very much Prof. Pawel Tambor, with whom I have worked before, for valuable discussions of philosophical nature on these topics.

Author details

Yves Gaspar
St. Edmund's College, University of Cambridge, UK

*Address all correspondence to: yves.gaspar@cantab.net

References

[1] Armas J. Conversations on Quantum Gravity. Cambridge, UK: Cambridge University Press; 2021

[2] Unger RM, Smolin L. The Singular Universe and the Reality of Time. Cambrigde, UK: Cambridge University Press; 2015

[3] Tumulka R. In: Chamcham K, Silk J, Barrow JD, Saunders S, editors. Essay Why Is there Something Rather than Nothing? In: The Philosophy of Cosmology. Cambridge, UK: Cambridge University Press; 2017

[4] Tegmark M. Our Mathematical Universe. London, UK: Penguin Random House; 2014

[5] Tourlakis G. Lectures in Logic and Set Theory. Cambridge, UK: Cambridge University Press; 2014

[6] Gaspar Y, Tambor P. The Laws of Nature and the Problems of Modern Cosmology. London, UK: Foundations of Science, Springer Nature; 2023

[7] Mumford D, Series C, Wright D. Indra's Pearls – The Vision of Felix Klein. Cambridge, UK: Cambridge University Press; 2002

[8] Wootters W, Zurek W. A single quantum cannot be cloned. London, UK: Nature; 1982;**299**(5886):802-803